Simulation Model Design and Execution

BUILDING DIGITAL WORLDS

Simulation Model Design and Execution

BUILDING DIGITAL WORLDS

PAUL A. FISHWICK

Department of Computer and Information Sciences
University of Florida

Prentice Hall International Series in Industrial
and Systems Engineering

PRENTICE HALL, Englewood Cliffs, New Jersey 07632

Library of Congress Cataloging-In-Publication Data

Fishwick, Paul A.
    Simulation model design and execution: building digital worlds
    Paul A. Fishwick.
        p.   cm.
    Includes bibliographical references and index.
    ISBN 0-13-098609-7
    1. Computer simulation.   I. Title.   II. Title: Simulation model
design and execution.
    QA 76.9.C65F57   1995   94-3675
    003′ 3--dc20   CIP

Acquisitions editor: Marcia Horton
Production editor: Bayani Mendoza de Leon
Copy editor: Zeiders & Associates
Cover designer: Rich Dombrowski
Buyer: Lori Bulwin
Editorial assistant: Dolores Mars

 © 1995 by Prentice-Hall, Inc.
A Simon & Schuster Company
Englewood Cliffs, New Jersey

Printed in the United States of America
10   9   8   7   6   5   4   3   2   1

ISBN 0-13-098609-7

Prentice-Hall International (UK) Limited, *London*
Prentice-Hall of Australia Pty. Limited, *Sydney*
Prentice-Hall Canada Inc., *Toronto*
Prentice-Hall Hispanoamericana, S.A., *Mexico*
Prentice-Hall of India Private Limited, *New Delhi*
Prentice-Hall of Japan, Inc., *Tokyo*
Simon & Schuster Asia Pte. Ltd., *Singapore*
Editora Prentice-Hall de Brasil, Ltda., *Rio de Janeiro*

Dedicated to my parents
John and Barbara Fishwick
for their philosophy
of free inquiry on any subject

Preface

This book started out as a disconnected set of lecture notes for two simulation classes taught at the University of Florida. Over the past six years, I have attempted to bring some order to the notes, and the result is what you find before you. When searching for simulation texts, I found many books on statistical methods and control engineering but hardly anything on *modeling methodology* or *algorithms for simulation*. Much of the material for the book has been sewn from many disciplines, as the reader can discern after skimming through the various examples. The interdisciplinary nature of the book fits well within the *systems approach* to studying phenomena. The first chapter covers, in some detail, the structure of the text and information for the instructor and for students taking a university course in simulation. I shall, therefore, spend the better part of this preface on acknowledgements to those numerous individuals who have helped to sculpt my thoughts, as well as the field.

My immediate family deserves a great deal of credit since they had to suffer through the actual creation of the book, which usually involved zombie-like periods of concentration and the strange case of me having conversations, out loud, with myself. I remain indebted to Martha (my wife) and Laurie (my daughter) for their patience and encouragement. My father (John Fishwick) read painstakingly through an earlier draft and red-marked many flagrant grammatical errors. I was often at a loss to justify my bizarre misuse of the English language. I would like to thank Joe Mize (Oklahoma State) and Noel Fabrycky (Virginia Tech) who provided many useful comments on a later edition of the manuscript. Many colleagues have provided me with feedback on my simulation thoughts. The first and foremost are the *Arizona Gang* composed of Bernie Zeigler, François Cellier and Jerzy Rozenblit (University of Arizona). I regard them as most-valued colleagues, and through my frequent correspondence with them they have helped me to shape this book. Bernie Zeigler, in particular, has helped more than anyone in helping to shape my written thoughts about systems and simulation. Ashvin Radiya (University of Wichita, Kansas) has used a draft in his simulation class and provided many useful comments. Jason Lin (Bellcore), David Nicol (College of William and Mary) and Richard Fujimoto (Georgia Tech) made substantial comments on Ch. 9. There were several anonymous reviewers of the book who pointed out a number of errors.

With the SimPack toolkit, I am grateful to both colleagues and students. Doug Jones (University of Iowa) donated his event set code for the queuing library. The following students have helped in extending SimPack: Hanns Oskar Porr (graphics and animation software); Daniel Hay (SimPack maintenance); David Bloom (miniGPSS compiler); James Bowman (DEQ); Brian Harrington (XSimCode) and Eric Callman (calender queue code), Victot Todd Miller, Jin Joo Lee and W. Dean Norris have made valuable suggestions regarding multimodel design.

There are several key people who helped me greatly during the preparation and production of the book. First, Marcia Horton had trust in me when the book was in an earlier form, and I thank her for the opportunity to publish this book through Prentice Hall. Marcia's assistant, Dolores Mars, always made my day by being optimistic and cheerful. Toward the latter stages of the book, I worked closely with Bayani DeLeon (production editor) at Prentice Hall and Anita Gupta at TechBooks. I want to thank them both for putting up with my never ending last minute manuscript changes. Also, even though I have never spoken with him, Rich Dombrowski did an excellent job on the chapter and book cover artwork from my flimsy stick drawings.

The text was composed and maintained in LaTeX on a Gateway 2000 Nomad notebook computer running the Linux operating system initiated by Linus Torvalds. Linux is the best thing since sliced (or unsliced) bread and I couldn't have maintained any kind of portability without it. I am indebted to many programmers who have created programs such as Xfig, Ghostview and dvips. All software used to create the book is freely available on the Internet.

Paul A. Fishwick

Contents

1 Introduction   **1**

1.1   Overview   1

1.2   Types of Simulation   2

1.3   Simulation Process   4

     1.3.1   *Phases, 4*
     1.3.2   *Model Engineering, 5*

1.4   Approach   6

     1.4.1   *Philosophy, 6*
     1.4.2   *Comparisons with Other Simulation Methods, 8*
     1.4.3   *For the Student, 9*
     1.4.4   *For the Instructor, 9*

1.5   Role of Simulation   12

1.6   Modeling Example   15

1.7   Object-Oriented Model Design   17

1.8   Book Organization   18

1.9   How and Where to Use This Book   19

1.10   Exercises   20

1.11   Projects   20

    1.11.1   How to Do Re-search, 20
    1.11.2   Where to Find Information for the Projects, 21
    1.11.3   Some Projects to Start Off, 22

1.12   Further Reading   23

1.13   Software   24

    1.13.1   SimPack Simulation Toolkit, 24
    1.13.2   Other Software and Online Simulation Information, 25

2   Foundations   **27**

2.1   Overview   27

2.2   Model Design   29

    2.2.1   Data: The Observables, 29
    2.2.2   Model Components, 40
    2.2.3   Model Definition and Basic Categories, 46

2.3   Serial Model Execution   51

    2.3.1   Time Slicing, 52
    2.3.2   Basic Event Scheduling, 53
    2.3.3   Structured Event Scheduling, 54

2.4   Behavioral Modeling   63

2.5   Financial Applications   64

2.6   Exercises   65

2.7   Projects   67

2.8   Further Reading   67

3   Conceptual Modeling   **69**

3.1   Overview   69

3.2   Text to Object Translation   70

    3.2.1   Nouns and Pronouns, 71

3.2.2   *Adjectives and Adverbs, 72*
3.2.3   *Verbs, 73*
3.2.4   *Prepositions, 73*
3.2.5   *Conjunctions, 73*

3.3   Picture to Object Translation   74

3.3.1   *Sketches: No Explicit Rules, 74*
3.3.2   *Concept Graphs: Some Explicit Rules, 75*
3.3.3   *Schematics: All Explicit Rules, 76*

3.4   Mapping Conceptual Models to C++ Code   76

3.4.1   *Class Structure, 76*
3.4.2   *Methods and Message Passing, 78*

3.5   Exercises   81

3.6   Projects   82

3.7   Further Reading   83

3.8   Software   83

4   **Declarative Modeling**   **85**

4.1   Overview   85

4.2   State-Based Approach   88

4.2.1   *Deterministic Automata, 88*
4.2.2   *Nondeterministic Automata, 99*
4.2.3   *Production-Based Models, 104*

4.3   Event-Based Approach   112

4.3.1   *Finite Event Automata, 112*
4.3.2   *Keyframe Animation, 114*
4.3.3   *Augmented Event Graphs, 116*

4.4   Hybrid State-Event Methods   118

4.4.1   *State-Event Graphs, 118*
4.4.2   *Inventory Control, 120*
4.4.3   *Petri Networks, 123*

4.5   Exercises   135

4.6   Projects   138

4.7   Further Reading   140

4.8   Software   141

5   Functional Modeling   **143**

5.1   Overview   143

5.2   Function-Based Approach   145

    5.2.1   Block-Models, 146
    5.2.2   Automatic Vehicle Control, 153
    5.2.3   Coupled Water Tanks, 157
    5.2.4   Encoding Functional Network Structure, 159
    5.2.5   Digital Logic Circuits, 161
    5.2.6   Queuing Models, 165

5.3   Variable-Based Approach   173

    5.3.1   Signal Flow Graphs, 173
    5.3.2   Kinetic Graphs, 173
    5.3.3   Pulse Processes, 175
    5.3.4   System Dynamics, 176
    5.3.5   Compartmental Modeling, 184

5.4   Exercises   189

5.5   Projects   190

5.6   Further Reading   192

5.7   Software   192

6   Constraint Modeling   **193**

6.1   Overview   193

6.2   Equation-Based Approach   195

    6.2.1   Discrete Delays in C, 195
    6.2.2   Fibonacci Growth, 197
    6.2.3   Difference Equation Coding with Circular Queues, 197
    6.2.4   Canonical Form Difference Equations, 198
    6.2.5   Differential Equation Algorithms, 200
    6.2.6   Predator/Prey Ecological Model, 206
    6.2.8   Harmonic Motion, 211
    6.2.9   Chaotic Behavior: The Lorenz System, 213
    6.2.10   Mathematical Expression Handling, 215
    6.2.11   Delay Differential Equations, 215

6.3   Graph-Based Approach   217

     6.3.1   Electrical Networks, 219
     6.3.2   Bond Graphs, 221

6.4   Exercise   230

6.5   Projects   231

6.6   Further Reading   232

6.7   Software   232

7   Spatial Modeling   **233**

7.1   Overview   233

7.2   Space-Based Approach   234

     7.2.1   Overview, 234
     7.2.2   Percolation, 235
     7.2.3   Cellular Automata, 236
     7.2.4   Ising Models, 244
     7.2.5   Partial Differential Equations, 246

7.3   Entity-Based Approach   252

     7.3.1   Overview, 252
     7.3.2   L-Systems, 253
     7.3.3   Gas Dynamics with Time Slicing, 260
     7.3.4   Gas Dynamics with Event Scheduling, 262
     7.3.5   Planetary Orbital Mechanics, 267
     7.3.6   Particle System Algorithms, 268
     7.3.7   Rigid Body Mechanics, 274

7.4   Exercises   278

7.5   Projects   281

7.6   Further Reading   281

7.7   Software   282

8   Multimodeling   **285**

8.1   Overview   285

8.2   Aggregation and Decomposition   287

8.3   Abstraction and Refinement   288

8.4   Abstraction as Homomorphic Simplification   289

8.5   Refining Aggregate Vectors   294

8.6   Discontinuity and Integration   296

8.7   Refining Declarative and Functional Models   297

8.8   Pendulum Phase Space   301

8.9   The Two-Jug System   303

8.10   The Dining Philosophers Revisited   307

8.11   Industrial Plant Control   308

8.12   Boiling Liquids   311

       8.12.1   Overview, 311
       8.12.2   Two Homogeneous Refinements, 312
       8.12.3   A Heterogeneous Refinement, 314
       8.12.4   Execution Results, 316
       8.12.5   Question Answering and Choice of Model, 318

8.13   Refining Petri Nets   321

       8.13.1   Two Flasks of Boiling Water, 321
       8.13.2   Robot Operations, 325

8.14   Exercises   328

8.15   Projects   331

8.16   Further Reading   331

8.17   Software   332

9   Parallel and Distributed Simulation   333

9.1   Overview   333

9.2   Architecture Styles   334

9.3   Suitable Model Types for Parallelization   336

9.4   Functional Models   338

       9.4.1   Causality, 338
       9.4.2   Synchronous Method, 339
       9.4.3   Conservative Asynchronous Method, 341
       9.4.4   Optimistic Asynchronous Method, 346

9.5   Spatial Models   350

          9.5.1   Space-Based Approach, 350
          9.5.2   Entity-Based Approach, 357

   9.6   Exercises   367

   9.7   Projects   368

   9.8   Further Reading   368

10   SimPack Toolkit   **371**

   10.1   Overview   371

   10.2   Copyright and License   372

   10.3   Event Scheduling   372

   10.4   Declarative Model Simulators   374

          10.4.1   FSA Simulator, 374
          10.4.2   Markov Simulator, 378
          10.4.3   Temporal Logic Simulator, 380
          10.4.4   Timed Petri Net Simulator, 382

   10.5   Functional Model Simulators   383

          10.5.1   Time-Slicing Block Simulator, 383
          10.5.2   Clock, 384
          10.5.3   Digital Logic Network Simulator, 385
          10.5.4   Queuing Model Library, 386
          10.5.5   Grocery Store I, 387
          10.5.6   Grocery Store II, 388
          10.5.7   Single-Server System, 389
          10.5.8   CPU/Disk System, 399
          10.5.9   Sample Traces for the CPU/Disk Problem, 401
          10.5.10 Blocking in Queuing Networks, 401
          10.5.11 General Network Simulator, 401
          10.5.12 Communications Network Simulator, 407
          10.5.13 Pulse Processes, 409
          10.5.14 Optimal Path Route Simulator, 410
          10.5.15 Mini GPSS Compiler, 413
          10.5.16 Xsimcode, 418

   10.6   Constraint Models   420

          10.6.1   Difference Equations, 421
          10.6.2   Ordinary Differential Equations, 423

   10.7   Spatial Models   426

10.7.1   Gas Dynamics, 426
10.7.2   Diffusion, 426

Bibliography   429

Index   441

Chapter 1

Introduction

Simulations can tell us things we do not already know.[1]

1.1 OVERVIEW

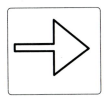

Computer simulation is the discipline of designing a model of an actual or theoretical physical system, executing the model on a digital computer, and analyzing the execution output. Simulation embodies the principle of "learning by doing"—to learn about the system we must first build a model and make it run. The use of simulation is an activity that is as natural as a child who *role plays* with toy objects. To understand reality and all of its complexity, we must build artificial objects and dynamically act out roles with them. Computer simulation is the electronic equivalent of this type of role

---

[1]H. A. Simon, *The Sciences of the Artificial* [207], p. 19.

playing. It is a highly interdisciplinary field since it is widely used in all aspects of industry, government, and academia. One can find the teaching of simulation in almost every academic department from economics and social science to engineering and computer science. Within the overall task of simulation, there are several possible foci: model design, model execution, or model analysis. Our primary goal in this book is to explain the *methods* in system model *design* and *execution* for computer simulation.

Before we embark on a discussion of different model design and simulation approaches, let's consider the word *model*. A model is something that we use in lieu of the real thing in order to understand something about that thing. Mach, in a lecture entitled "The Economical Nature of Physics," states [137] (p. 193):

> The communication of scientific knowledge always involves description, that is, a mimetic reproduction of facts in thought, the object of which is to replace and save the trouble of new experience. Again, to save the labor of instruction and of acquisition, concise, abridged description is sought. This is really all that natural laws are.

While Mach spoke of economy of physical law, it is natural to extend this concept to the broader area of system modeling. Our purpose in creating different model types is to promote economy in our information representation for dynamical systems. We want to understand how time-varying phenomena operate in the real world, and so we *model*.

1.2 TYPES OF SIMULATION

The task of executing simulations provides insight and a deep understanding of physical processes that are being modeled. A simple example of a simulation involves the tossing of a ball into the air. The ball can be said to "simulate" a missile, for instance. That is, by experimenting with throwing balls starting at different initial heights and initial velocity vectors, it can be said that we are simulating the trajectory of a missile. This kind of simulation is known as analog simulation since it involves a physical model of a ball. A *computer* simulation, on the other hand, is a simulation where the model is a computer program. We build a computer-based missile simulation by constructing a program that serves to model the equations of motion for objects in flight. If we regard the actual, physical system under study as being exact, we can create approximations of that model as being at different levels of abstraction. The scale in Fig. 1.1 demonstrates different types of simulation and gives an example of each type in the missile domain. Let's discuss the types in Fig. 1.1. The physical model is the system under study—the missile in flight. That is, we seek to study a missile in flight by creating a simulation of it. One way of simulating this system is to create a smaller (or scaled) version of the system. A bullet is a kind of miniature missile—therefore, we can study missiles in flight by shooting bullets instead. Bullets are, after all, much cheaper than missiles when it comes to experimentation. We can then answer questions about missile simulation by following this general procedure:

1. Obtain the question to be answered or the problem to be solved in a given domain—such as ballistics. For instance, we can ask "How far does a 120-mm missile go horizontally when the gun turret is placed 30 feet above the ground and aimed at a 40 degree elevation from the horizontal?"

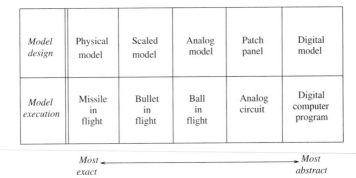

Model design	Physical model	Scaled model	Analog model	Patch panel	Digital model
Model execution	Missile in flight	Bullet in flight	Ball in flight	Analog circuit	Digital computer program

Most ◄──────────────────────────────────────► Most
exact   abstract

Figure 1.1   Various missile models up to the physical object.

2. Translate the question and problem into the simulation model. We then translate the above problem into a small-scale model using a bullet fired from a rifle. This involves correctly scaling down all aspects of the problem. This is not trivial, and many physical phenomena are not invariant to the scaling operation.

3. Fire the rifle.

4. Measure the horizontal distance of the bullet after it has hit the target.

5. Scale the answer back to the missile domain.

6. Provide the answer to the original question.

These six steps reflect the general method of simulation as applied to problems of *prediction*. The analog model takes some liberties with respect to modeling the physical system. Scale invariance is one such liberty. For missiles, any object can really be used to simulate a missile simply by throwing that object. Therefore, a tennis ball might serve as a reasonable model. Already we can see the need for picking a model that serves to correctly *validate* the real system behavior. Balls might not be so good in the end, since balls may have greater drag than missiles. Also, there are many other factors such as the amount of propellant to use, that will help us to faithfully model the original system.

   The first electronic computers were *analog* as opposed to *digital*. The primary use of an analog computer is to simulate a physical system that can be modeled as a set of differential equations. Electronic devices such as operational amplifiers—when used in conjunction with resistors and capacitors—function as integrators and differentiators. This means that one can simulate a set of equations by "patching" a panel (i.e., creating a program) on an analog computer. The most abstract type of simulation model is the digital model in which an computer program permits us to execute the model. Our interest is in digital models since they are the most flexible and can be programmed easily.

   In this text, we will often use the term *simulation* synonymously with computer simulation. Simulation is an applied methodology in that we describe the behavior of complex systems using mathematical or symbolic models. Our model serves as a "theory" or "hypothesis" for how a system really behaves over time. We may also simulate nonreal systems by varying parameters, initial conditions, and assumptions about our model. Modeling missile flight on a planet other than Earth, for instance, would involve a different gravitational constant and different parameters for the planet's atmosphere. Rules can be changed so that, for instance, trees grow sideways or on top of clouds. Fictional simulations are very popular in the form of role-playing simulation games, and they permit the simulation user to learn about a new environment by exploring it interactively. The concept of *learning* is paramount in simulation. We learn about an environment in an extremely effective way and modify rules while seeing the effects of our interaction.

1.3 SIMULATION PROCESS

1.3.1 Phases

Simulation is a tightly coupled and iterative three component process composed of 1) *model design,* 2) *model execution,* and 3) *execution analysis* as shown in Fig. 1.2 along with the relevant subareas and book chapter numbers. The bold lines in Fig. 1.2 are to show our emphasis in this text: *model design* and *model execution.* The third area of *execution analysis* already has broad coverage, and references to good textbooks will be mentioned shortly. While simulation is normally broken down into even more detailed subcomponents, the three components in Fig. 1.2 characterize the research field and the way that simulation is taught and used in science. The first step to building a simulation model of a real system is to gather data associated with that system. Data can be in either symbolic or numeric form. Typically, numerical data are obtained using either physical sensors (such as a thermocouple or stress gauge) or human sensors (as with eyesight or tactile feedback). Nominal data are obtained from humans using interviewing methods or knowledge acquisition techniques developed to obtain qualitative knowledge. From the data and knowledge of past experiments with similar systems, we formulate a model. Models can be created without knowledge of the data; however, we should eventually be concerned with deciding on the accuracy of our model. The model will usually contain parameters which must be estimated (i.e., initialized to some specific value). The model is usually termed "mathematical" in that we aim to specify any model clearly. Ambiguity is best removed using formal semantics. Models

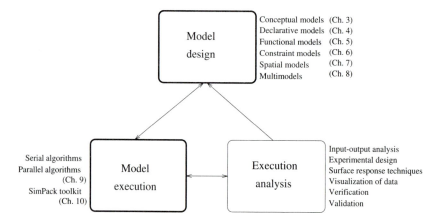

Figure 1.2   The study of computer simulation.

must be converted to algorithms to run on a digital computer. Verification is the process of making sure that the written computer program corresponds precisely to the mathematical model. Validation, the next step, is the process of making sure that the model's output accurately reflects the behavioral relationships present within the original real system data. The model, when simulated, should be able to produce the same sorts of data and input-output[2] relationships that were gathered initially. Next, we perform tests on the data generated from the model by running specific analysis such as Fourier analysis (to gain an understanding of frequency response) or statistical analysis (to gain an understanding of the qualitative nature of the data, such as dispersion and clustering). The most basic type of analysis is simply to "look" at the data and to make inferences based on this examination.[3] For instance, if the output data specifies a time-dependent state variable value that demonstrates a damped oscillation, we deduce that the system will gradually achieve a steady state after fluctuating for some initial time period.

1.3.2 Model Engineering

Perhaps the hardest general problem in simulation is determining the exact method that one should use to create a model. After all, where does one begin? Just as the discipline of software engineering has emerged to address this question for software, in general, modelers also have a need to explore similar issues: *How do we engineer models?* While there are many modeling techniques for simulation, we are often in a quandary as to which model technique to use and under what conditions we should use it. Our approach is depicted in Fig. 1.3 along with the associated chapter references where each modeling method is defined. Here, we have a rough guideline to building and executing models. First start with a concept model of whatever dynamic system is being investigated. Break the system into a hierarchy of abstractions (multimodel) and then choose models to represent those abstraction levels. Some levels may be declarative or functional in nature, and some might combine these elements (on the same level) using the appropriate hybrid modeling technique. Other model abstractions may be of a spatial nature. The final solution to a model will be a multimodel of the system since no one model type will be sufficient to describe a system except in only the most elementary circumstances. A multimodel is a collection of individual models—each characterizing an abstraction level—connected together in a seamless fashion to promote level traversal. After the model has been designed, it needs to be executed. Serial model execution methods are discussed in each of the model design chapters. That is, the way that one executes a model depends on the type of model. Parallel simulation methods, while generally applicable and far reaching, are still used on special machine architectures (of which there are numerous examples). In Ch. 9 we cover several key algorithmic approaches. Finally, as an appendix, we discuss the SimPack toolkit, which provides a software laboratory for simulation experimentation.

---

[2]Behavioral model of system.

[3]Validating a model through an informal means by making sure that the output "looks right" is called *face validation*. This is performed routinely in computer animation since, here, detailed analysis is less important than an entertaining video sequence.

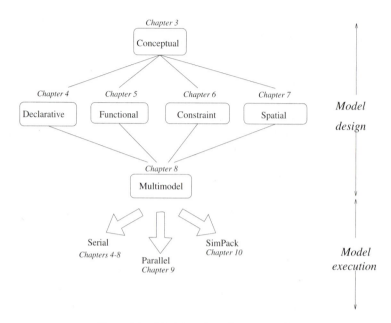

Figure 1.3   Model engineering progression.

1.4 APPROACH

1.4.1 Philosophy

The purpose of this book is to discuss different type of model designs and algorithms for simulating those designs. It was sometimes difficult to decide on what to include and what to leave out of the text. I chose to discuss those model types that have gained posterity after decades of employment. While some books focus on one *monolithic* modeling technique, I believe that no single technique by itself can be used to model a system successfully, although it is certainly true that many modeling methods are *powerful enough*[4] to model *any* system. The issue in modeling is not complexity as much as it is expressiveness and orientation. Automata, for instance, are useful for representing high-level discrete control of a system, or for representing a system from an abstract perspective. Each model type is like a different-colored lens. When you look at the world with a rose-colored lens (a Petri net, for instance) you can model in a certain kind of way. The rose-colored lens, though, might block out an understanding of a subsystem that is better understood with a green lens. It is not enough to learn one method of modeling any more than there can be one way to model with clay or paint a picture. The area of modeling is diverse and the student should become familiar with key classes of models. Considering the variety of models, we introduce Eq. (1.1) as an example model:

$$X(t) = X(t-1) + U(t) \tag{1.1}$$

---

[4]"Power" is synonymous with computational expressiveness so as to satisfy Turing machine equivalence, for instance.

Now, consider the model in Fig. 1.4. This model and the equation model in (1.1) provide exactly the same semantics. That is, either model can be translated into the other model. However, these are *different* models because each of them reflects a different way of looking at the physical system. The way that you like to view systems will influence your decision as to which model to choose. A strictly formal view of this situation would dictate that the model in Fig. 1.4 is merely a "diagram" representing the model in Eq. (1.1) since the use of algebra and difference equations predates—and is more refined than—the graphical block model. However, this view is narrow since the diagram in Fig. 1.4 uses a graphical language that is common to engineers who model machines and devices. The graphical model is far more than a mere diagram. The analyst *thinks* that using this language and this action makes the block model technique more valuable than the equational technique for systems where blocks map structurally to physical things. These represent two different ways of reasoning about systems. Block models are convenient for representing systems with distinct human-made physical components which interact with each other. If the physical system being modeled is an electronic component creating discrete delays, then the block model in Fig. 1.4 is appealing since the rectangular block is homologous to a component or integrated circuit performing the delay. Conversely, if the system being modeled is the population growth of ants over a short period of time, the equational model in (1.1) might be more appropriate since there is no clear structural analogy between an ant growth process and a block network.

   Let's specify the differences between *model design, model specification,* and *model execution*. A model is a high-level specification. The term *specification* is used most often to describe a formal methodology often grounded in predicate logic or a mathematical theory. Models are what scientists use to communicate about system design. The semantics of a model are assumed to be known and common when scientists employ common model types. For instance, if I create a model of a Petri net, I do not write down the semantics because it should be clear to my audience how Petri nets are executed. Even though models are the key to successful systems thinking and understanding, one must also have a *rule book* in case there are disagreements or questions about how transitions fire and how time is incremented. The rule book is where specifications come into play. Even though as modelers we do not talk in terms of detailed specifications (due to an adherence to Mach's principle of economy in modeling), it is vital to have such a specification in case arguments arise as to the semantics of certain models. Specifications can be written in many ways; our method in the book will be to use pseudocode in defining algorithms. The algorithms formally specify how the models behave and are simulated over time. Model execution is different from model design because one model design may be executed in many different

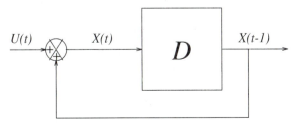

Figure 1.4   Block model with positive feedback.

ways. Consider a queuing model. This can be executed using a fixed-time quantum, an event scheduling approach, or a distributed simulation approach over multiple processors.

I have written the text while including certain commonly used components:

1. *Structure.* I have made an attempt to create a taxonomy for model design and execution that transcends the usual discipline-specific barriers which cause our understanding of systems to become fragmented.

2. *Examples and Figures.* There are numerous examples in the text because I am a believer in the phrase "a book cannot have too many examples." The same can be said of figures. As educators, we must sometimes be reminded that the purpose of all writing—for research, instruction, or otherwise—is to inform and educate. The use of multiple knowledge cues is, therefore, critical to a proper understanding of scientific material.

3. *Formalism and Density.* I have made an attempt to mix formalism with modeling practice with an emphasis on the latter. This mixture is difficult to create because, on the one hand, there is a need to specify unambiguously what one is trying to say and, on the other, one needs to engage the reader so that a deep understanding takes place. I was trying to avoid writing either a book of prose or a reference encyclopedia.

4. *Algorithms and Code.* Algorithms, when presented in a kind of pseudocode style, specify a procedure necessary to achieve a result; code unambiguously implements an algorithm and serves as the "proof" (at least as far as simulation is concerned). There are no existing standards for presenting pseudocode, so I opted to use fragmented English and commonly used words such as "while" and "if-then" instead of limiting the pseudocode to describing a more formal manipulation of variables. I did this since the actual code implementing the pseudocode is available in the *SimPack* toolkit described in Ch. 10.

1.4.2 Comparisons with Other Simulation Methods

This book stresses a taxonomy based on the way that scientists think and do modeling. There are many existing taxonomies in simulation; however, I found them to be incomplete. First, it is rare to find books that have coverage of both discrete and continuous modeling approaches, let alone describing how to fit them together. I feel that discrete and continuous phenomena should be seamlessly integrated in any discussion of modeling. Whereas continuous and discrete methods in mathematics are generally covered under separate umbrellas, many system *modeling* methods are invariant with respect to issues of the continuous nature of the input signal. The relationship between queuing models and functional block models is one such example: the way of thinking about these models (and simulation algorithms to support execution of the models) is no different—there are flows/signals into and out of boxes often with a transformation of local object state. Therefore, we try to avoid terms such as "discrete event simulation" and "continuous time simulation." Instead, we have distinctions according to the modeling flavor: conceptual, declarative, functional, constraint, and spatial. In addition, we have an approach to glue these types together: multimodeling. These distinctions cover discrete and continuous components while focusing on the kind of classifications that modelers use. Some taxonomies cover a very narrow class of model and are consequently inappropriate for a general theory of model design. Taxonomies based on "discrete event" modeling (process, entity, or activity approaches) fall into this category. Other approaches to categorizing models which are based on the complexity of system definition (1) start with behavioral systems; (2) introduce the concept of state, then system; and (3) proceed in this fashion to create a hierarchy of modeling. While this is mathematically pleasing and well-structured, there are two problems with it: 1) scientists who model do not think in terms of system complexity and so they would continually have to map between the world of modeling and formal definitions of models, and 2) apparent system complexity based on components can be deceiving.[5] Even though we stress model design rather than formal specification, we demonstrate the relationship between model types and the canonical system definition (Sec. 2.2.3). In general, formalism and modeling work together at *different levels of system description*—the models being what analysts use to think about systems, and the formalisms serving the need for more unambiguously describing the semantics of models.

1.4.3 For the Student

As the reader of this book, what will you gain from reading it? You will obtain a thorough treatment of model design and execution for simulation. In the coming decades, computer simulation will have a tremendous influence on all facets of science and engineering since it is often too costly or prohibitive to build real physical systems. At the very least, physical systems will be augmented with simulations to help you to make decisions and pose hypothetical scenarios. A key activity of applied scientists is *to model* some aspect of reality, and the computer plays a central role in modeling. For this reason, most employers will want students who have a strong expertise in simulation. Computer simulation will play a central role in computer science curricula, as technology makes it easier to create digital worlds in which we immerse ourselves. Already, the computer science community has strong movements in computational science, high-performance computing, and experimental computer science. All of these areas rely upon computer simulation as a fundamental methodology. Through reading this book, and performing exercises, the student who majors in a field such as computer science, systems engineering, or mathematics will come to understand simulation methodology and how it is applied *broadly* to solve real-world problems. For students in a specific application area such as mechanical engineering, ecology, or biomedicine, there will be many techniques in the book that are not normally associated with your application field. By studying alternative approaches, you will develop a broader information base about modeling.

1.4.4 For the Instructor

Since there are several books on computer simulation and systems modeling, it is natural for the reader, and especially the instructor, to question the utility of this book in relation to

---

[5] An example of this deception is found when studying artificial neural networks or fuzzy logic controllers. Both of these modeling approaches are *behavioral* in that they represent input-output associations without referring to explicit changes in state; however, both methods are quite powerful and are being employed increasingly in system design and control.

other simulation books. I will answer this question by highlighting the following differences between this book and other available simulation texts:

1. *Movement toward Heterogeneity in Modeling.* Most modeling approaches that people espouse are *monolithic* in nature. That is, the people advocating a particular modeling approach stay with that approach; if the approach is found too limiting, an extension is made to accommodate the imperfection. The structure of most simulation languages falls into this category; you either *buy* the whole methodology or else use something completely different. This approach is a little like settling on brick as the only possible method of building a house. All houses are constructed from a variety of materials, including building blocks such as bricks, wood, concrete slab, and metal frame. The same is true for modeling. It is better to choose a variety of well-utilized and proven modeling methods and then search for ways to glue them together to yield a *multimodel* rather than always to view the world to be modeled through a single-model colored lens perspective. Certain modeling approaches are appropriate for certain abstraction levels, and there is actually little to be gained by trying to force every analyst into using modeling method XYZ. In a more general sense, the approach to multimodeling parallels the ongoing distributed, heterogeneous approach to network computing: learning how to connect and interface distinct systems.

2. *Unifying Discrete Event and Continuous Modeling.* The only reason why the concepts of "discrete" and "continuous" have separate treatments in the mathematical literature is because each approach comes with a set of static rules and methods specific to each class. Such rules are commonplace in static, noniterative, approaches. With simulation these differences are less pronounced, and a comprehensive simulation modeling textbook should focus on both discrete and continuous models as integrated facets. I cover both in a seamless fashion. The strength of this approach is seen in Ch. 8, where complex models must contain both discrete and continuous components, often representing different levels of overall model abstraction. The more commonly referenced method of *combined modeling* is viewed as a special case of *multimodeling* where two specific model types (discrete and continuous) are integrated.

3. *New Taxonomy of Models.* Many disciplines require models to represent dynamical systems. While there is some agreement in modeling types in software engineering and artificial intelligence, in computer simulation we are split. The use of *world views,* for instance, provides a good taxonomy for discrete event orientation but is not general enough to cover continuous and spatial models. Our approach attempts to create a taxonomy that better reflects a more global view of modeling. *Process orientation* is seen as representing a functional modeling approach; *activity scanning* is representative of a declarative approach; and *event scheduling* is a phenomenon that we characterize as having both a sense of model execution (i.e., using event set scheduling algorithms) and a sense of model design (using event-oriented models such as event graphs or a logic of events). Modeling within computer simulation has tended to be divided along the lines of analysis instead of design, hence a breakdown of models that involve discrete events or continuous behavior. In contrast to this, we focus on model design since design should drive the way that we represent a model for a system, execute that model, and then finally analyze the model's output. Functional models are useful for modeling systems composed of distinct physical objects where there is directionality in the flow among the objects. It does not matter if the thing that flows is discrete or continuous as far as the model design is concerned, and this is why we must not make divisions between "discrete event" and "continuous" models. It is not the model that is discrete event in form; it is the execution method utilized once the model has been designed. A perceived split between discrete event and continuous modeling methods exists only because of our persistence in thinking about modeling from the perspective of how we are required to analyze and execute models before they are fully designed. Consider a water pipe which is segmented into strips, where each strip is joined by a water processing box. Each box may serve as a capacitor for water, or it may contain a filter or pump. This scenario suggests the use of a functional model where each box acts as a function processing water as it flows.[6] It makes no difference to the model design whether we decide to execute this model by treating water in discrete chunks (i.e., using discrete event simulation) or whether we use a continuous technique. The high-level design is the same in either case: a functional model. As it turns out, we may choose to subdefine the functional model, at the top abstraction level, into different types of submodels depending on the granularity desired but the top-level model remains functional in design. We begin with the design of the model (and multimodel) and later use execution methods that are appropriate for the model. For multimodels, several execution algorithms must be synchronized so that all abstraction levels are executed properly.

4. *Focus on Modeling.* The simulation community has produced some excellent textbooks on the *analysis aspect* of computer simulations. While we tread lightly in this area, our focus is on the modeling problem, or more specifically, engineering models and executing them. For instructors with an interest in teaching a combination of modeling and analysis, I recommend using this text in conjunction with a text that focuses on analysis, such as:

> *Simulation Modeling and Analysis,* Averill Law and David Kelton, McGraw-Hill, Second Edition, 1991, or
>
> *Discrete Event Simulation,* Jerry Banks and John Carson, Prentice Hall, 1984.

Also, I do not focus on one specific application area. Instead, I am interested in how widely different disciplines such as economics, computer system architecture, microbiology, and physics relate to one another under the common umbrella of system simulation and analysis. The emphasis on methodology does not imply that we will ignore applications. I present many different application examples so that the methodological foundations are placed into perspective.

5. *Focus on Algorithms and Execution.* In modeling systems, one must know how to write computer programs that execute models in serial and parallel. There are many algorithms and programs that are of key interest; however, there are few simulation texts that include this vital component. In the case of parallel algorithms for simulation, the situation is much worse. Despite a plethora of journal literature on this subject, there is still no textbook coverage. This situation will surely change as more simulation texts focusing on algorithms

---

[6]A spatial model might also be appropriate if we do not wish to abstract out pieces of the pipe as transfer functions.

and model design come forth. The text by M. H. MacDougall (*Simulating Computer Systems,* MIT Press, 1987) is an exception in that the author presents C code (SMPL) along with problems in modeling computer-based systems. Other exceptions include texts by W. Kreutzer (*System Simulation: Programming Languages and Styles,* Addison Wesley, 1986) and by R. Davies and R. O'Keefe (*Simulation Modelling with Pascal,* Prentice Hall, 1989). Note that while many texts include references to well-known simulation languages such as GPSS, SLAM, and Simscript II, there is little formal discussion of exactly how to construct such a language using freely available software tools. This book is meant to satisfy the simulation reader who wants to know about the implementation knowledge instead of only a description of a simulation language. Many algorithms in this text represent programs or code libraries in C and C++, all located within a freely available toolkit called *SimPack,* which is available over the Internet (refer to Sec. 1.13 and Ch. 10). SimPack contains code in C and C++ for executing many different model types. Without accessible simulation software, the state of the art of model engineering will become stagnant. Fortunately, many freely available simulation tools are becoming available on a wide scale. It is vital that students learn the details about simulation algorithms and code, not just "how to use simulation language XYZ."

These reasons point out the key differences between my approach and many existing approaches. With the advent of inexpensive hardware on which to execute models, system simulation is becoming the *modus operandi* for computational science. With this new responsibility, simulationists must present modeling solutions that are generally applicable and can handle very large scale system representations. This book is a step in achieving this goal.

1.5 ROLE OF SIMULATION

Science proceeds with spurts of intense research activity often taking the form of a new thrust with a new name. The "tug of war" between the traditional approaches and the new approaches takes place in conference sessions and journal articles. Even though the term "simulation" is old, it reflects the way in which a good deal of science will be done in the next century. Scientists will perform computer experiments in addition to creating scientific hypotheses and performing experiments on empirically obtained data. When one says *simulation,* many different ideas may come to mind, depending on the discipline. I suggest that simulation represents a fundamental discipline in its own right regardless of the specific application. Let's take some of the key focus areas of the past decade and view them in light of computer simulation. Figure 1.5 illustrates the central role that simulation plays with regard to several other areas described below. Basically, my goal here is to shed light on the potential or existing role that simulation plays in each area:

1. *Computational Science.* Simulation is the *lingua franca* of computational science. Computational science involves the use of computers (especially "supercomputers") for visualization and simulation of complex and large-scale phenomena. While it is the increasing throughput of ever-faster computers that aid us in studying complex phenomena, it is the discipline of simulation that provides the conceptual basis for such study. The

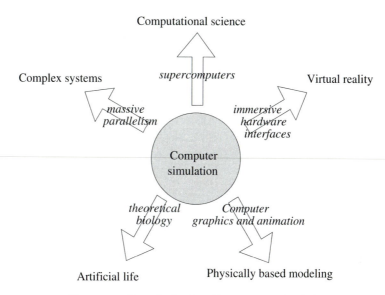

Figure 1.5   Central role played by computer simulation.

emphasis on supercomputing is really a focus on simulation, which happens to require fast computing hardware. Studies involving N-body simulations, molecular dynamics, weather prediction, and finite element analysis are often quoted within the general thrust of computational science. Hopefully, we will see curricula and academic departments devoted to computational science, and simulation will play a central role in these formations. Just as computer science has its basis in computability theory, then computational science has as its basis computer simulation.

   2. *Chaos and Complex Systems.* The idea that one can observe complexity within a structurally simple deterministic model is of fundamental technical interest and concern. Furthermore, most models exhibiting chaotic behavior do not yield to static forms of analysis; simulation must be used to provide a more detailed view of the system in question. Qualitative topological phase space features of linear systems may be determined analytically, but simulation must be used for nonlinear systems. Iterative methods must work in tandem with determining special features such as fixed points, eigenvalues, and hyperbolic points. In the broad view, simulation languages of the future will work closely with symbolic manipulation routines and languages so that a system can be analyzed with every available tool (both analytic and iterative). Combined analytic-iterative approaches in simulation fall under the heading of *hybrid simulation.*

   3. *Virtual Reality.* The push within virtual reality research is to *immerse* the analyst within the simulated world through the use of devices such as head-mounted displays, data gloves, 6 degree of freedom sensors, and force-feedback elements. Although virtual reality is often seen as being synonymous with human-machine hardware interfaces, the technology must incorporate methods for building digital (or virtual) worlds in order to be effective. The construction of dynamic worlds is precisely what computer simulation is all about, so there will be a marriage of these two fields as both progress. This union will yield better simulation environments and easier ways for analysts to grasp the simulation-produced information.

4. *Artificial Life.* The relatively new area of *artificial life* is an offshoot of computational science in that it challenges how we define the term *experiment.* An experiment in artificial life is one where a computer program is written to simulate artificial life forms, often carrying along metaphors such as genetic reproduction and mutation. A basic issue arises with the study of artificial life and its relationship to simulation. First, some quick background material—most simulations are *validated* by comparing real-world behavior against that of the model execution output. This comparison is statistically based where we claim, within a confidence interval, the fidelity of the model. But what if we are simulating non-real-world behavior? Can it be said that our simulation is reflective of a scientifically sound research method? For the same reason that we pick up an insect and study its natural behavior, we should also acknowledge that the computer can act as a generator of life, diversity and complex behavior. There are those who argue that the two instances are different—that is, the insect was not engineered by a human. However, when we construct a complex model of artificial behavior by engineering it, are we completely confident in our ability to infer the insect's behavior just because we engineered the computer on which the simulation model is executed? After all, natural selection is a mechanical process as well. Clearly there is much room for debate on this topic; however, we should explore every avenue with regard to artificial life based on the philosophy of computational science.

5. *Physically Based Modeling and Computer Animation.* Within computer graphics, there has been a noticeable push in the direction of physically based modeling.[7] In actuality, the movement is more due to an increased attention on *system modeling,* per se, rather than on physically based modeling. To understand this, consider the way the computer animations have been performed in the past. The first computer animations were drawn on individual sheets of paper or celluloid. When a series of sheets is transferred to film and run at a speed of 24 frames per second, an illusion of motion is perceived. From a simulation perspective, the sheets reflect a declarative type of model that is similar to a finite state automaton—one state per sheet. Therefore, early animation models were finite state automata. The next generation of animation programs allowed users to create in-between states automatically by specifying certain key states (key frames) and then using geometric smoothing methods such as a Bezier or cubic spline. Now, given this knowledge, we can see how progress has been made in computer animation. Specifically, we can create more cohesive models by basing the animation on more complex model types—such as event modeling, functional modeling, or constraint modeling. The term "physically based" normally refers to constraint-based, equational models derived from physical laws; however, the movement called physically based modeling is a step in the direction of using simulation models to replace the older automata-type models—inherent in frame-based simulation—with more comprehensive system models.

---

[7]Application of physical principles to computer animation.

1.6 MODELING EXAMPLE

To get an idea of what modeling entails, we consider a system involving waiting in a line for service. One way of thinking about a queuing system is from a *declarative* perspective. Using a declarative approach, we specify the system *state* and then make transitions or rules to tell us what the state looks like at a later time. The given states represent the number of entities in the system (either located within the queue or being served); arrivals and departures serve as external and internal events that cause the system to change state. Another declarative approach is to focus on the dual of state (event). An event-based approach defines *arrival* and *departure* events using a *scheduling* or *causal* arc. The events *arrival* and *departure* are actually partitions of event space where *arrival* refers to the set of all arrivals and *departure,* the set of all departures. The event relation is strictly from the arrival of token $i(a_i)$ to the departure of token $i(d_i)$ since there is no such general relationship among events whose token identifiers differ. Figure 1.6 shows four classes of model for the single-server queue. Figure 1.6(a) represents a first crack at a model for the system—a conceptual model. The conceptual model is quite informal; however, it communicates the basic nature of the queuing process. Also, the model provides us with a vocabulary for the system, although it is too ambiguous for simulation. The declarative model [Fig. 1.6(b)] defines four possible states for the system, where a state is identified by the number of people waiting in line. At first there are no people waiting, so the system remains in state s_0. When a person arrives into the system, the system moves to state s_1. If someone then departs the system immediately, the state transition is back to s_0. The event which causes the change in state for this system is an *arrival* or a *departure*. The arrival is an external input to the system (external event) and the departure is an internal input to the system (internal event). External events come from outside the system, whereas internal events come up from lower abstraction levels. This type of model is a finite state automaton. We can also create a change in state by placing probability values on the arcs (i.e., instead of events); however, this type of model would not involve any inputs. For the simulation to be of interest, we have to specify how long the system spends in each state (this temporal specification is easily made part of the model by putting it inside each state). Another, more compact way of declarative modeling is to specify that an arc exists between state i and state $i + 1$ when $i \in \{0, 1, 2\}$. The returning arcs are defined in a similar manner. This is the production rule (or logic) approach where the state specification need not be constant, but can use patterns.[8] A Prolog-like logic rule can be created: `state(A)` `:- state(B), 0 <= B < 3, A is B+1`. The *implication* symbol `:-` is treated as a state transition for the purpose of simulation. Figure 1.6(c) illustrates the event-based approach to declarative modeling. Events and states are dual concepts.

Instead of viewing the system from the perspective of state or event, we can treat the server and queue as a kind of "black box" [see Fig. 1.6(d)] which takes as input a discrete signal defined by customer arrivals over time and produces as output a delayed signal of customer departures. The delay is what causes the overall simulation time, for a functional model, to advance; simple delays in functional block models are either discrete (i.e., as performed when solving a difference equation) or continuous (i.e., when solving

---

[8]The patterns are shorthand for functional mappings—mapping a pattern to a subset of state space.

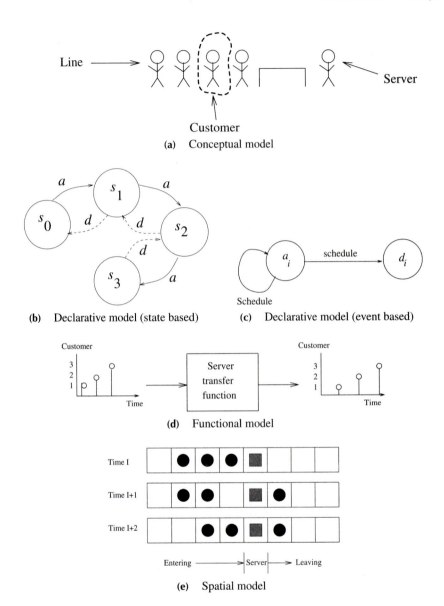

(a)   Conceptual model

(b)   Declarative model (state based)   (c)   Declarative model (event based)

(d)   Functional model

(e)   Spatial model

Figure 1.6   Different ways of modeling a single-server queue.

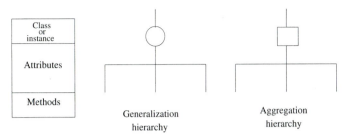

Figure 1.7   Object-oriented model design notation.

a differential equation). The functionality of models such as Fig. 1.6(b) and (d) is easily integrated—using declarative models inside functional blocks. The model in Fig. 1.6(e) represents yet another different way of modeling. Figure 1.6(e) is a spatial (or geometrical) view of the system as if looking down on the queue from overhead; entities are drawn as filled circles and the server is a filled box. When there is a line in front of the server, there are two phases associated with the behavior of the system: (1) the server processes the first customer, and (2) the remaining customers move up one place in line.

1.7 OBJECT-ORIENTED MODEL DESIGN

I adopt an object-oriented (OO) perspective to modeling by first introducing the fundamental model categories, listing the key model types within each category and finally specifying how to proceed from real-world objects to *programming objects*. Sometimes, the main contribution of OO methodology is seen as being *code reuse* or *inheritance*. For simulation, the key contribution is the *mapping* between real world and digital world. The fact that one can map physical objects to computer-based objects is the most powerful concept for simulation. We will use a simple graphical notation without getting carried away with graphical OO symbology. Figure 1.7 specifies the notation to define a *class hierarchy*.[9] A *class hierarchy* is composed of classes only. Classes are connected to form hierarchical structures using two possible connectors: *generalization* or *aggregation*. An *object* is an instance of a class. Each object will have *attributes* and *methods*. Attributes refer to properties of an object while methods refer to the functions and procedures that operate upon the attributes. Attributes can be one of two types: *static* and *dynamic*. A static attribute is one that does not change over time (i.e., such as the color of a material object), while a dynamic attribute generally represents a component of the state vector associated with the object. Simulation is chiefly concerned with dynamic attributes but we want to set the stage for having an OO database (i.e., persistent objects) containing all kinds of information about the system. Such a database allows simulation to be fully integrated with other activities, such as explanation, diagnosis, and information retrieval. A *model* is created by constructing objects and connecting them through message passing.

   When we label an object, we can specify the object name `engine1` or we can specify its class so that we know how it was constructed: `engine(engine1)`. In general,

---

[9]Class hierarchy is also called a *class library,* especially from an implementation point of view.

we will use the notation `class(object).method()` when we want to express all three concepts: class, object, and method. If we do not explicitly denote the class, we will use just `object.method()`. This is close to C++; however, in C++ there is just `class::method`; the tie between an object and the class from which it was constructed is specified in the program text using a *constructor*. Message passing deserves some discussion within the context of the C++ programming language. Objects communicate by sending messages to one another. This approach can be used to handle both discrete and continuous signals. Continuous signals are discretized first. Discrete signals are mapped into the OO paradigm by having each message hold a time-stamp along with information. Some OO languages provide the illusion for message passing; however, in C++ message passing is accomplished in the following manner. Suppose that object A sends messages to object B. The output stream of object A is directed to the input stream of object B. Sending a message in C++ is encoded as B selecting the appropriate method based on input that it receives from A. One could imagine several possible ways that B could process such a "message." Inside B, there can be a method called `solve()` which takes the current state (an attribute of B) and the input from A and generates a new state for B. The generic linear formulation $\dot{\mathbf{X}} = M\mathbf{X} + N\mathbf{U}$ expresses a possible way that `solve()` could work. `solve()` might also use a switch (case) statement based on the input (U) to determine how to update the state (X) of B. In no instance should one object update the state of another object. Each object updates its own state.

We treat the object-oriented approach as a lowest common denominator among all presented model types. It provides an infrastructure in which models can be coded. We do not suggest that all models begin with an object-oriented design, but that the object-based design be used to relate different modeling methods into a more unified representation. Sometimes it may be convenient, for instance, to model part of a large system first as a functional block network or Petri net and then later consider how these models can be unified within an object-oriented methodology.

1.8 BOOK ORGANIZATION

We begin by discussing the nature of data obtained from the real-world and basic components that are part of every model. Chapter 2 deals with a foundation for further discussion on model design and execution. This will provide the reader with a basic systems vocabulary. We begin with the fundamentals of model design, including ways of expressing data (and its extension in the form of probability density functions and fuzzy sets), and then finish with key approaches to sequential model execution. The rest of the book is a discussion on modeling techniques. The discussion of each model is broken into two parts: design and execution. Occasionally, we will elaborate on analytic features of a model and its output, but this is not of primary concern in this text. Chapter 3 begins the modeling enterprise with the concept model. A concept model defines a physical system at a very high abstraction level. We employ the object-oriented approach when building concept models—a knowledge base is first constructed before we can design dynamic models. Chapter 4 discusses declarative models, that is, models where the design concentrates on the form of the current system state and a subsequent system state after a state transition occurs. Declarative models deemphasize the actual functions that cause the state changes.

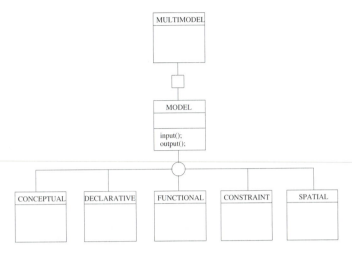

Figure 1.8   Model hierarchy.

The functional approach (Ch. 5), on the other hand, focuses on functions that transform input into output while, normally, keeping track of a state vector along the way. Constraint models (Ch. 6) present models as collections of state variables that are related to one another in a constraint network. The usual form of a constraint network is a set of equations; however, there are other graph-based forms such as electrical networks and bond graphs. Spatial models (Ch. 7) focus on relating geometry directly to the dynamic model. Such models divide the world into many small pieces which interact in a geometrical fashion. In Ch. 8, I discuss the definition of "multimodel," which is a hierarchy or network of models. This chapter serves to unify all other model design chapters since multimodels are composed of individual models whose types may be functional, declarative, spatial, or constraint. In Ch. 9, I present parallel and distributed simulation algorithms and approaches for batch simulations and distributed simulations where humans are part of the digital world interaction loop. Chapter 10 serves as an appendix for discussing the available SimPack code written in C and C++. Figure 1.8 displays the model hierarchy representing each model design chapter. One creates a multimodel which is composed of models which are of specific types. For instance, an FSA-controlled multimodel might be composed of a finite state automaton (FSA) at the high level of abstraction and a functional block model at the low level. The FSA is a subtype of declarative model and the block model is a subtype of functional model.

1.9 HOW AND WHERE TO USE THIS BOOK

This book is meant to satisfy a wide range of readers interested in simulation model design and execution. In this sense, the book can be used as a professional reference for those having the need to model systems, and by students primarily at the upper-level undergraduate or introductory graduate level. Earlier drafts of this text have been used at the University of Florida in our computer simulation classes for the past four years. The book will appeal to those teaching simulation courses whether in engineering, computer science, or any other science where system model design and computer experimentation is important. University departments that have traditionally offered simulation instruction include (1) computer science, (2) industrial and systems engineering, (3) statistics, and (4) decision and information sciences. In places where *analysis* is of primary import, the book should be used in conjunction with one of the several available aforementioned texts which focus on analysis. This will give a broader view of simulation but will also require more lecture material and time.

1.10 EXERCISES

In all chapter exercises where you are asked to design a model, you should clearly define objects and classes, while detailing the design. This is especially important when you design multimodels, since there can be many refining levels for model components.

1. Where have you heard the term "simulation" in your everyday experience? Most have seen television commercials or advertisements with the statement "this is a simulation" or "simulated XYZ." Can you identify others?

2. Consider the following real-world systems, and state what a possible *analog* or *scaled* model would look like.
   (a) A grocery store checkout counter
   (b) The growth of a tree
   (c) A telephone conversation
   (d) A sailing ship in the ocean
   (e) A waterfall onto rocks

3. What differentiates a simulation program from an ordinary computer program?

4. Make the argument that simulation is a third way of "doing science." Consider the first two ways as "theory" and "experiment." Is this a valid viewpoint? Label the pros and cons.

5. Discuss the notion of a physical item being a "simulation of itself." Does this have any meaning? Can you imagine any examples in the real world where this is true? Is it useful for a device to simulate itself?

6. Under what circumstances could an algorithm for sorting a set of numbers be considered a computer simulation? That is—does (or could) the procedure of sorting reflect physical behavior?

7. How would you determine how detailed your simulation should be? What criteria would you use? Consider the missile simulation as one example and invent another.

8. Consider two models: (1) the block model depicted in Fig. 1.4, and (2) the difference equation in Eq. (1.1). Explain under what circumstances an analyst might choose each model.

9. The primary part of an "expert system" in artificial intelligence is composed of a collection of IF-THEN rules—such as "If the mushroom is yellow, then the mushroom is poisonous." Expert systems exist for many different domains, such as medicine, mineral exploration, and chemistry. Speculate how an expert system could be used to help someone determine which modeling method to use for a given physical system.

1.11 PROJECTS

1.11.1 How to Do Re-search

Research means to search and search again—digging for information wherever you can find it. Each chapter will contain a *Projects* section which provides research exercises related to computer simulation. Each written project report should include the following sections:

1. *Abstract.* Tell what you did and what results were obtained.
2. *Introduction.* Overview what is to be done and give background of who else has done related work.
3. *Information Trace.* Specify precisely where you obtained the information. Show this in the form of a trace—that is, show every step along the way during your information hunt. Examples of information sources: library, Internet, CD-ROM, personal reference, electronic mail.
4. *Research.* Tell what you did and show your results.
5. *Conclusions.* What did you learn from your research?
6. *References.* List references.

1.11.2 Where to Find Information for the Projects

General.   There are many ways to locate information and the ways are continuing to change as technology improves. I will list some of the ways that have helped me:

- Always do *parallel* research. This means that you should use a *sledgehammer* approach to research and seek information in all possible ways and in parallel. Do not make part of your research process dependent on the output from another source. Some aspects of the search process are unavoidably sequential but avoid being sequential at all costs. And do not assume that because you found relevant information in one place, you need to stop your search process.
- Start with your library. Do a keyword search of all relevant on-line or card catalogues. Try different keywords because massive amounts of information are indexed and you must provide the right word that happened to be used to create the original index table.
- Use CD-ROMs or videotapes if they are available.
- Exploit the massive Internet resource by using the information tools. There are many tools for searching *Gopher* space, the *World Wide Web* (WWW), and FTP *file transfer* space.
- Post a query to electronic news groups. Again you must do an effective *search for keywords* before doing the keyword search.
- Send electronic mail to anyone who might be able to provide pointers.

This is not an exhaustive list but it should help get you started in completing the research-oriented projects listed in each *Projects* section.

Specific information resources.   At the University of Florida, there is a large archive for simulation-related information available through a variety of mechanisms. The simulation archives are: (1) Simulation Digest, (2) Frequently Asked Questions, and (3) Simulation Software. Here are the methods of access:

FTP

```
ftp ftp.cis.ufl.edu
binary
cd pub/simdigest (tools are under /tools)
ls (do a list of files)
get (particular files)
quit
```

Gopher

```
gopher gopher.cis.ufl.edu
Go to: "CIS Information" and then "Simulation Digest"
```

Alternatively, to make a direct link:

```
Name=Simulation Digest
Type=1
Port=70
Path=1/cis/simulation
Host=gopher.cis.ufl.edu
```

World Wide Web (WWW).   From within Mosaic or another WWW client program, open the URL `gopher://gopher.cis.ufl.edu:70/11/cis/simulation`. A general CIS Department Hypermedia Document and on-line introduction to the Internet "Tunneling Through the Internet" can also be obtained by opening the URL `http://www.cis.ufl.edu/welcome.html`.

1.11.3 Some Projects to Start Off

1. If you were going to construct a *digital world,* what exactly would you need to build it? This book focuses on one aspect of building digital worlds: the dynamic models necessary to simulate time-dependent activity of objects. In addition to simulation model design and execution, you would need:
   - A geometry building tool that would allow you to construct the computer-aided design (CAD) aspects of your model.
   - A good human-computer interface. You might choose existing approaches or *immersive* technology as described in the *virtual reality* literature.
   - A good set of analysis tools for viewing the results of a simulation in some graphical form.
   - A tool to help you to make decisions about which model types to use.

   Write a report using some of the above concepts, and specify what is needed to build *digital worlds.* Include citations to the literature that you find over the Internet and in your library.

2. Construct an analog model out of *real materials* (solid, liquid, or gas) for one of the following systems. What are the limitations of your model? Observe the system and determine the *error.* Does the error vary with time, input, or state variable? Choose one more example of an already existing device that simulates one of the given systems and use a picture of it with an explanation of how it performs its function.
   - **(a)** An integrator. It takes an input x and produces $\int_0^t x\,dt$.
   - **(b)** A differentiator. It takes an input x and produces dx/dt.
   - **(c)** A multiplier. It multiplies two inputs x and y to produce xy.
   - **(d)** A sorter. It sorts n inputs and produces an ordered sequence.
   - **(e)** A server. It takes a discrete set of inputs into a queue, provides a service on the first in the queue, and then takes the next input.

3. Take each of the publications listed in Sec. 1.12 and summarize how each of them define the following terms using pictures and exposition.
   - **(a)** System
   - **(b)** Model

(c) State

(d) Event

Explain in your report how the authors differ in their definitions. Is there an overall agreement on what the terms mean?

4. Video arcade games with 2D displays can "immerse" the user in an artificial environment. Visit an arcade and focus on one particular game that strikes your interest. To what extent can these be considered simulations of a real or potential system? Pick one in particular, and report on its possible use as a training vehicle. How might the game be improved? Discuss the visuals and human-machine communication devices.

5. Interview three people at your location who use simulation models in their work. Tell them that you need at most twenty minutes of their time. Your goal is to document the model types that they use, and why they have chosen these types over other methods. Obtain the following information.

(a) What model type(s) do they use?

(b) Is there another method that they would like to use if given different circumstances (i.e., if they had a different position in the company, or if they had more money to buy a particular simulation package)?

(c) How do they decide when to use a specific type? Do they use rules of thumb (heuristics), or do they have a more specific decision criterion?

(d) How do they like simulation approaches versus analytic approaches that employ closed-form methods? Do they mix the two methods? When do they use one method over another?

(e) Do they seem to be biased toward one particular modeling approach?

After obtaining this information (the data), write a report detailing the results of your study. What did you learn from the interview experience?

6. Take one of the areas in Sec. 1.5 which rely on simulation on a discipline and write a short paper describing how simulation can help in that discipline. The comments in the section will help you to get started, but you should use the library and seek out references in that area.

7. You will find that many people have their own definitions for the term *simulation*. Take a large sample of people (at least 20) outside your class and ask them to define "simulation." Then try to categorize their answers into meaningful groups. Some may say "Simulation is something that is not real" or "Simulation is role playing." Regardless of the type of answer, this project is interesting and you should obtain a reasonable characterization of how people view simulation. A more complex, group project is to target certain *groups*: people who work in a hospital, people who work in a law firm, or computer programmers.

1.12 FURTHER READING

The formal framework for model types in this text are based on the canonical systems definition defined by Padulo and Arbib [165]. Zeigler [233, 234, 235, 238] and Klir [123] provide similar frameworks for the theory of modeling founded upon systems theoretical underpinnings. Systems researchers [20, 123] have stressed the importance of the "systems approach" to problem solving, and a relatively small number of simulation researchers have stressed a methodology for simulation based on systems theory and science. Zeigler [234], Oren [162, 163], Elzas [53, 54], and Nance [156] have advanced the state of the art in simulation methodology; however, few existing simulation books concentrate on simulation *modeling* as a methodology. Most simulation texts have extensive treatments of the statistical analysis of discrete event systems [14, 132, 204]. Pritsker's text [179] has a better treatment of modeling than others, and it is oriented toward modeling and analysis using the SLAM simulation language. Here is a partial chronological list of books which contain a broad coverage of system modeling:

An Introduction to Cybernetics, W. Ross Ashby, John Wiley & Sons, 1963.

Systems Theory: A Unified State Space Approach to Continuous and Discrete Systems, Louis Padulo and Michael A. Arbib, W.B. Saunders, 1971.

Theory of Modelling and Simulation, Bernard P. Zeigler, John Wiley & Sons, 1975.

Multifacetted Modelling and Discrete Event Simulation, Bernard P. Zeigler, Academic Press, 1984.

Architecture of Systems Problem Solving, George J. Klir, Plenum Press, 1985.

Dealing with Complexity: An Introduction to the Theory and Application of Systems Science, Robert L. Flood and Ewart R. Carson, Plenum Press, 1988.

Structures of Discrete Event Simulation: An Introduction to the Engagement Strategy, John B. Evans, Ellis Horwood Limited, 1988.

Alternate Realities: Mathematical Models of Nature and Man, John L. Casti, John Wiley-Interscience, 1989.

Object Oriented Simulation with Hierarchical, Modular Models: Intelligent Agents and Endomorphic Systems, Bernard P. Zeigler, Academic Press, 1990.

Continuous System Modeling, Francois E. Cellier, Springer-Verlag, 1991.

The object-oriented method has been used in number of simulation studies. The first use and design of object-oriented methodology is presented within the Simula language [22]. Interestingly enough, this language was developed with simulation studies in mind even though it is a general-purpose programming language. Zeigler [238] has an object-oriented design methodology based on DEVS and SES [234, 235] formalisms. Graham [92] has a good explanation of object-oriented methods as they apply to programming languages, databases, and artificial intelligence. C++ [215] is our preferred language of implementation for pragmatic reasons—it is compatible with C and is efficient. The texts by Booch [27] and Rumbaugh et al. [195] have good high-level presentations of object-oriented systems design.

1.13 SOFTWARE

Material in this section of each chapter is likely to become out of date. If you are not able to locate the software as it is listed, try using `archie` or `xarchie` to search on the name.

1.13.1 SimPack Simulation Toolkit

SimPack is a set of tools to accompany this book. The SimPack toolkit supports experimentation in computer simulation with a variety of model types. The full version is written in the C programming language, and a subset (SimPack++) is written in C++. The purpose of SimPack is to provide researchers and educators with a starting point for simulating a system. The intention is that people will view what exists as a template (or "seed") and then *grow a simulation program*. SimPack tools have been used for several years of teaching computer simulation at the senior undergraduate and introductory graduate levels at the University of Florida. Many complete programs are included to illustrate how the routines are employed.

At the end of each chapter there is a discussion on relevant software products that are freely available on the Internet. SimPack is a free product that is copyrighted by the author and covered under the GNU Free Software Foundation General Public License referenced in Ch. 10. In a nutshell, the tools can be used freely for research, teaching, or private use but not sold for commercial profit without free distribution of the complete source code. Feel free to distribute SimPack to whomever might have an interest in exploring the use of simulation to understand real or hypothetical environments.

1.13.2 Other Software and Online Simulation Information

There are many other free (or nominal cost) software packages for simulation that will be of interest to readers, so they are mentioned at the end of each chapter. One program that cannot be overemphasized is `archie` or `xarchie` for searching for software in general. Also, the reader should be aware of the Gopher hierarchy at the University of Florida, which can be accessed at gopher.cis.ufl.edu. Under "CIS Information" one finds the archives and a Frequently Asked Questions (FAQ) list for *Simulation Digest*—which was compiled over a six-year period beginning in 1987. In this information tree, there are numerous pointers to simulation topics as well as freely available simulation software.

## Chapter 2

Foundations

Events and intervals are labeled positive and negative. One is a condition of the other.[1]

2.1 OVERVIEW

To *model* is to abstract from reality a description of a dynamic system. Modeling serves as a language for describing systems at some level of abstraction or, additionally, at multiple levels of abstraction. A *system* is a part of some potential reality where we are concerned with space-time effects and causal relationships among parts of the system. By identifying a system, we are necessarily putting bounds around nature and the machines that we build; we define a system to be a "closed world" where we clearly separate items that are integral parts of the system (i.e., they are

[1]B. Krone, *Logic and Design in Art, Science and Mathematics* [131], p. 52.

within it) from items that can affect the system from the outside. Models are used for the purpose of communicating with each other—the alternative being that two people wishing to discuss dynamics would be forced to work with the real system under investigation. A model permits us to use devices such as equations, graphs, and diagrams. If I wish to discuss the hydrodynamics of a fluid line, I can draw straight lines (instead of connecting real pipes), arrows (instead of pumping fluid), and circles and boxes (instead of installing valves and tanks). The model reflects the language or metaphor that I will use to talk about a system. I also need to have some *consistent* way of describing dynamics by specifying the semantics of graphical models.

Modeling is a way of thinking and reasoning about systems. The goal of modeling is to come up with a representation that is easy to use in describing systems in a mathematically consistent manner. There isn't a single modeling language any more than there is a single natural language; instead, there are numerous instances of model languages, each good at representing a certain type of system or a particular aspect of dynamics. All models are constructed to answer a set or class of questions and so, not only are there many modeling approaches, there are also levels of abstraction within each approach. If someone wants linguistic answers about system dynamics, it may suffice to construct a high-level qualitative model; conversely, quantitative answers will require lower-level models. We will see that finite state automata, for example, are good at representing discrete control systems with a small number of states. Functional models are suited for representing systems that have coupled subcomponents—such as machines and plants. Petri nets are good at representing situations where one has resources and contention. Constraint modeling, often in the form of equations, is appropriate when the underlying dynamics are based on conservation of some property such as force, mass, or energy. One can also extend the basic modeling approach by *multimodeling*. With the multimodeling approach, we glue together models of the same type (homogeneous decomposition of models) or models of different types (heterogeneous decomposition). Various technical communities have nourished their own favorite modeling techniques and, through posterity, the modeling community uses a number of well-defined techniques. These "well-defined" techniques have withstood the test of time and have been found to be useful for describing systems, and so it is these techniques with which we will construct models and multimodels.

In this chapter, we will overview the fundamental aspects of system modeling. As with most scientific study, we begin with *data*. Our use of the word "data" refers to values obtained either through observation or arbitrary assignment of values to model components. The engineer will be familiar with the *real* and *complex* spaces since the state of most engineering products are most naturally defined with real numbers (i.e., 2.3 ft/s, 55 psi), and the composite unit containing frequency and phase information uses a complex number representation. Social scientists and artificial intelligence researchers must often deal with nominal and fuzzy data that does not come from mechanical measurement, but from humans communicating natural language. *Model components,* which serve as fundamental building blocks for models, take on the data values. Sample components include: state, event, input, output, parameter, and time. Such components are coupled together using declarative and functional perspectives to form *models.* Moreover, models can be coupled together to form *multimodels,* each of which can be composed of a set of models.

2.2 MODEL DESIGN

2.2.1 Data: The Observables

Data classes. In most computer languages, one has a variety of basic types, such as integer, floating point, and enumerated. The following are convenient abbreviations for certain sets: $\mathcal{Z}$ for the set of integers and $\mathcal{R}$ for the set of reals (i.e., floating point). Certain subsets are delimited by the endpoints of the set such as $\mathcal{R}_0^+$ for the set $[0, \infty)$.

The following scales are useful in characterizing a "higher level" of data type:

1. *Nominal.* Symbols of some type are used to label or classify data.

   (a) Reflexivity: either $A = A$ or $A \neq A$.
   (b) Symmetry of equivalence: $A = B$ iff $B = A$.
   (c) Transitivity: $A = B$ and $B = C \rightarrow A = C$.

2. *Ordinal.* We introduce the concept of the $<, >$ operations.

   (a) Irreflexivity: $A \not< A$.
   (b) Symmetry of equivalence: $A = B$ iff $B = A$.
   (c) Asymmetry of order: $A < B \rightarrow B \not< A$.
   (d) Transitivity: $A > B$ and $B > C \rightarrow A > C$.

3. *Interval.* Attributes in the real world are ranked and there is a fixed interval between ranks. The choice of the zero point is arbitrary.

   (a) Symmetry of equivalence: $A = B$ iff $B = A$.
   (b) Asymmetry of order: $A < B \rightarrow B \not< A$.
   (c) Commutativity: $A, B \in \mathcal{R} \rightarrow A + B = B + A$ and $AB = BA$.
   (d) Association: $A, B, C \in \mathcal{R} \rightarrow (A + B) + C = A + (B + C)$, and $(AB)C = A(BC)$.
   (e) Substitution: $A = B$ and $A + C = D \rightarrow B + C = D$; and $AC = D \rightarrow BC = D$.
   (f) Uniqueness: $A, B \in \mathcal{R} \rightarrow A + B$ and AB produce a single real number, respectively.

4. *Ratio.* The ratio scale is associated with attributes having interval properties as well as a natural zero point. The natural zero ensures that any two intervals on a ratio scale have comparable magnitudes determined by the number of times that one contains the other.

Examples of these scales are (1) nominal: marital status, occupation, color; (2) ordinal: formal education, class; (3) interval: biblical time, Fahrenheit temperature scale; and (4) ratio: height, age, distance, cost. Marital status, for instance, is a variable of type *single* or *married*. One cannot relate these two values except to compare them as being equivalent or nonequivalent. Therefore, marital status is a nominal measure. Fahrenheit is an interval, but not ratio, measure since 6 degrees is not twice as warm as 3 degrees; the zero point is arbitrary and does not reflect a "true zero." However, if an item costs 6 dollars—this is twice as much as an item costing 3 dollars, so *cost* is a ratio measure; the zero point naturally falls at the zero dollar amount and cannot be arbitrarily placed.

Random variables. Probabilistic measures are important to simulation since it is often possible to define the frequency associated with some dynamical process; however, it is not presently possible to define a precise relationship among state variables using a graph or set of equations. Also, probabilistic models can serve as high-level abstract (or lumped) models for processes. Random variables (r.v.) are the key elements in probability. A random variable is unlike a real variable in that real variables have one specific value such as $X = 2.45$; a random variable can have a range of possible values where a value is called a *random number* (RN). The value is obtained by *sampling* the distribution that defines the random variable. Random variables, like real-valued variables, can be either discrete or continuous. A *population* is the total number of items in a group: the population of people in the United States, or the population of motor casings being forged in a plant. A *sample* is a subset of a population. We usually operate on samples because a population is too large and not accessible. How does a random variable differ from a variable within a computer program? The answer to this question is seen when we realize that variables have two virtual operations associated with them: *get* and *store*. With *get,* we obtain the value of a variable, and with *store* we put a new value into a variable.

Operation	Variable	Random variable
get	$X \rightarrow$ value	sample $X \rightarrow$ value
store	$X \leftarrow$ value	$X \leftarrow$ function with value range

Generating random numbers (RN). An important aspect of simulation is the ability to create "close-to" random numbers. Let's work with a uniform distribution which is defined as a probability distribution function that is "flat" or constant across the range of values that a random variable will assume. For instance, to generate random numbers in the range $[0, 1)$ for a uniform distribution, we would expect any number in $[0, 1)$ to be as likely as any other number when a number is picked. The "picking" of a number from a distribution is known as *sampling* the distribution. Other ranges can be chosen also. If we know how to generate random numbers between 0 up to, but not including, 1 ($u \in [0, 1)$) and we desire a uniform random distribution between A and B (that is, $y \in [A, B)$, where y is a random number between A and B), we can generate y from u where $u \in [0, 1)$ by the following:

$$y = A + (B - A)u$$

If we have a random number generator (RNG) that provides numbers from 0 to 1 and we want to pick a number from the range 20 and 25, we first sample the uniform distribution between 0 and 1 to obtain, say, 0.4 and then solve for y:

$$y = 20 + (25 - 20)0.4 = 20 + 2 = 22$$

So, we now use 22 in our calculations. Continued generation in conjunction with this linear association will produce any number of values uniformly between 20 and 25.

Here is one way to generate random numbers starting with a *seed* value of r_0:

$$r_{i+1} = r_i a \,(\text{mod } m)$$

This is a multiplicative congruential generator, and there are many other variations. a must be relatively prime to m, and m should be defined to be the largest integer that can be supported. A short integer may be defined as $m = 2^{31} - 1$ or $m = 2^{63} - 1$ for a 64-bit word (recall that the first bit in each word defines the sign of the number). There are many ways of checking the competence of the generator. These methods include simple visual inspection by plotting generated values, autocorrelation, and the Fourier transform to detect hidden periodicities. The above generator will have a specific period since no number can be greater than $m - 1$; once a number is generated that was generated previously, the sequence will repeat. With random number generators, we try to make the period as long as needed.

For generating nonuniform random numbers, we can use other methods. One well-known technique is called the *inverse transformation method*. Here are the steps associated with this method:

Algorithm 2.1. Inverse Method for Nonuniform RNs
Procedure Main

> *1a. Specify the nonuniform distribution*
> *1b. Discretize distribution if it is continuous*
> > *(using a sampling method)*
> *2. Create a cumulative distribution (CD) from the discrete distribution*
> *3. Calculate the inverse of the CD function either using a table (T)*
> > *or a direct solution method*
> *4. Use a uniform distribution random number generator to generate*
> > *a number (N) between 0 and 1*
> *5. Use N as the probability for the CD probability and locate the*
> > *corresponding range value in T*

End Main

Figure 2.1 shows us each step of this process for a continuous distribution f. As you can see, we are "taking the inverse" of the cumulative distribution function to sample nonuniform distributions. This is, then, a numerical method for inverting a function. If the inverse function can be calculated analytically, then this is superior to the iterative numerical method. If the density function does not have a closed form solution for the inverse then the numerical approach is warranted.

Another solution involves applying Monte Carlo simulation to determine whether a sample point falls under the curve defining the probability density function. If we generate a point $(r1, r2)$ in the two dimensional space of the density function, then the point either falls under the curve, or over the curve. If the point falls under the curve then we take this value as our random number, otherwise we sample another point in the 2D space. An algorithm for this approach follows:

Algorithm 2.2. Rejection Method for Nonuniform RNs
Procedure Main

> *1a. Specify the nonuniform distribution f*
> *1b. Discretize distribution if it is continuous*
> > *(using a sampling method)*

2. Let f be bounded by a maximum value M
3. Generate a pair of uniform random numbers (r1, r2)
3a.   r1 is generated on a finite interval [A,B] (horizontal axis)
3b.   r2 is generated on a finite interval [0,M] (vertical axis)
4. If r2 < f(r1) then accept r1 and output r1
*   Else reject r1 and go to step 3*

End Main

Figure 2.2 demonstrates the use of the rejection method for an arbitrary distribution. Note how point $(r1, r2)$ falls outside the curve (and is rejected), but $(r1', r2')$ falls inside the curve (and is accepted). In this case, if $r1$ was the first random number to be generated, it would have been rejected and the algorithm would have proceeded to the second generated value, $r1'$, which would have been output.

Discrete distributions. If X is a discrete random variable, then $p(x_i) = P(X = x_i) =$ the probability that X equals x_i. Random variables also have conditions that govern their definitions — $\forall i (0 \leq p(x_i) \leq 1)$ and $\sum_{i=1}^{\infty} p(x_i) = 1$. We call $p(x)$ the *probability mass function* (pmf) of X. Figure 2.3 displays an example pmf for a six-sided die. We

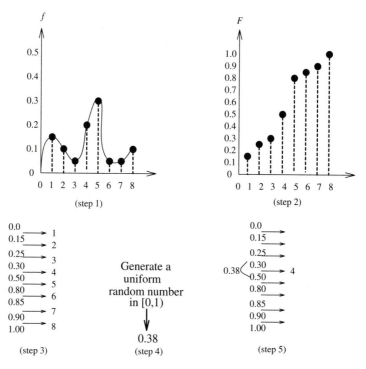

Figure 2.1   Generating nonuniform random numbers (inverse method).

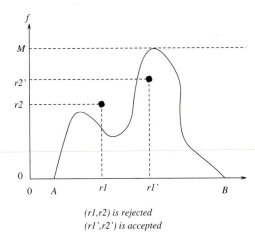

(r1,r2) is rejected
(r1',r2') is accepted

Figure 2.2   Generating nonuniform random numbers (rejection method).

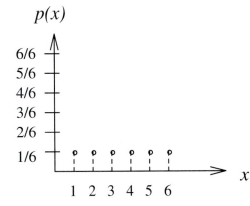

Figure 2.3   Discrete random variable for die tossing.

see that there is a 1/6 chance that the die will land with its top side equal to any one of the numbers 1, 2, 3, 4, 5, or 6. A die that was "loaded" would have a differently shaped pmf—with certain numbers having a greater chance than others of being rolled.

The die-tossing example shows a *uniform distribution,* which means that $\forall i, j\,(p(x_i) = p(x_j))$. This value was 1/6 for tossing the die. The uniform distribution can be used to model many processes where queuing is nonexistent, such as the distribution of pip number on the face of a die—dice do not have to wait in line and it is equally likely to roll a 6 as it is to roll a 3. Uniform distributions model processes that have no (or negligible) biases toward any specific outcome.

Nonconstant distributions are plentiful—we will briefly consider the *Bernoulli, binomial,* and *Poisson* distributions. The Bernoulli distribution is defined as a two-event distribution where the outcome is either a success or a failure. Imagine flipping an unbalanced coin where there is a 1/4 chance of getting heads and a 3/4 chance of getting tails. In this case, we broadly interpret heads as a success and tails as a failure. Then we define the distributions as follows:

$$P(X = heads) = p(heads) = 1/4 \tag{2.1}$$

$$P(X = tails) = p(tails) = 1 - p(heads) = 3/4 \tag{2.2}$$

A Bernoulli process is a stochastic process that involves an experiment with n trials. The n trials (each trial being one flip of the coin in our example) performed in sequence define a Bernoulli process as long as the trials are independent. The independence criterion is an important one in stochastic simulation and says that one trial is not dependent on any past trial. Each Bernoulli trial will have one of two outcomes: success or failure. Let p be the probability of a success and q be the probability of a failure.

The binomial distribution will be defined in terms of Bernoulli trials. A p designates the probability of a *success,* and q specifies the probability of a failure. The binomial distribution is defined by an r.v. that denotes the number of successes in n Bernoulli trials:

$$p(x) = \begin{cases} \binom{n}{x} p^x q^{n-x} & \text{for } x = 0, 1, 2, 3, \ldots, n \\ 0 & \text{otherwise} \end{cases}$$

where $\binom{n}{x} = \dfrac{n!}{x!(n-x)!}$ (the number of ways of taking x items from a total of n items without replacement). As an example of the definition $\binom{n}{x}$ consider the following: there are 5 items numbered 1, 2, 3, 4, 5 and we want to find out how many ways we can take 2 items from this collection. The answer is $\binom{5}{2}$, ways, which is $\dfrac{5!}{2!3!}$ or 10 ways:

$$\{(1, 2), (1, 3), (1, 4), (1, 5), (2, 3), (2, 4), (2, 5), (3, 4), (3, 5), (4, 5)\}$$

For our biased coin example, the probability of 2 successes (heads) is $p(2) = \binom{5}{2}(.25)^2(.75)^3$, which evaluates to $p(2) = \dfrac{5!}{2!3!}(.0625)(.422) = .264$. Therefore, the probability of having two heads come up is .264.

The Poisson distribution is defined as follows:

$$p(x) = \begin{cases} \dfrac{(\lambda t)^x}{x!} e^{-\lambda t} & \text{for } x = 0, 1, 2, \ldots \\ 0 & \text{otherwise} \end{cases}$$

$p(x)$ is the probability of x events occurring in time t, where λ is the mean number of events occurring per unit time. There is a story that a Poisson distribution was used to model the number of deaths of Prussian officers by horse kicks. In such a situation, we have a large number of officers and only a few deadly horse kicks. The Poisson distribution is the limiting form of the binomial distribution when there are a very large number of trials but only a small probability of success with each of them.

Continuous distributions.   A continuous random variable is defined as follows:

$$P(a \leq X \leq b) = \int_a^b f(x) \, dx$$

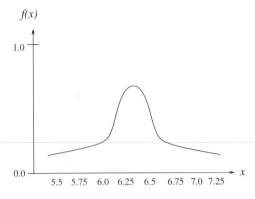

Figure 2.4   Continuous random variable for player height.

with the following conditions: $\forall x, (0 \leq f(x) \leq 1)$, $\int_R f(x)\,dx = 1$, where R is the total range of values that X can assume. If x is not in R, then $f(x) = 0$. $f(x)$ is called the *probability density function* (pdf). These conditions are similar to those for the discrete r.v. with the natural extensions for continuity. Figure 2.4 displays a sample pdf describing the height of a basketball player if sampled from a population of people.

I will define the uniform, triangular, exponential, and normal distributions. The uniform distribution is just like the discrete case except that the pdf, $f(x)$, defines a continuous line representing a constant value. Here is the uniform distribution:

$$f(x) = \begin{cases} \dfrac{1}{b-a} & \text{for } a \leq x < b \\[2ex] 0 & \text{otherwise} \end{cases}$$

The uniform distribution is widely used in simulation, primarily as a basis for generating other nonuniform distributions.

The triangular distribution has one more piece of information than the uniform distribution: a *mode*. With a triangular distribution, we know that a variable may take on a value within a specific range, and we know the value that occurs with the greatest frequency (the mode). If we consider the random variable to have the range $[a, b]$ with a mode at value $c \in [a, b]$, we can calculate the frequency at mode c by considering the definition of the cumulative distribution F. We know, for instance, that $F(a) = 0$ and $F(b) = 1$, and knowing that the area of a triangle is calculated as $\frac{1}{2}$base $\times$ height, we can calculate $F(b) = 1 = \frac{1}{2}(b - a)H$, where H is the height of the triangle. Then $H = 2/(b - a)$. So $f(a) = 0$ and $f(c) = 2/(b - a)$, and for values, x where $a \leq x \leq c$, f is defined by the location of x with respect to a and c.

$$f(x) = \begin{cases} \dfrac{2}{b-a}\dfrac{x-a}{c-a} & \text{for } a \leq x \leq c \\[2ex] \dfrac{2}{b-a}\dfrac{b-x}{b-c} & \text{for } c \leq x \leq b \\[2ex] 0 & \text{otherwise} \end{cases}$$

The exponential distribution is perhaps the most used in queuing models where event scheduling methods control the time increment. The distribution is defined as

$$f(x) = \begin{cases} \lambda e^{-\lambda x} & \text{for } x > 0 \\ 0 & \text{otherwise} \end{cases}$$

Exponential distributions are good at representing the facets of many different processes. In simulation, we can use the exponential distribution to model the distribution of interarrival times of entities entering a queue for service. Most entities will arrive within some given interval of the the preceding entity; however, it is still possible to obtain a few very large interarrival times.

The normal distribution is defined by

$$f(x) = \frac{1}{\sigma\sqrt{2\pi}} \exp\left[-\frac{1}{2}\left(\frac{x-\mu}{\sigma}\right)^2\right], \qquad -\infty < x < \infty$$

and represented as $N(\mu, \sigma^2)$. μ is the population mean and σ^2 is the population variance. The normal distribution models a random variable that is a combination of other random variables. Say that a student's total score in a class is based on the total sum of five tests. The tests are 20 points maximum and 0 points minimum. Let the actual score on each individual test be a random variable with any distribution. Therefore, we are interested in finding the distribution for $Y = X_1 + X_2 + \cdots + X_n$, where $X_1, \ldots, X_n$ are the individual test scores, and Y is our final course score. The Central Limit Theorem states that for a large enough sample, Y will be normally distributed (in the limit). If we try some simple examples using a discrete example, we can convince ourselves of this. Consider when $Y = 2$ and when $Y = 10$—clearly, there are more possible ways (combinations of individual test scores) to get a total course score of 10 than to get a total course score of 2. This means that $P(Y = 50)$ will be larger than $P(Y = 10)$, and $P(Y = 25)$ will be somewhere in between. Similarly, $P(Y = 100)$ will be as small as $P(Y = 0)$ since there is only one possible set of combined scores for each of these two cases ($0 + 0 + 0 + 0 + 0$ for a total score of 0, and $20 + 20 + 20 + 20 + 20$ for a total of 100).

Cumulative distributions. A *cumulative distribution function* (cdf) defines a non-decreasing function that says whether a r.v. is less than some given value. For instance, the chance that we would throw a die numbered 1, 2, or 3 is 50%; for any number less than or equal to 5, the probability would be $5/6$: one needs to simply add (or integrate) all possibilities for numbers ≤ 5. The cdf is defined as $F(x)$. For the discrete random variable we have

$$F(x) = \sum_{\forall x_i \leq x} p(x_i)$$

For the continuous case, we have

$$F(x) = \int_{-\infty}^{x} f(t)\, dt$$

With a cdf, F is nondecreasing [$a < b \Rightarrow F(a) \leq F(b)$].

Discussion on probability distributions. With the discussion of distributions and their associated definitions it is not at all clear when we should use, say, an exponential distribution instead of another distribution. With each distribution, we have discussed an example application of that distribution. The distributions that we've outlined are important because they *recur* in a large variety of different applications. You could easily come up with your own distribution, but it would most likely fall into one of two categories:

1. Your distribution would be an approximation to an existing well-known distribution, or

2. Your distribution would reflect your specific data, and the qualities of those data (such as dispersion from the mean) which do not correspond "across the board" to other completely different applications.

Of course, new distributions are not in any way ruled out; it is important, though, to realize these two caveats when considering entirely new distributions. One way of choosing a distribution is to simulate your process over time and collect data and statistics. This gathered information can be used to match the data with a distribution. Another method is simply to choose a distribution based on the kind or class of application that you have; it is known that certain scenarios require certain distributions. This second method is especially useful when it is not possible (or too expensive) to perform data gathering.

What are probability distributions, really? With regard to the simulation of systems, a probability distribution is simply a high-level abstraction of a process. A pdf provides a *model* for a process which would otherwise require a very complicated constraint (Ch. 6) or spatial (Ch. 7) model if we were to model the process in great detail with inputs, outputs, and state variables. The method of choosing distributions, therefore, permits us to model certain aspects of a process (interarrival time of entities into a queue) or entire processes. Usually, a simulation model will have both probabilistic methods combined with graph-based or equational models to make an effective combination which accurately validates the model at reasonable cost in terms of model size and speed. There are many philosophical issues regarding whether nature actually operates on deterministic or stochastic basis (in quantum mechanics); however, this does not change our use of probability distributions for differing levels of abstraction in a model. We see that even our so-called "random" numbers are generated from a deterministic difference equation using a modulus.

In summary, probability distributions help us to specify lumped, abstract knowledge and incorporate this knowledge into our models. A complex system with observable outputs can be quite complicated to model; however, a probability distribution representing the frequency of observations can serve as an abstract substitute for the system being modeled. Also, in many cases, a detailed system model is unknown; however, it is possible to gather data sufficient enough to define a probability distribution. There are discrete as well as continuous probability functions, and the cumulative distribution function provides us with a nondecreasing function with a range between $0 \leq F(x) \leq 1$.

Monte Carlo method.   In the discussion of the rejection method for generating random numbers, we saw our first example of *Monte Carlo simulation*. This method is widely used in the analysis of model output. Monte Carlo simulation is the process of simulating a process in order to estimate a probability density function. With the rejection method, we *simulated* the picking of horizontal and vertical values to determine whether these were located in the area under the density function curve. Where is the simulation in this? Consider that a person rolled dice for each x and y direction and then placed a marker on a large piece of paper on which the density function was located. By playing this game with many markers, the player will eventually succeed in closely estimating the area, and likewise can generate nonuniform random numbers. Therefore, the simulation that we do is the rolling of the die and the player's act of setting down markers even though this *virtual* physical process is deemphasized or not mentioned at all.

This situation becomes more interesting when considering the use of the Monte Carlo method *within* a computer simulation. Consider the simulation of an electronic network composed of a voltage source and a network of resistors, inductors, and capacitors. Resistors are marked with certain tolerance values such as 5% or 10%, meaning that the ohm rating for a resistor can vary plus or minus a percentage of the nominal resistance. Now, getting back to the simulation, we can form a model of the network and perform a simulation. In this simulation, we must choose a specific value for the resistance of each resistor. We pick this value by sampling from the known distribution for the resistor based on its tolerance rating. After running the simulation, we obtain state variable output and then perform an analysis of some sort. What we would really like, though, is to sample repeatedly from resistor density functions and rerun the simulation many times. From such repeated experiments, we obtain more accurate state variable values. What we have done is to run one simulation *inside* another. As in the rejection method, imagine that when we sample from a resistor density function, we are actually picking out physical resistors from a box of 100 resistors. Our electronic network simulation, then, simulates the current flow through the network while also simulating picking and choosing resistors from a box. This process of having one simulation inside another can be taken to the nth degree. For example, the electronic network may be part of a chip that is surface mounted onto a printed circuit board with many other chips. If the original chip feeds its output into another chip's input, then by doing multiple simulation runs of the original chip's functionality we estimate the relative frequency of input value to the second chip.

The importance of Monte Carlo simulation is that we can *abstract* the behavior of part of a model and replace it with a probability density function. For the electrical network, we can abstract away the complicated state relationships necessary for the first network (for a given input signal to the first chip) and replace them by a single density function representing the chip's output. When we want to simulate the second chip and we require input from the first, we could sample the first chip's density function.

Fuzzy data. What simulation model components can be made fuzzy? It is not too surprising to find that most nonfuzzy mathematical structures can be made fuzzy simply by extending the appropriate definitions to encapsulate fuzzy, not crisp, sets. The *extension principle* permits the transfer of existing mathematical methods to incorporate fuzzy semantics. Here is a sample of how this relates to simulation. We can make fuzzy:

- A state variable value. This includes both initial conditions and values at a specific time.

- An event variable value.
- Parameter values. Time variant systems can use fuzzy functions for parameters.
- Inputs and outputs.
- Model structure.
- Algorithmic structure.

In the last item, we note that a simulation model is really just an algorithm at the lowest semantic level, and therefore methods in fuzzy algorithms can be utilized.

Considering the wide variety of applications of fuzzy set theory to simulation, we have designed and implemented procedures to deal with fuzzy-valued variables. We term a first order *fuzzy number* to be one that equals a single real value. As we increase the order, we obtain the following:

- An interval of confidence. Ex: $\mathbf{X} = (1, 3)$, where $\mathbf{X}$ is an interval fuzzy number with $\mu(1) = 1$ and $\mu(3) = 1$.
- A triangular number. Ex: $\mathbf{X} = (1, 1.2, 3)$, where $\mathbf{X}$ is a triangular fuzzy number with $\mu(1) = 0$, $\mu(1.2) = 1$, and $\mu(3) = 0$.
- A trapezoidal number. Ex: $\mathbf{X} = (1, 2.5, 3, 4)$, where $\mathbf{X}$ is a trapezoidal fuzzy number with $\mu(1) = 0$, $\mu(2.5) = 1$, $\mu(3) = 1$, and $\mu(4) = 0$.
- General discrete fuzzy number. $\mathbf{X} = \sum_{i=1}^{n} \mu_i / x_i$.

A fuzzy number is defined as a fuzzy set that is both convex and normal. The simple types of fuzzy numbers are piecewise linear, so they can usually be abbreviated using the interval notation just delineated. With generalized fuzzy numbers we must, though, write down each domain value and its corresponding confidence level. The general form of the discrete fuzzy number (as shown above) is

$$\sum_{i=1}^{n} \mu_i / x_i$$

and the continuous fuzzy number has the associated denotation for a continuous domain:

$$\int_X \mu(x)/x$$

These two types of fuzzy numbers are similar in concept to the discrete probability mass function and continuous probability density function found in the theory of probability. It is natural to assign lexical values to fuzzy numbers, so that we can assign *high* to (200.0, 400.0), *low* to (10.0, 200.0), and *hardly anything* to (0, 10.0) for a given application.

Figure 2.5 defines a *fuzzy variable,* also known as a lexical variable since it can take on several natural language words as values. The fuzzy variable is *temperature,* and its possible values, using triangular fuzzy numbers, are (1) cold, (2) cool, (3) ambient, (4) warm, and (5) hot. It may be appropriate to use fuzzy numbers when there are few available data from which a probability distribution can be induced or when there are enough data,

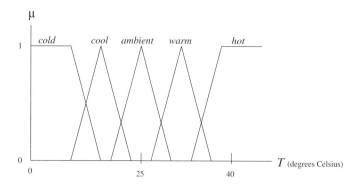

Figure 2.5   Fuzzy numbers for temperature.

but obtaining the data is costly or difficult. Fuzzy sets are a useful bridge between expert
systems knowledge and quantitative knowledge since there is a direct mapping from quality
to quantity.

2.2.2 Model Components

States and events.   States, events, and time are three of the most fundamental concepts in system modeling. States and events are dual components in that a state change in
a system occurs as a result of an event occurrence.[2] A *state* describes the system for an
interval of time. An *event* is a point in time that designates a change in state. An event is
an instantaneous state, so an event is just a special kind of state which has no time duration.
A state is usually defined as a simple tuple $\langle s_1, \ldots, s_n \rangle$, where n is the number of components in the state vector and s_i are the state's components. An event is a time-tagged state
$\langle t, s_1, \ldots, s_n \rangle$. The time t represents the instant of time that is associated with the event.
This definition of event was first used in physics, and is defined in systems theory. While a
system is often in a state for an interval of time that interval will be equal to a point value if
the system state is continuously changing. Furthermore, an event is a point that is relative to
the *level of abstraction* for which the system has been defined. At lower-level descriptions,
the event is a function that includes additional state-to-state transitions (thereby forming
subevents). For example, if the circus "arrives in town," then this is an important *event*
even though the arrival itself may be associated with an interval of time. If we construct a
model in which the circus arrival process is deemed unimportant—at our desired level of
detail—we treat the event as a point in time.

   Later, we will create models at multiple abstraction levels, and each of these levels
will have states and events. While states and events exist in every model, it is sometimes
convenient to use other terms, depending on the context. A *phase* is simply a state. When
term *phase* is used, however, it most often implies that there is a lower level of abstraction

---

[2]The ancient Chinese symbol called T'ai-chi T'u, incorporating, *ying* and *yang,* provides a somewhat more
exotic visualization for the interplay between state and event.

model that contains a state space and that the phase is a subset of this state space. The word *cold* represents a phase of water during a heating process. Of course, *cold* is nothing more than a state in a model, and it is termed a phase within a lower (more complex) state space for which *cold* serves as a partition. For instance, *cold* might refer to the situation where temperature (T) is less than 50 degrees. This inequality ($T < 50$) partitions the lower-level state space, including T as a subcomponent. Therefore, while a phase is just a state, one can see the reason for using the word *phase* in a multimodel. We mentioned, also, that an event can be considered either as a *point* or as a *function,* depending on the abstraction level. From the context of a given abstraction level, all state changes result from an event occurrence. If we are considering only models at a single abstraction level, all events occur due to a change in input. (*Note:* The change in input might be further specified using a condition such as "input becomes greater than 5 volts.") These inputs to a model are called *external events*. An *internal event* is also an input to the model; however, the input comes from a lower-level abstraction model and not from outside the system. The existence of internal events implies that multimodeling is in effect since it implies that there are state transitions where a specific state transition causes an output to change.

When making tea, I boil water, insert the tea bag in the hot water, and then wait a while until it steeps. Therefore, states for this system might be: (1) heating water, (2) inserting bag in water, and (3) steeping tea. Each of these activities is assumed not to overlap in time, and has a time duration so that the system is in just one of these states for a given length of time. We can let state 1 be active for 6 minutes, state 2 for 0.1 minute, and state 3 for 4 minutes. The amount of time spent in a state depends on the input trajectory. So, if state "(1) heating water" is active for 6 minutes, there needs to be a six minute input trajectory segment which corresponds to this duration. Consider Fig. 2.6(a): this is a model for our dynamic tea system. When we execute this model, we obtain a trajectory shown in Fig. 2.6(b). The model in Fig. 2.6(a) has been created by *coupling* three states together. Coupling is the joining of system components to form a network or graph. What is the "input" that causes the change in state? If our system is to be modeled only at a single abstraction level, we might assume an arbitrary input signal (or string) that facilitates state transition. However, we will see that a detailed modeling of the causes for state transition will require more than one model. All events occur because of a change in input; sometimes the input can be viewed as being defined at the same level (i.e., an *external event*) or a lower level of abstraction (i.e., an *internal event*).

In natural language, states are most often represented by participial phrases for a verb, such as "emptying," "placing," and "boiling." Event names can easily be formed from state names by prefacing the participle with "start" or "end," such as "start boiling" or "end emptying." Let's consider some examples:

1. A pendulum that swings back and forth can be characterized using any number of states. Let's pick three possible states: (*left*), (*center*), and (*right*). Then the system can be said to be in any of these three states depending on its current position for a point in time. A typical event would be (12.23, left) meaning that at time 12.23 in the simulation, the state of the system equals *left*. In this case we have one state variable (position) and

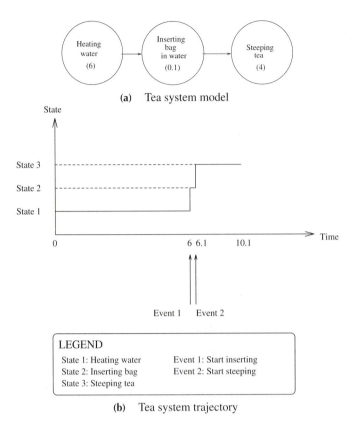

(a)   Tea system model

(b)   Tea system trajectory

Figure 2.6   Tea model.

three possible values for that variable to assume. We could just as easily have picked the numbers 0, 1, and 2 to refer to these states. The nominal values are more amenable to human interpretation. Now, suppose that we take the pendulum example and decide that the pendulum can swing starting at some specific angle (with respect to the vertical) and reach its maximum position (before swinging back again) at another angle. Then we might want to augment our state definition to include an angle in radians: (position, angle). An example state might then be (left, 2). In the latter case, our state vector is composed of two state variables: position and angle.

2. A train is moving at constant velocity. In this modeling scenario, it will be important for us to know whether the train is near (within one mile of) a station. We want to include this information in our simulation model. We can let the state be defined as (location, station-flag), where location is a real number denoting displacement along the track, and station-flag is a boolean value (i.e., *true* or *false*) denoting whether or not the train is within one mile of a railroad station.

3. Consider the following program in the C programming language:

```
main()
{
  int i,j,in,A[10],B[2];
  B[0] = 2; B[1] = 3;
  printf("Enter an Input: ");
  scanf("%d",&in);
  for(i=0;i<10;i++) A[i] = i + in;
  for(i=1;i<10;i++) {
  A[i] = A[i-1] + B[0]*B[1];
  for (j=0;j<10;j++) printf("%d",A[j]); printf("\n");
  } /* end for */
}
```

We assume that this program is a simulation of some unspecified process for a total of 10 time units (0 to 9 inclusive). What would be the state for this program/simulation? We see that array A is changing, whereas array B is not. We might, therefore, call B a parameter array (i.e., $B[0]$ and $B[1]$ are parameters of the system), and the actual state vector is $(A_0, A_1, \ldots, A_9)$, where A_i is a short notation for the C array element $A[i]$. The system also has an input identified by the variable in. Here is the output for the program:

```
Enter an Input: 4
4 10 6 7 8 9 10 11 12 13
4 10 16 7 8 9 10 11 12 13
4 10 16 22 8 9 10 11 12 13
4 10 16 22 28 9 10 11 12 13
4 10 16 22 28 34 10 11 12 13
4 10 16 22 28 34 40 11 12 13
4 10 16 22 28 34 40 46 12 13
4 10 16 22 28 34 40 46 52 13
4 10 16 22 28 34 40 46 52 58
```

This example shows us that, broadly speaking, *any* algorithm can be considered as a kind of "system" with inputs, states, outputs, and parameters.

4. A puck slides on an ice rink during a hockey game. Its location over time at any point in the game is of concern to us, so we decide to form a state vector with two variables x and y representing coordinates of a two-dimensional space. A typical state of the game might be (10, 3), where units are in meters. At coordinate points where the puck enters the goal cage, we will create events, so that if the puck enters the goal cage at time 39.0 and the puck's location is (15, 2), the event is (39.0, (15, 2)).

5. A person waits in one line (queue) and proceeds directly to another line. This system, therefore, has two queues. The state of a queuing system is the number of entities in each queue at any one time.[3] For each queue, we have a state variable, so the state of the system is (x_1, x_2) at any one point in time, where x_i represents the number of people in queue i. Arrivals into a queue, and departures from the server, are potential events.

Even though an event is formally defined as a state vector value at a given time, many systems researchers identify discrete events as being synonymous with events. It is often assumed that an event naturally accompanies a change in state. This, however, does not correspond to our definition—an event is nothing more than a state at a specific time. If the state in a trajectory changes value does this mean that there is an event, or do events apply only when a cognitive or lexical association can be identified with the event? Does a state of $\langle 2.3, 4.5, 1.1 \rangle$ have a lexical mapping to a natural language word or phrase such as "warm," "cold," or "large"? It might not; however, it is still a state. Likewise, events need not have a "name tag," but they often do in the literature simply because identifying anonymous events is not as helpful in our attempt to gain system understanding. A spot in the air 3 feet above a floor most likely has no cognitive or linguistic mapping; therefore, we will not naturally think that an event has occurred at this location even if an object passes through this spot during a motion trajectory. Although if the ball hits the floor, then this serves as an event since we have a strong cognitive mapping for "floor." To make the discussion of events clearer, the term "discrete event" is normally associated with events that cause changes in state. The discrete events for a model are those events that have lexical or cognitive mappings. For instance, an "arrival" or "departure" in a queuing model are two examples of discrete events. To keep with common practice, we will often use the term "event" in the traditional sense meaning "discrete event." The following is a list of heuristics that can be employed to identify events.

- An event can occur when a significant (or marked) change occurs in one of the following trajectories: input, output, state. Consider an input trajectory that is a square wave. Leading and trailing edges are good potential locations for events. In continuous systems, state variables specify orders of derivatives. The sign change for a state variable denotes a potential event location.

- Form a list of questions to be asked of the model. If objects are part of the question, they should be represented in the simulation model, and interaction with the objects are possible events. In general, actions are applied to objects and have a beginning and end—these are possible locations for events.

- Consider all *boundary conditions*. With respect to collisions between geometrical objects, the points of collision indicate events. Example events are when a projectile hits a target or when an articulated body comes into physical contact with its environment.

- Just as programs have *context switches* (i.e., subroutines), models can have *model context switches*. When a model is deliberately composed of separate submodels, each submodel can be seen as a *phase* where events demarcate phase boundaries.

---

[3]We can also add to the state description the entities getting service.

Our systems specifications will involve variables of types: scalar, vector, or matrix. We prefer to extend these basic mathematical types to include the type *record*. A record in a computer language such as C or Pascal is a repository of data and knowledge for a variable defined as a record. In Pascal the record is called `record` and in C, the record is called a `struct` (short for structure). For instance, consider a system where a human arm is moved. The arm is treated as an object in the simulation; however, the state of the arm at any time involves several variables, each of different types. We can create a C record called `arm` with type `StateType` as follows:

```
#define LEFT 0
#define RIGHT 1

typedef struct {
    float x,y;
} CoordType;

typedef struct {
    int side = LEFT;
    CoordType upperarm;
    CoordType lowerarm;
    CoordType hand;
} StateType;
```

Then we can create a system model that updates the state of the arm by setting the appropriate fields of a variable that is declared to be of type `StateType`. The general field of data structures provides what is necessary to define the state of a system. In this sense, computer science —within the subfield of data structures—provides a rich environment for the creation of structures for state values.

A final point about states and events. It is important to characterize the state of a system using the minimal number of state components while still satisfying the requirements of the model to answer the appropriate questions. This implies that we want state components to be *orthogonal* to each other. The state of a robot arm needs to include only those degrees of freedom at the joints (for revolute motion) and at the interface (for translational motion). Why is this so? Why don't we include points somewhere along the link or at the base of the robot arm in the state description? At the base, the points do not move in space so there is no need of tracking them, and points along a link can easily be calculated from the joint positions since the link points are simple algebraic *functions* of the joint positions. If the links were to deform or vibrate in time, and this vibration was to be modeled, then our state vector would need to grow in size. In the definition of state, we should consider both necessity and sufficiency.

Input and output.   The definition of *input* is relative to the particular system being described. That is, an input is simply a state that has a controlling influence on a system which does not contain the input state. So, in a general way, an input is just another kind of state except that it permits us to place boundaries around what is considered to be "inside" and "outside" of a system. An oven knob is an input if the system to be modeled is a pot of boiling water; however, the oven knob status is just part of the state description if the system, instead of being the pot, is the kitchen. In the tea system, there is an implicit input—representing an action by the human—that permits us to change from state to state. An input that is constant (i.e., non-time varying) is a *parameter*. An *output* is a function of the system state and input. For the tea system, we might consider the output to be equal to the current system state. The input for a system is something that controls the system's behavior and the output is an observable (see Fig. 2.7). Some systems do not have input or output; systems without an input are sometimes termed *autonomous* systems. We must be careful, though, with our definition of autonomy since one would argue that any robot would require input to interact intelligently with its environment.

   Time.   A basic concept of systems is *time*. Time is denoted by either an integer or a real number, although there is nothing preventing us from making time a nominal variable, such as *early, late,* or *before-lunch*. Time for discrete systems will jump in irregular or regular intervals. For continuous systems, time is incremented by very small time intervals (which can also vary in interval size). The chief difference between discrete and continuous time is that with continuous time, the time step can be made *arbitrarily* small—the only practical limitation is that of the computer word size. With discrete time systems, time cannot be subdivided in an arbitrary manner since the given time intervals are an inherent part of the physical system description.

2.2.3 Model Definition and Basic Categories

   Formal definition.   Before discussing specific modeling methods it is best to over-view a unifying formalism that serves to represent a wide variety of system models. A deterministic system $\langle T, U, Y, Q, \Omega, \delta, \lambda \rangle$ within classical systems theory is defined as follows:

- T is the *time* set. For continuous systems $T = \mathcal{R}$ (reals), and for discrete time systems, $T = \mathcal{Z}$ (integers).
- U is the *input* set containing the possible values of the input to the system.
- Y is the *output* set.
- Q is the *state* set.

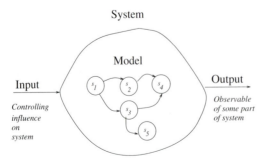

Figure 2.7   System with input and output.

- Ω is the set of *admissible* (or *acceptable*) input functions. This contains a set of input functions to use during system operation. Often, due to physical limitations, Ω is a subset of the set of all possible input functions $(T \to U)$.

- δ is the transition function. It is defined as $\delta : Q \times T \times T \times \Omega \to Q$.

- λ is the output function, $\lambda : Q \to Y$.

A system is time invariant if its behavior is not a function of a particular interval of time. This simplifies δ to be of the following form: $\delta : Q \times \Omega \to Q$. Here, $\delta(s, \omega)$ yields the state that results from starting the system in s and applying the input segment ω.

This formalism, although concise, is quite general. For structural reasons we employ techniques such as inter-component coupling, interlevel coupling, and hierarchy to make the overall system more manageable; however, we first identify some key pieces within the above definition. Q is also known as the *state space* of the system. An element $s \in Q$ is termed the *state* of the system, where the state represents a value that the state components can assume. The state of an ice hockey puck would be (x, y), whereas the state of a two-queue system would be (q_1, q_2), where q_i represents the number of entities in queue i. Normally, the state has a structure of an n-tuple: $\{s_1, s_2, \ldots, s_n\}$ however, it is best generalized as a *data structure*—a tuple being one type of data structure. Superstates provide flexibility in describing some model forms. A *superstate* is a subset (rather than an element) of state space; therefore, "goal" is the subset of the hockey puck state space representing the geometrical region where the goal net is located. A pair consists of a time and a state (t, s), where $s \in Q$ is called an *event*. Events are points in event space just as states are points in state space. Event space is defined as $T \times Q$. Events normally represent values of state that correspond to definite cognitive or lexical mappings. For instance, in a queuing model we identify an event as an "arrival," but we may not have words to represent the values of other states whose values do not correspond to a cognitive or lexical association. *State* and *Event* are critical aspects of any system and by focusing on one or the other, we form two different sorts of models: *declarative* models that focus on the concept of state, and *functional* models that focus on the concept of event. States represent a "snapshot" of a system, while events, even though they occur at points in time, are naturally associated with functions (i.e., routines, procedures). A change in state is associated with an event, and vice versa; so there is a duality between states and events.

Figures 2.8 and 2.9 show two sample models that can be represented using a (1) state transition model and (2) block model, respectively. Both of these formalisms are subclasses of the system formalism. Declarative models such as Fig. 2.8 focus on state transitions either by explicitly specifying state or by using a pattern which matches a part of state space. All FSAs such as this one should be thought of as built within a single functional block; so the idea of *function* is not absent from this representation; it is just that a declarative orientation is used. One might think that state transition networks should always be subservient to functional block networks; however, we will see with multimodeling that this is not always true. Specifically, we can also have a single state which is refined into a block network. In this example, outputs Y_i are associated with states S_i. Input conditions involving U_i provide determinism where there are multiple branches from a state.

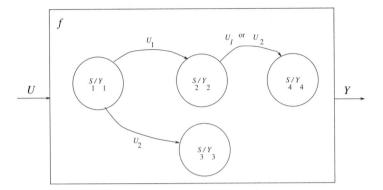

Figure 2.8   Declarative transition orientation: many states, few functions.

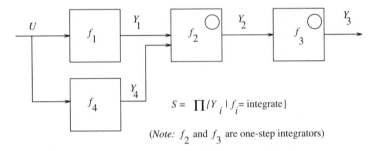

$$S = \prod \{ Y_i \mid f_i = \text{integrate} \}$$

(*Note:* f_2 and f_3 are one-step integrators)

Figure 2.9   Functional network orientation: many functions, few state descriptors.

In Fig. 2.9, transfer functions are represented as boxes with inputs and outputs connected to each box. This is termed a *block model* in systems engineering or a *data flow graph* in software engineering. The state vector for this system is a subset of the outputs Y_i. Specifically, the outputs connected to boxes with memory (i.e., integrator or delay boxes) make up the system state ($\prod$ represents a set-theoretic cross product):

$$Q = \prod \left\{ Y_i \mid \delta_i \equiv \int_0^t dt \right\}$$

Two of the functional blocks, f_2 and f_3, have circles within them. This is to designate that they are integrators which contain state transitions defined by the method of integration.

Declarative and functional modeling.   Imagine someone in a car asking you for directions. Assuming that you have a map, you have two distinctly different approaches to use for an answer:

1. *Declarative Approach.*   Show the driver a map of the area. Identify where the driver is located at the moment as *A* and the driver's destination at *B*. What you are, in fact, saying is "Here is where you are now (*A*), and there (*B*) is where you want to be. In between, here is what the map looks like." With this approach, you focus on describing the key states that the driver needs to know, including the start and end states.

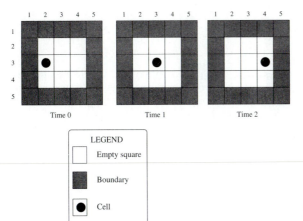

Time 0   Time 1   Time 2

LEGEND

Empty square

Boundary

Cell

Figure 2.10   The system with a horizontally moving cell.

2. *Functional Approach.* Tell the driver a procedure for getting from *A* to *B*. You might say something of the sort "... go to the third traffic light, turn left and go for ten minutes until you get to the interstate, then head east...." With this kind of explanation, you focus on providing a function[4] to the driver. The driver is the signal (or data) that is "flowing" through functions that will ultimately transfer the driver to the destination. The signal is actually composed of state variable values that denote things that change over time, such as the *X, Y* position of the driver.

A focus on state provides a *declarative* view of modeling, while a focus on event provides a *functional* view. It is convenient to split two fundamentally different approaches to simulation modeling along the lines of declarative versus functional modeling. In declarative modeling, we build models that focus on state representations and state-to-state transitions. In functional (or procedural) modeling, we focus on the system as a coupled network of functions each of which takes inputs and produces outputs. There are hybrids of these two approaches; however, we will initially focus on these basic modeling techniques. These two approaches—declarative versus functional—can be found in other disciplines, such as software engineering and artificial intelligence. In software engineering there are two distinct views of large software systems: an imperative data flow view (the functional view) and a state transition view (the declarative view). In artificial intelligence, there are declarative versus functional approaches to knowledge representation and symbolic programming. Lisp, for instance, is a programming language based on the lambda calculus (a functional calculus), while Prolog is a language based on predicate calculus (a declarative calculus).

To illustrate the basic difference in the declarative versus functional model classes, consider Figure 2.10. Here we have a simple 2D board where each board square can take one of three possible values: empty, boundary, or cell. The dynamic system is defined by the cell, moving until it strikes a boundary. For simplicity, we assume that the cell moves horizontally, although this is not a necessary condition. Figure 2.10 illustrates the system for two time steps. To model such a system declaratively, we must specify what the system "looks like" from one time step to the other. If we assume that the only possible cell motions are horizontal where the cell begins in rows two, three, or four, then Fig. 2.11 provides two

---

[4]A procedural algorithm with input and output.

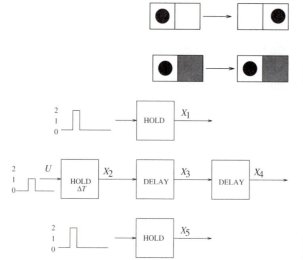

Figure 2.11   Declarative model: two rules yielding horizontal motion.

Figure 2.12   Functional model: block model for cell motion.

local rules. The advantage of the local rules—rather than rules involving pictures of the entire board—is that the board may be made arbitrarily large and the rules remain the same. The two square templates shown in Fig. 2.11 should be thought of as a pattern. To simulate the system, one first scans the board in its first state to see if a rule pattern exists. The first pattern in Fig. 2.11 is seen to match the part of the board in Fig. 2.10, where the cell is located. When the first rule is applied to the initial board, the board appears as it does at the second time location in Fig. 2.10. This continues until the cell encounters the boundary, at which time it stops (where the stopping is facilitated by the second rule in Fig. 2.11); when this rule is applied, the state of the board remains the same.

For a functional model, we do not focus on state transition; instead, we focus on what *operations* or *functions* must be executed to get the cell from its initial location to its final location at the other end of the board. Let x be defined as an array where the state of the board is kept. A block model equivalent of the code is presented in Fig. 2.12. We simplify the block model by assuming that there is a variable x that represents a one-dimensional vector (with elements $x_1, \ldots, x_5$) representing the third row in Fig. 2.10. In Fig. 2.12, the *HOLD* function accepts a digital pulse and indefinitely holds the output at the pulsed input value; the *HOLD* ΔT function holds the input for some period of time ΔT and then the signal returns to level zero. The *DELAY* function takes the input and effects a specific delay corresponding to the time that the cell is to remain in a square. The behavior of Fig. 2.12 is shown in Fig. 2.13.

Principle of locality.   There is an assumption so basic to modeling that we hardly realize it when constructing models. We will term it the *principle of locality*.[5] In artificial intelligence, there is the *frame problem,* which describes the difficulty in representing change from one situation to another. In our terminology, situation means *system state*.

<hr />

[5]The principle of locality holds in most physical domains, except for nonlocal interactions expressed within quantum mechanics [32], pp. 142–153.

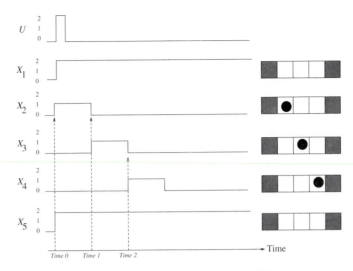

Figure 2.13   Trajectories for input and state variables $x_{1,...,5}$.

To define how one transits from one state to another, we must carry along all of the state values that do not change, as well as those that do change in the transition. There are several ways around the frame problem from a practical perspective, but the frame problem demonstrates that modelers need to find ways of expressing the change without creating unduly large state descriptions. The use of patterns and unification in production systems and logic programming languages serves to help our efforts in making declarative models. We create lumped states based on the *locality* of base states; base states cluster together to form groups. In this way, we partition system spaces for more expressibility in declarative system modeling. For functional models, we couple functions together whose blocks represent physically adjacent or nearby system components. Another example of locality is with respect to spatial models and processes such as diffusion and convection. Certain objects have a local effect on other objects (namely, those with whom they are connected), not a more global effect. With several recent solutions to the N-body problem, similar approaches are used—to localize the effect that one particle or body has on others by virtue of its distance from them. By using the principle of locality, we may create much simpler models. Without it, we would be forced to contain the entire state of the system when expressing any state transitions or we would have to maintain fully connected graphs for representing functional models or the interactions among particles.[6]

2.3 SERIAL MODEL EXECUTION

There are two primary approaches to executing a system model on a sequential, von Neumann-style architecture: (1) time slicing and (2) event scheduling. Parallel approaches to simulation are covered in Ch. 9. Time slicing involves incrementing time using a global clock by adding a certain amount to the clock value over time. The increment need not

[6]The naive $O(N^2)$ solution to N-body simulations is an example of this type of connectivity.

be the same for each update of the clock; the key attribute of time slicing over the event scheduling method is in the *way* that time is incremented: through addition. While addition, at first, may seem like the only way to increment a clock, there are other methods, such as event scheduling. Event scheduling is more general and subsumes time slicing in that time slicing is easily incorporated into event scheduling by scheduling a new event at a time that would normally be set using addition. Event scheduling can be likened to a "calendar" or an "alarm clock" in that events are scheduled to occur at a specific date or time. For a calendar, events are recorded on the day when the event is to occur; the calendar keeps track of an ordered sequence of events. For an alarm clock, one must imagine a clock with multiple alarms capable of being set for different times.

The two methods of time advance serve a single purpose: *to provide the illusion of parallelism on a serial architecture-based computer.* Most systems that we will want to simulate have many things going on in parallel—a grocery store has many lines, all of which may be occupied simultaneously, and a molecular dynamics simulation involves the simultaneous motion of particles each with a different velocity vector. Time slicing and event scheduling give us clear approaches to simulating parallelism. On parallel architectures, other methods can be used since we can partition pieces of the distributed architecture to map to pieces of the model (or system to be simulated).

2.3.1 Time Slicing

In time-slicing model execution, a *virtual clock* is updated in intervals of some arbitrary size which can change throughout the simulation. The system is normally defined as a group of functions each of which must be evaluated at the end of each interval. This approach is very similar to round-robin scheduling in operating systems, where a time quantum plays the role of the time interval, permitting each "process" to gain access to the CPU.

Algorithm 2.3. Template for Time-Slice Execution
Begin Main
    Initialize state variables
    Loop through all functions
    While not end of simulation
        UpdateFunctions
    End While
End Main
Procedure UpdateFunctions
    For all functions
    Switch on function type
        Case type 1: apply function type 1 to state and inputs
        Case type 2: apply function type 2 to state and inputs
        $\vdots$
        Case type n: apply function type n to state and inputs
    End For
  Increment time using time slice
End UpdateFunctions

The time-slicing method is a straightforward way to simulate a functional network or a set of equations; however, it can be inefficient in circumstances where the values of inputs and outputs do not change regularly. Consider the case, for instance, where we have a functional block that performs a logical *and* of its two inputs. Furthermore, suppose that the two inputs change very rarely. Our discrete time-slicing simulator would check this gate continually at the lowest level of time resolution (i.e., the interval) to provide its current output. But since the output is not changing (except for rare instances), we are wasting precious computation time in the extra checking. Therefore, for continuous systems with continually varying inputs, state values, and outputs, the time-slicing method is appropriate and usually necessary; however, for discrete time systems such as digital logic circuits, we should explore other methods if we are to have an efficient system. Another problem with the time-slicing method is the problem of encoding a delay into a functional element. This can be handled by using event scheduling (to be discussed in Sec. 2.3.2) or a circular queue (as discussed in Ch. 6). Many functional elements will have a "memory" or propagation delay built into the device. So, for instance, an *and* gate may have a 10-ns delay between the time the input changes and the time that the new output reflects this input change.

2.3.2 Basic Event Scheduling

Event scheduling is defined as executing code that contains an *event loop* where events are posted and then checked according to minimum time. That is, the event on the future event list with the minimum time tag executes first. While most event scheduling methods employ a data structure such as a linked list or heap, it is not absolutely necessary to maintain such a structure in order to schedule and execute events.

Consider a single-server queue—there are two events of importance: an *arrival* into the system and a *departure* from the system. Arrivals come according to a probability distribution. Departures are slightly more complicated. Let's consider an implementation where times are generated on the fly using a random number generator.

- The next arrival time is always relative to the last arrival time.
- The next departure time depends on whether there is an entity already in the system besides the entity that is departing.

  — If there is no other entity, the next departure time is relative to the next arrival time.
  — If there is another entity, the next departure time is relative to the last departure time.

This is defined as follows:

$$D_0 = 0$$

$$D_i = \max(D_{i-1}, A_i) + \sigma_i$$

D_i refers to the departure time of the customer i (i.e., the output of the server block), A_i is the input to the server (i.e., arrivals), and σ_i represents the time to service customer i. Let's consider the example of a single-server queue and five customers where the interarrival times are 2, 8, 3, 9 and the service times are 3, 4, 4, 2, 1. Table 2.1 displays the times. A simple piece of code for basic scheduling follows. Note the following variable definitions:

TABLE 2.1
BEHAVIOR OF A
SINGLE-SERVER
FUNCTION

i	A_i	σ_i	D_i
1	0	3	3
2	2	4	7
3	10	4	14
4	13	2	16
5	22	1	23

ua and *us* are mean times for interarrival and service; `ta` and `td` are time of next arrival and time of next departure; `te` is "end of simulation" time; `time` is the global simulation clock time; `n` is the number of entities in the system; and `dist()` is a random number generator for an arbitrary probability distribution.

```
float ua,us,te,ta,td,time;
int n;
n=0; ta=0.0; td = ta + dist(us); time=0.0;
while (time < te) {
if (ta < td) {
/* arrival event */
time = ta; n++; ta = time + dist(ua); }
else { /* departure event */
time = td; n--;
if (n==0) td = ta + dist(us);
else td = td + dist(us);
} /* end if */
} /* end while */
```

To reflect on the above implementation for a single-server queue, we see that only two variables are needed (`ta` and `td`) even though there may be a number of events that require scheduling. For instance, if *m* entities arrive in a queue, it is not necessary to store the times of all *m* entities. It is enough, for this particular application, to store the latest arrival time and the next departure time. This constraint is not general, though, and applies only because of the first-in, first-out nature of the single-server queue that we have defined. Now, let's proceed to a more general solution to the event scheduling problem.

2.3.3 Structured Event Scheduling

Event scheduling refers to a model execution approach where time is incremented through the interaction with a global event data structure. The event methods can help us to simulate systems while overcoming the above two deficiencies associated with discrete time simulation. In the event scheduling method, a global future event data structure *controls* the simulation, while individual block elements submit events to occur to the data structure. A central switch statement continually checks the data structure and causes the event with the least time to occur. The "occurrence" of an event means to execute an event routine which handles the semantics associated with that event. The type of data structure used in event scheduling is called a *priority queue,* and it involves the following key operations:

1. *Enqueue.* Insert a time-tagged event into the data structure so that the structure maintains its ordered sequence (also known as *insert*).
2. *Dequeue.* Delete the event in the data structure that has the minimum time (also known as *delete_min*).
3. *Cancel.* Cancel the event in the data structure with a specified time or event identification (also known as *delete*).

There are a number of possible data structures that may be used to implement a priority queue, and we will cover a subset of them. When we consider each key operation, certain optimal structures come to mind. Take the *enqueue* operation. Clearly, a hash table is a perfect structure for enqueuing events; however, the *dequeue* operation can be costly. For *dequeue* a list structure which maintains a sorted order is convenient; the least time event is always located at the head of the list. The problem is trying to find one data structure that can handle both *enqueue* and *dequeue* operations in a reasonable amount of time. We will consider three major categories of data structure: (1) list, (2) tree, and (3) table.

Linear list method: linked list.   There are two types of linear list implementations: (1) linked and (2) array. In the first case, a list composed of items and pointers is created, and in the second, a contiguous array is implemented. Linked list operations are included below and graphically depicted in Fig. 2.14.

Algorithm 2.4. Linked List Operations
Procedure LinkedEnqueue (X)
    Start with the head of the list and locate Y
      whose time is greater than X
    Insert X directly before Y in the list
End LinkedEnqueue

Procedure LinkedDequeue
    Remove X from the head of the list
End LinkedDequeue

Procedure LinkedCancel(X)
    Start with the head of the list and traverse to locate X
    Delete this item from the list by:
      (1) Letting X be between items A and B
      (2) Creating a new variable T to point at B
      (3) Removing X from dynamic storage
      (4) Making A point directly to B using Ts value
End LinkedCancel

The general next_event ↔ schedule cycle is depicted in Fig. 2.15.

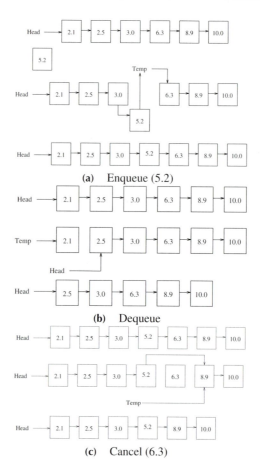

(a)   Enqueue (5.2)

(b)   Dequeue

(c)   Cancel (6.3)

Figure 2.14   Linked list operations.

Linear list method: array.   An array may also be used instead of the linked list approach. With an array, *dynamic data storage management* must be performed by the person writing the simulation rather than by the operating system or language compiler. Here are the advantages and disadvantages of using an array:

Advantages

- Enqueue and dequeue operations are usually faster than with linked lists. Since array elements can be located contiguously on the disk, array accesses are faster than with dynamic pointers to areas of memory that might require paging from secondary storage.
- Having not to deal with pointers can ease the burden of debugging pointer-based operations. With an array, the indices are integer and therefore easy to work with.

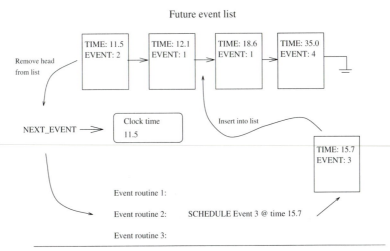

1. *An event is removed from the head of the future event list*
2. *The routine corresponding to this event is now in "control"*
3. *The event routine will schedule other events*

Figure 2.15   Basic cycle for an event scheduling.

Disadvantages

- You must do your own dynamic storage management. This is often something that is easier if left to the compiler and operating system. Arrays may become too large for the available memory, in which case the programmer must also deal with external storage and procedures such as swapping to disk. If an array becomes fragmented (after many deletions), it may be necessary to institute garbage collection to group the "holes" back together.

In short, a linked list approach is preferable since storage management issues are left to the compiler and operating system. The only case where an array should be considered if the extra speed gained is worth the tradeoff, or if the programmer does not have knowledge of pointers. If speed is of a primary concern, however, linear list approaches are the worst since they have $O(N)$ average complexity despite their limited appeal due to their simple structure. *SimPack* contains a linked list option primarily for educational purposes: it provides a very understandable mental model for the event scheduling approach.

Tree method: binary search tree.   A binary search tree is a tree where every node has from zero to two possible children. If a node has zero children, it is a *leaf node,* and if a node has either one or two children, it is an *internal node.* Given an internal node p with datum x, the search tree is organized so that

$$p.\text{left} \neq \text{NIL} \rightarrow p.\text{left}.x < p.x$$

$$p.\text{right} \neq \text{NIL} \rightarrow p.x < p.\text{right}.x$$

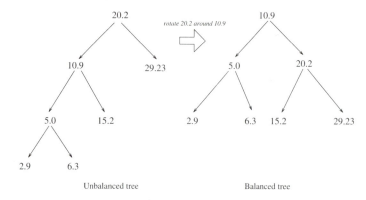

Figure 2.16 Result of the operation $rotate(10.9, 20.2)$.

The above two requirements are termed *invariants*—they list the constraints that define a binary search tree. The enqueue operation is achieved by inserting a node into the tree in order that the above invariant is maintained. The minimum node value is always found at the lower left part of the tree (i.e., by traversing the left links, starting at the root node, until a leaf is found). Therefore, a dequeue obtains the minimum value node and deletes the node. This repeated deletion of the leftmost leaf nodes causes a *right skew* in the binary tree, therefore making it somewhat undesirable from the perspective of event set manipulation. The problem is that skewing results in an unbalanced tree which further results in fairly long access times for tree elements since every search begins at the root, and the search must proceed as if the tree were a linear linked list since skewing results in long linked list chains.

We can correct the balancing problem of the binary search tree if we incorporate a self-organizing tree balancing technique. That is, when we find that the binary tree is becoming unbalanced, we do something to make it balanced. The most common action is *rotation*. The AVL tree was an early and effective structure made to combat the balancing problem, and it uses rotation to maintain reasonable times for search operations. The AVL tree is a binary tree that contains nodes with data and a balance factor. The balance factor for tree node p measures the difference in height between the left substree, *p.left,* and the right subtree, *p.right*. For a full binary tree (with 2^i nodes), all balance factors are zero; however, with incomplete or skewed trees the balance will increase (or decrease depending on the direction of the skew). When the absolute value of the balance factor is greater than 1, a rotation takes place to rebalance the tree. A rotation is defined as $rotate(X, Y)$, where X and Y are adjacent tree nodes. Node Y is "rotated" around node X. Figure 2.16 illustrates the process of rotation for an example binary tree.

The rules of rotation are as follows for the operation $rotate(X, Y)$, rotate Y around (or with respect to) X, where Y is the root node and X is one of Y's children:

1. If X is a left leaf node, after rotation it will have Y as a new right subtree (similar case if X is a right leaf node).
2. If X has one left subtree but no right subtree, Y will become its new right subtree after rotation (similar case for a singular right subtree).

3. If X has two subtrees, as in Fig. 2.16, let's take the case where X is the left child of Y. X will maintain its current left subtree; however, its right subtree will become a new left subtree of Y after rotation (similar case when X is the right child).

Tree method: implicit heap.   A *min-heap* is a complete binary tree where all child nodes contain larger times than the parent node time. The definition for a complete binary tree (T) of height (h) is:

1. T is empty, or

2. The left subtree of T is a full tree of height $h - 1$ and the right subtree of T is a complete tree of height $h - 1$, or

3. The left subtree of T is a full tree of height $h - 1$ and the right subtree of T is a full tree of height $h - 2$, or

4. The left subtree of T is a complete tree of height $h - 1$ and the right subtree of T is a full tree of height $h - 2$.

A good way to envision heaps is to consider drawing a binary tree by starting at the root and successively filling in nodes from left to right in a breadth-first manner. If this process is used, each tree drawn will be complete.

Given an internal node p with datum x, the heap is organized so that

$$p.\text{left} \neq \text{NIL} \rightarrow p.x < p.\text{left}.x$$

$$p.\text{right} \neq \text{NIL} \rightarrow p.x < p.\text{right}.x$$

Heaps are either *explicit* or *implicit*. An explicit heap is organized with pointers, whereas an implicit heap is implemented in a contiguous array. The array-based (implicit) heap implementation is normally quite fast, so we will concentrate on this approach while displaying heaps using both pointer and array illustrations. The adjective "implicit" refers to a minimal space approach; heaps with links require storage for the links, whereas a contiguous array stores only the data. We do not have fragmentation problems, as we would with a linear list array because the heap is always a complete tree; items are added to the end of the array and a reorganization through swapping is used to move data around. In the array implementation, we note the following:

1. A parent of any child is determined by taking the child's index in the array and dividing by 2.

2. The left child of any parent node is determined by multiplying the parent's index by 2. The right child (if it exists) is determined by adding 1 to the left child's index.

Figure 2.17 illustrates the enqueue, dequeue, and cancel operations on a heap-based priority queue.

Algorithm 2.5. Heap Operations
Procedure HeapEnqueue (X)
        Insert X in the heap as a new last heap item
        Bubble X up the heap to its correct location

(done by swapping child with parent while moving
up the tree)

End HeapEnqueue

Procedure HeapDequeue

> *Remove the item from the root with the minimum time*
> *Swap the root item with the last item*
> *Delete the last tree item*
> *Reorganize the heap by swapping parent with child*
>> *where necessary until heap order is preserved*
>> *(done by bubbling item down the heap)*

End HeapDequeue

Procedure HeapCancel(X)

> *Find X to delete in the heap by searching breadth first*
> *Swap X with last heap item*
> *Reorganize heap*

End HeapCancel

Table method: dynamic hashing.   A hash table is composed of a sequence of rows or buckets. Each bucket will contain either a single value or a pointer to a data structure containing values. It is convenient to think of a hash table, for purposes of simulation, as a *year*. Then each bucket is a *day* in the year and a day has a specific range associated with it. The *current date* is simply the global clock simulation time. A hashing function is what is used to take a time-based *key* and convert it directly into an address within the hash table. For instance, if we wanted to store a simulation event time within a hash table that has 16.0 time units per year, we can create one as follows:

$$h(k) = k \bmod 16.0$$

where h is the hashing function and k is the key (or event time).

Hash tables are data structures that can operate as effective priority queues under special conditions. The regular hash table, while it supports the enqueue operation without any difficulty, has a difficult time with the dequeue operation because of the sequential nature of dequeue—that is, dequeue depends, not on an absolute data location, but on a relative location to the current location (namely, the next-lowest time for events stored in the hash table). Where, exactly, will the next-lowest event be located? In a large hash table, the search for the next-lowest event time might take an exorbitant amount of time. On the other hand, a small hash table will involve "horizontal" searches through the data structures stored in each bucket. For this reason, dynamic hashing is the only real alternative. With dynamic hashing, the table changes its size depending on the number of stored events.

Figure 2.18 shows an example hash table (or calendar) that contains events (only the event times are shown in the table). This hash table represents the current calendar after some period of system simulation. The current time (or date) is 86.22 on day 3 of this calendar year. The year is composed of 8 days, where each day is 2.0 time units. As we expect, there can be no time less than 86.22 in the calendar since a dequeue would have occurred previously on such an item. Let's consider a series of dequeue operations without

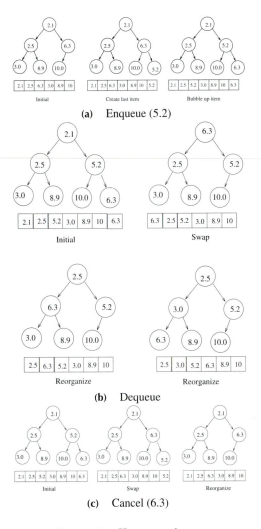

Figure 2.17   Heap operations.

an enqueue. Let's begin at the current time. The node adjacent to the one containing 86.22 contains 87.8. The algorithm first checks to see whether 87.8 is in the same calendar year as the current one (year: $[80.0 \rightarrow 96.0)$). In this case, we confirm that $86.0 \le 87.8 < 88.0$. Then we continue searching the following day (day 4). Since day 4 has no entries, we continue to day 5. Two successive dequeues continue from the nodes containing 91.2 and 91.9. However, 171.5 is clearly not in the day range for the current year. This means that we leave this alone and resume our search with day 6. The last two nodes in day 6 are not in the current year either, so we leave them alone. We are now at the end of this year. For the following year, we can continue to dequeue nodes that remain for days 0, 1, and 2. Then, since days 3 and 4 are now empty, we continue to day 5. On day 5 we see that 171.5 is still

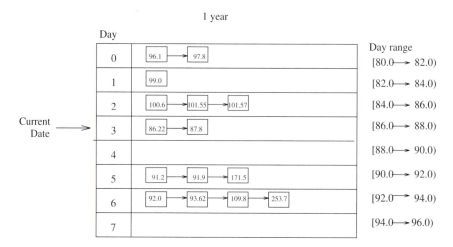

Figure 2.18   Hash table containing events.

not within the current year, so we leave it alone and review day 6. The node with 109.8 can be dequeued but 253.7 cannot. Now, something interesting happens—as we proceed to the following year and sweep all of the days, we find that there are no events that can be dequeued; the remaining events in days 5 and 6 are due to occur in the future. Specifically, 171.5 occurs 5 years since the very beginning, and 253.7 occurs in 10 years. A brute force algorithm would simply scan year after year until it found events for that year; however, we can use the following logic to speed things up: *if the scan of a year yields no events scheduled for that year, dequeue the next stored event.* The next event is 171.5, and the dequeue operation proceeds in this fashion.

Enqueuing events is more straightforward. Hashing is an ideal way to enqueue events because a functional hash on the key (the event time) directly produces the location of the event. Actually, the event may be located at a location close to the hashed value because of a *collision*. A collision occurs when a value or other data structure is already located at the hashed location.

- By attaching a data structure to each bucket, collisions are handled by storing the event in a linked list, tree, or other structure. A secondary (or *n*-ary hash table could also be used.

- Through probing, events can be stored in a bucket that follows the hashed bucket. The "probing" process looks for an empty bucket and stores the event in that location.

Algorithm 2.6. Dynamic Hashing Operations
Procedure HashEnqueue (X)
           Hash to the appropriate bucket using a modulus
                and a hashing function applied to X
           Double the hash table size if the number of events
                is greater than 2 ∗ number of buckets
End HashEnqueue

Figure 2.19   Behavioral model as a black box with inputs and outputs.

Procedure HashDequeue
> *Search all buckets for the event with the minimum time*
>> *(make sure that time is within the current year)*
>
> *If no events are found for this year(a year of empty*
>> *buckets), then take the minimum time event*
>
> *Remove the item from the table*
> *Halve the hash table size if the number of events*
>> *is less than 0.5 ∗ number of buckets*

End HashDequeue

Procedure HashCancel(X)
> *Hash to appropriate bucket by applying the hashing*
>> *function to X*
>
> *Locate X and remove it from the bucket list*

End HashCancel

2.4 BEHAVIORAL MODELING

With the exception of the models in this section, all of the modeling methods discussed in this book are symbolic and state-based. They are also economic in that we attempt to express a system's behavior with a minimal number of components. Whether we are building a model representation from a declarative, functional, or spatial perspective, we can associate pieces of our model with pieces of the world. While many models are characterized as being composed of events and states, all models need not be created in this fashion. A *behavioral model* is one where we model a physical phenomenon by associating inputs to the system with outputs that arise as a result of the inputs. That is, a behavioral model is a repository of (input, output) pairs without regard to what states or events may have transpired for the system to take the current input and produce the current output. Figure 2.19 displays the essence of the *black box* behavioral model. The black box is named as such since one *cannot see inside* to describe the behavior through local state transitions and interfunction coupling. We still need to come up with a formalism that allows us to associate inputs with outputs. An example of a behavioral model is one where an input vector $\mathbf{u}(t)$ is related to an output vector $\mathbf{y}(t)$ through a simple linear relationship in Eq. (2.3):

$$\mathbf{y}(t) = A\mathbf{u}(t) + B \qquad (2.3)$$

where A and B are matrices. However, the simplest method of defining a behavioral model is to enumerate all outputs corresponding to inputs in a table and then present the table as the behavioral model. In this sense, tables that provide digital logic gate behaviors for combinatoric networks are behavioral models. However, this is very uneconomical. The best behavioral models are ones that require a concise model description that can interpolate (or *memorize*) a large number of (input, output) pairs. The following model types are often used to create behavioral models:

1. *Linear or Nonlinear Regression Models.* Arbitrary functions whose parameters are estimated while attempting to match the behavior against the data to be *stored* in the model.
2. *Neural Network Models.* Neural networks with hidden layers can associate output with input.
3. *Fuzzy Logic Models.* Input/output fuzzy logic models have been used for a variety of control operations.

A back-propagation neural network, for instance, can associate input with output by feedback of the error of the neural net output response, the output selected by the trainer. Behavioral models are *generic* in that the model structure is generally used regardless of the physical domain.[7]

Why are behavioral models employed? If the underlying local state transitions and functional decomposition are unknown or too costly to obtain, behavioral models are the method of choice. Determining detailed models of the kind discussed in this book can be time consuming, and behavioral models provide a *first cut* at modeling the empirical data obtained through experimentation. In the area of automatic control, neural networks and fuzzy logic controllers demonstrate that it is not essential that humans know a more exact state-based model in order to control a device.

2.5 FINANCIAL APPLICATIONS

Models are useful for several reasons, two of which are (1) to gain a better understanding of a system from a research perspective, and (2) to use a model to determine the cost of running a system which has either not yet been constructed or not yet extended from its present configuration. The topics of finance and cost are relegated to this chapter since they are orthogonal to modeling in general and, therefore, foundational in nature. That is, all models can be made to have an associated cost.[8] Whenever there are the attributes of time, or states and events that occur over time, cost values can be assigned to these variables. This demonstrates that simulation is an essential tool for cost modeling and financial analysis. Figure 2.20 shows the basic setup where a system is simulated, and the cost associated with the system is a function of the system output. A few examples of cost determination are

---

[7]One must not confuse a neural net model of brain function with the use of neural nets for associating generic input with output. The former would be a simulation model whose internal neuron states (for a memory-based neural net) characterize validated local state transitions, whereas the latter model is used only for a generic mapping of input to output.

[8]For a greater emphasis on cost modeling, it will be most constructive for the instructor to assign exercises, with the goal of having the students create a cost function for any model design if one is not already provided.

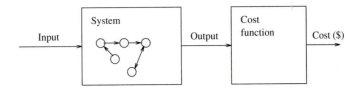

Figure 2.20   Assigning cost to an arbitrary system.

given in this book, but let's first go over each major model class to see how costs may be applied.

1. *Declarative Models.* A finite state machine which models phase transitions for a daily schedule for a cotton mill. Phase transitions, in this case, represent shifts. The change in shift occurs as a result of changing input to the FSA, and a cost variable can be assigned to each FSA state specifying its weighting with respect to the other states. For instance, if day-shift workers make less than night-shift workers, the cost for the day-shift state (or phase) will be higher. The output of the FSA represents state trajectories which are fed into the cost function.

2. *Functional Models.* A queuing network which models the flow of raw material through a lathe, with the resulting part being routed to other workstations and machines. If the server in the queuing network is a human, one must compute the labor cost associated with the time taken to process parts. If the server is a machine, the machine can be depreciated over the time taken to use it. For every server in the network, there is a cost that can be assigned so that the simulation reports the operational costs.

3. *Constraint Models.* Bond graphs are models that focus on the exchange of power in a system and the accumulation of energy by applying power over time. Energy utilization is directly related to cost. Consider the regional utilities from which we get our water, electricity, and other energy-based consumables. The accumulation of energy in the form of capacitance, for instance, can be output to a cost function.

4. *Spatial Models.* Spatial phenomena can be associated with cost by relating the state of the system (i.e., the spatial configuration) and time to cost. State and time are combined in the form of the *state trajectory*. A cellular automaton for forest fire will include rules that take a tree through various local states (normal, onfire, burnt) by using surrounding state information. The cost of a forest fire is in direct proportion to the area covered by the fire at any given time.

5. *Multimodels.* Multimodel costs are dictated by the costs of the individual layers and the costs of the models defining those layers.

2.6 EXERCISES

1. Consider a barbershop system where you enter the shop and have your hair done.
   (a) provide an automaton model with states and transition arcs.
   (b) Decompose the automaton model into a lower abstraction level so that states such as "sitting in chair" (for instance) are broken down into lower-level states.

2. Specify the type (nominal, ordinal, interval, ratio) for the following values.
   (a) The date 2500 B.C.
   (b) 23.5 years
   (c) The word "hot"
   (d) 2 tomatoes
   (e) the interval [2.1, 3.43]

3. You are asked to model the activities during a fire escape from a thirty-story building. Under what circumstance would you model the "ringing of the alarm" as an event—as opposed to treating it as a subprocess with substate changes?

4. Draw fuzzy sets for the following lexical variables.
   (a) Speed of a car
   (b) Height of a person
   (c) Size of a house

To the instructor: Once all students have handed in their solutions to this exercise, provide a statistical report to the students defining how the individually chosen sets were different. Did most students choose similar curves? Were there only one or two outliers in the distribution, or was there more variety?

5. Specify at least four alternating states and events for the following activities, and for each activity subdefine one state and one event to subdivide for a more detailed view of the system.
   (a) A baseball game in progress
   (b) Eating dinner
   (c) A phone conversation
   (d) Buying groceries at the store
   With the specification of events and states, try to pick single words rather than to use instruments such as *begin XYZ* (for an event) or *doing XYZ* (for a state).

6. In the two-level aggregation of the algebraic circuit of Fig. 8.2, the top level is dependent on the bottom level for the precise semantics. Note that we could isolate level 1 from level 2 by storing the input/output behavior (say, in tabular form) of level 1 inside its block. What are the advantages or disadvantages of this approach versus simulation at the lowest level?

7. What are the differences between simulation of a model (specified in the form of a program) and execution of that model?

8. Specify the following system components (input, output, state, event, and parameter) for each of the following systems.
   (a) An operating grandfather clock with three weights
   (b) A moving herd of sheep on a farm
   (c) A game of chess (or checkers) being played
   (d) Stomping one foot on the ground
   (e) An executing bubblesort algorithm
   (f) A glass of milk being drunk using a straw

9. Specify a declarative and functional model for the process defined by getting $\frac{1}{2}$ lb of sliced turkey from the supermarket delicatessen counter. This process includes both the customer and the supermarket worker.

10. Consider the binary search tree to represent an event set. Provide a sample sequence of operations and a complete tree of height 3, using *enqueue* and *dequeue* so that the tree remains balanced after the operations. What types of operations (and in what order) are likely to skew the tree?

11. The reorganization of a heap is an $O(\log(N))$ operation given the height of the heap. Provide the specific circumstances in which the heap never requires readjusting when an *enqueue* operation occurs. What is the worst-case scenario for heap readjustment given (1) an *enqueue* and (2) a *dequeue?* That is, describe the situation that would require the most work of each given operation in terms of tree size and organization.

12. If a simulation uses only integral time values (such as the natural numbers), suggest an improvement in efficiency for the dynamic hashing scheme. Assume that the order in which operations occur, for events with the same time, is unimportant.

13. The assignment of a cost function to a model may not seem to apply on small-scale models such as the growth of bacteria in a petri dish. It would appear that a small-scale model would imply a small operational cost. Is this always true? Give a specific example to justify your answer.

2.7 PROJECTS

1. Do a study on random number generators. There are quite a few methods for generating random numbers and different methods that can be used to *seed* them. What are the key issues in generating random numbers? Does it make any difference in the type of application in which they are used?

2. Probability density functions are abstractions for lower-level processes which are left undefined for reasons of computational cost. Pick three *pdf*'s from different sources and describe, in words and diagrams, what kind of more detailed model could be used in lieu of the pdf given that a lower level of detail was desired.

3. States and events are ubiquitous. Take a room anywhere (in a building, apartment, or house) and list states and events that can be associated with objects at that location. Specify the abstraction level that you choose. Is it high or low?

4. There are many types of event scheduling methods implemented in the SimPack queuing code. Do a performance analysis on one of the models already provided within SimPack. While varying parameters, run each simulation for a given time and then time how long it takes for the simulations to execute. The key things to do here are to **(a)** find a small set of variables or parameters to vary and **(b)** run the simulation for two or three significantly different time intervals.

5. Build a software tool that will allow users to witness a set of FEL data structures as they change over time with the simulation. This has already been done for the linked list option of the queuing library (using `trace_visual()`), so you will need to provide it for the other data structures.

2.8 FURTHER READING

Systems science literature [75] provides a description of the practice of the "systems approach" to problem solving. Basic notations of state and time, while found more informally in general physics literature [185], are more fully developed in systems theory [118], while Padulo and Arbib provide an expository account of basic systems theory with the inclusion of events [165]. Zeigler [234] specifies a number of formalisms—most notably, DEVS—that build upon the classical canonical systems definition and elaborates upon the key definitions of internal versus external events. Nance [156] provides a discussion of time and state for simulation model definition. The concept of "superstate" is a consequence of the system homomorphism and is also discussed in the software engineering literature [42]. Combined simulation models were the first form of multimodel and are discussed in [33, 174, 179]. The use of fuzzy numbers in simulation model execution is discussed by Fishwick [64, 65]. Jones [117] has a good discussion of data structures for event scheduling. Brown [32] introduced the concept of *calendar queues,* which are characterized more generally as dynamic hash tables. The texts by Banachowski et al. [13] and Horowitz and Sahni [109] serve as good introductions to some advanced data structures appropriate for event scheduling. Kosko [128] provides a good description for using neural networks and fuzzy logic maps as behavioral models.

## Chapter 3

Conceptual Modeling

In general, we mean by any concept nothing more than a set of operations; the concept is synonymous with the corresponding set of operations.[1]

3.1 OVERVIEW

Models containing components that have not been clearly identified in terms of system-theoretic categories such as state, event, and function are called *conceptual models*. The first type of simulation model that we want to create is a conceptual model that emphasizes objects and their relations to one another. From such a model, we can gradually progress to more system-theoretic constructs. The creation of a concept model is an early model engineering design stage. The topic of *model engineering* (or how to design, create, and revise models over time) is of central importance

---

[1]P. W. Bridgman, *The Logic of Modern Physics* [31], p. 25.

to systems study just as software engineering is a key aspect of computer science. Our first descriptions of systems will usually be in one of two forms: natural language text or pictorial. Always build a conceptual model before constructing other model types. The conceptual model serves as a knowledge base for not only the dynamic aspects of a system, but all knowledge about the system.

Conceptual models can be reformulated in terms of object-based models. Even though there are several different ways to encode conceptual models, including semantic networks or arbitrarily drawn graphs, the object-oriented modeling approach fits the purpose of allowing the construction of large repositories of static and dynamic information about a system, thus setting the stage for information retrieval and simulation within the same context.

The statement "John starts his automobile and travels to the store" can be considered a very high level system model—a text model. If we are to simulate such a "system," we will need to be able to translate from vague linguistic passages and pictures to derive compact mathematical models that permit us to reason about system behavior. Consider the following statement:

> John threw the baseball through the window which shattered onto the walkway and into the room. It struck a bedpost and left a narrow mark.

This seemingly innocuous statement causes many problems in interpretation:

- Which window was struck? We assume that the bedroom window was struck since the bedpost was hit and the bedroom window is near the bedpost.
- Where did the glass window shards land on the walkway and in the room? What portion of the shards landed in each area?
- To what does the pronoun "it" refer? Does "it" refer to the baseball or to a window shard? We have to resort to extra knowledge that tells us that a window shard is more likely to leave a narrow mark since the shard has sharp edges.
- What kind of narrow mark did the glass shard leave?

The more abstract a model, the more ambiguities are present and the less precise and unique our system behavior will be. However, we should not infer from this that building conceptual models is unimportant. To the contrary, if we want to understand model engineering, we have to start with vague, ambiguous models and proceed gradually to more refined versions. We see that for this kind of model a lot of *a priori* knowledge is required to answer some of the questions and, therefore, to simulate the model. This extra knowledge is stored in a knowledge base or database to which the natural language processor has access.

3.2 TEXT TO OBJECT TRANSLATION

The *class hierarchy* is a precursor of functional and declarative specifications of a program's behavior. The purpose of the class hierarchy is to store basic relationships and attributes associated with the domain problem. From the object model, we can "extrude" models defining behavior. The following are some heuristics to aid in creation of the object model from a textual description:

- *Make nouns classes or instances of a class.*
- *Use adjectives to make class attributes, subclasses, or instances.*
- *Make transitive verbs methods which respond to inputs.*
- *Use intransitive verbs to specify attributes.*

Let's consider some example sentences.

1. *The Ford truck accelerated down the highway.* We have the nouns "truck" and "highway," which serve as classes. The word "Ford" modifies the noun "truck" and can be made a subclass or instance of "truck" (that is, a "Ford" is one kind of truck). The verb "accelerated" provides a method for "truck" which will produce a change to truck attributes (such as position and velocity) as well as producing an output that is input to object "highway."

2. *The steam rises from the outlet valve at the top of the boiler.* All nouns are made classes or instances: "steam," "valve," and "boiler." If "valve" is treated as an instance, we make "outlet" an attribute for "valve"; if it is treated as a class, "outlet" serves to identify a type (i.e., subclass) of "valve." The phrase "at the top of the boiler" serves to identify an attribute of position for "valve." The verb "rises" serves as an association or relationship between "valve" and the outside of the system which is implied to exist even though it is not stated in the sentence. The verb "rises" can also specify an attribute of "steam."

3. *The gyroscope spins at a fixed rate.* The word "gyroscope" is the main object. The verb "spins" specifies a *reflexive* relationship from the gyroscope to itself since it is the only object and there is no direct interaction between two or more objects. Alternatively, we could create an additional implied object that must have started the action, such as a person or motor. "Fixed rate" is a link attribute which could also serve as an attribute to gyroscope; however, it relates mainly to "spinning" rather than to an object.

Using these heuristics, we create a class hierarchy for the ball scenario in Fig. 3.1. Note the use of (1) generalization (three types of balls), (2) aggregation (window is composed of shards, room and bed are composed of subitems), and (3) relationships encoded as methods (throw, strike, fall, mark). Normally, a class hierarchy will show only the hierarchical generalization realtionship (and not aggregation links); however, we include both in the same figure when there is sufficient room. This is by no means a unique way of representing all of the objects; however, it serves to permit us to build a knowledge base consisting of physical objects involved in the system. We now present heuristics based on grammatical categories.

3.2.1 Nouns and Pronouns

Nouns can be classes or instances; however, nouns may also be encoded as class attributes. Consider our previous example using a Ford truck: *Ford* is a type of *truck*, but it could be considered to be an attribute of the truck as well. That is, the truck class can have an attribute called "type." The suggested method to use whenever considering attributes that are "types" is to use a subclass or instance rather than an attribute. This is because our

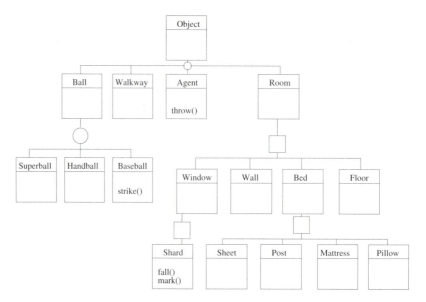

Figure 3.1   Class hierarchy: ball scenario.

object-based methodology has a specific structure for types (namely, the class hierarchy). Now, should *Ford* be a subclass or an instance of truck? The difference between class and instance is a subtle one. Make an object a class if you plan to develop your knowledge structure to the extent that there are levels below this object. Upon reading this, it seems almost too obvious; however, that is all there is to it—if there are several instances of a *Ford* truck, make *Ford* a subclass. Alternatively, if the word *Ford* clearly identifies a unique truck object in the representation, make *Ford* an instance of *truck*.

3.2.2 Adjectives and Adverbs

An adjective modifies a noun and therefore serves as an attribute value within the object for the noun. A "wet rag" refers to a rag object that has an attribute called *saturation level*, for instance. Note that an adjective gives us the *value* of the attribute field—not the attribute name. To obtain the attribute name, we must decide what category might include the adjective. For instance, if something is "wet," it might also be "dry" or "damp," since each of these adjectives forms a category called *saturation level*.

In systems, an attribute is either passive or active. A passive attribute is one that is unchanged by any of the methods stored in the object. An active attribute's value will change in accordance with the semantics defined by the object's methods. For instance, consider the statement "John hit the ball." We have two objects (John,ball) and one verb; however, there is an implicit attribute connected with the ball object—its new position as a result of the hitting action. In two dimensions, there would be two attributes X and Y. These attributes are changed by the object's methods (defined below).

Adverbs are very similar to adjectives except that they modify verbs instead of nouns. Therefore, an adverb specifies a link (rather than object) attribute value. The link attribute name is found in the same way as for adjectives—one forms a category to include the value and uses the category to define the link attribute name.

3.2.3 Verbs

In addition to attributes in an object, we can have methods. Methods define actions to be taken on the object's attribute values. When an attribute specifies a state of an object, a method explicitly defines how the state will change as a result of the relations that point to the object. Therefore, there is a strong connection between relations of arcs pointing to an object and the methods stored in that object; namely, the methods specify the semantics of how the attributes change over time.

Intransitive verbs such as "to be" suggest attributes (like adjectives); there is no object (in the sense of subject/object in grammar) associated with an intransitive verb. The statement "the cup is red" suggests an object or instance called *cup* and an attribute name called *color* (i.e., the category for red) whose value is "red." Transitive verbs connote an action associated with a subject and object. "The palette struck the automated guided vehicle (AGV)" provides us with a subject noun "palette," a verb "struck," and an object noun "automated guided vehicle." We make *palette* and *AGV* two objects and then form a relationship between them called *struck*. This relationship is encoded as a method `strike()` or `collide()` located in the `palette` object.

3.2.4 Prepositions

Prepositions operate within phrases, including nouns and pronouns, and they specify different types of information. Consider the following examples:

- "The valve is turned to the right." Here "to the right" modifies the link verb "turned," so we would create a link attribute under *turned* called "angular direction" containing the value "clockwise."
- "The valve was elevated by the cam." We can turn this sentence around to read "The cam elevated the valve," thereby eliminating the prepositional phrase.
- "The algae are in the lake." The use of "in the lake" serves to identify a location of the algae and so can be an attribute value within an object *algae*.

3.2.5 Conjunctions

Independent clauses, created from conjunctives, form the basis for a hierarchy in conceptual modeling. If "Jack turned the wheel while the water was boiling," we have two independent clauses: "Jack turned the wheel" and "the water was boiling." Moreover, both of these clauses are occurring simultaneously. For graphically based concept models, one way of handling conjunctions is to (1) consider each clause separately, (2) create separate object models, (3) create new objects that represent each object model, and (4) relate the new objects using a Petri net. Petri nets are discussed in Sec. 4.4.3.

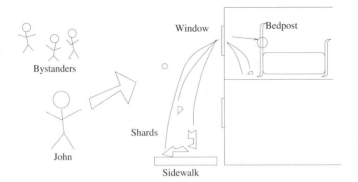

Figure 3.2   Picture model: throwing a ball.

3.3 PICTURE TO OBJECT TRANSLATION

Drawing pictures is another basic method for system description. It is the oldest representational method known for communication. Pictures from cave dwellings to oil paintings serve to show us some aspect of reality. Pictures are not just for artists; scientists use them to better understand systems and to communicate their system concepts to their colleagues. Most often, pictures are a representation for what our visual senses tell us. This means that when we draw conceptual models of systems, our first pictures will represent the *geometry* and *topology* of a system, not its *dynamics*. Graphical portrayals of *state* and *event* are far too abstract compared against pictures of the physical structure that we see with our eyes. For these reasons we need some ways of going from pictures to models.

As in the translation process from text to objects, we have similar problems when taking a pictorial depiction of a dynamic system to break it into subcomponents. We can ask questions such as "Where are the boundaries between different figures in the picture?" or "At what abstraction level should the object model be defined?" The following sections describe heuristics for identifying objects and classes.

3.3.1 Sketches: No Explicit Rules

We define a sketch to be a completely informal drawing of a system—meaning that we have used no particular rules in depicting each system component. Even though explicit rules have not been used, the artist or scientist has usually used implicit rules such as "draw a boundary around an object of interest" and "draw lines and arrows between interacting objects." Also, many objects, such as people, machines, and buildings, can be identified through some rather complicated pattern recognition process. Consider the ball-throwing scenario shown in Fig. 3.2. Humans can clearly identify physically separated objects such as the (1) people, (2) ball, (3) window shards, and (4) bed. Some objects, such as the bed, have subobjects which are drawn as smaller geometric objects (the posts and mattress). There are no explicit rules followed and we have been lackadaisical in our drawing of arrows and people—some people are smaller than others and two different types of arrows are drawn. What do they mean? Are the bystanders far away from John (the agent of the action) or are they controlling him (which could be inferred from the fact that they are drawn on top of

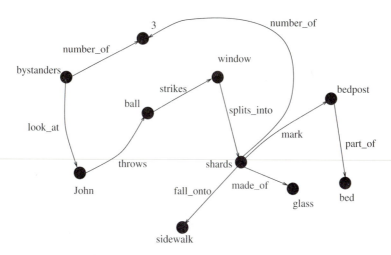

Figure 3.3   Conceptual graph: throwing a ball.

him)? Why are two types of arrows drawn? Does the thicker arrow denote greater velocity or is the thickness irrelevant?

3.3.2 Concept Graphs: Some Explicit Rules

Frequently, a graph (composed of nodes and arcs) is used to represent causality and time flow in a system. Since graphs are loosely structured, they serve a key role in building conceptual models. All graphs are guaranteed to have uniformity in that they all have vertices (nodes) and edges (arcs); this structure is a first attempt at creating some kind of order in defining the system. The next question relates to the semantics attached to each node and arc. That is, does each arc carry the same type of meaning? Are the semantics of each node grouped cleanly into categories? If we are very rigid about our graph formulation rules, we can specify dynamic models in a characteristic language such as finite state automata—where all nodes are states and all arcs are transitions to states. For a conceptual graph, though, we relax these conditions. For instance, it may be that some graph nodes represent events while others represent states. While there exists the overall graph structure, there is little formal structure in categorizing vertex and link semantics. Furthermore, some arcs may contain probability values, while others do not. To someone who is bound to implement a simulation, such a conceptual graph may appear frustrating since it lacks the cohesive structure necessary to increment time and update state; however, the conceptual graph plays an intermediate role in making it easier for people to communicate their system thoughts to one another. Figures 3.3 and 3.4 display example conceptual graphs. Conceptual graphs divide the world into rough categories and relationships. Some relationships will be dynamic, such as *throws* in Fig 3.3, and others will be static, such as *number_of*. In Fig. 3.4, some nodes are of type "physical object" (such as controller or steel), and others have to do with an activity (such as swinging phase). Our goal here should be to use the natural language text heuristics to categorize the text on each node and arc; we can use the existing arcs to make it easier to identify relationships between objects that are created. We will,

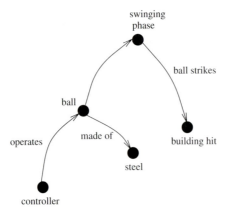

Figure 3.4   Conceptual graph: building demolition.

therefore, make the following objects: (1) controller, (2) ball, and (3) building. "Steel" is an attribute of ball and "swinging phase" represents a state value (or active attribute value) of the ball where the attribute would be "swinging" or "position."

3.3.3 Schematics: All Explicit Rules

If the picture represents a *schematic*, there are rules that have been followed in laying out components and icons. This helps considerably in creating an object model. Specifically, one should map the schematic's icons directly to objects and specify object attributes as those normally identified with the physical component represented by the icon. Standard symbols for resistors, capacitors, inductors, and transistors, for instance, enable a straightforward creation of objects. Moreover, straight lines drawn between two components become the relationships between objects. Generally, schematics of the physical objects engaged in dynamical activity are created by designers in computer-aided designs (CAD). It is possible that the schematic is a picture of the dynamical model itself (such as a finite state automaton, Petri net, or block diagram). In this case, we are indeed fortunate, since there is a straightforward translation of these icons to an executable simulation, as will be seen in the remainder of the book.

3.4 MAPPING CONCEPTUAL MODELS TO C++ CODE

3.4.1 Class Structure

C++ is an object-oriented language that is an extension to the C language. It is upwardly compatible with C, so that if code is written in C, C++ is a logical choice if object-oriented methods are to be employed. We have seen how to create object models for the "ball throwing" scenario. Now, we will study how to take these object models and translate them into C++ code. First, it is useful to think of all constructs in C, such as `int i,j,k;` or `struct card deck[52];`, as being object constructions instead of data declarations. In the first example, we are constructing (i.e., building) three objects of an integer "class." In many programming languages, such as Pascal and C, we have a mental model that defines a program as a list of data structures followed by separate code which is normally hierarchically organized into subprocedures. In object-oriented programming languages, we must abandon this model in favor of object creation and destruction—where objects encapsulate both data and function. For simulation, the object-oriented approach is quite natural since physical objects that are being modeled are efficiently mapped to objects in our program. When we map a concept to be modeled to an object, it is most useful to think of all of the object's attributes as being the data (inside the object) combined with methods which manipulate and update that data. For simulation, part of the data will be the object's state description. Therefore, to create an object called a "deck" (i.e., of cards), we will include data describing information on each card, but also actions that are associated with the deck, such as "shuffling," "dealing," or "throwing," each of which changes the deck's state information. Objects are instances of a class, so when we review the pictorial form of an object shown in Fig. 1.7, we can create the following C++ code:

```
//
// Piece of code declaring a C++ class along with
// two of its methods
//
#include <stream.h>
class ClassName {
private:
   int attribute1, attribute2;
   float attribute3;
public:
   ClassName(int) {attribute1 = 3;}
   ~ClassName(void) {}
   void method1(int);
   int method2(void) {attribute1++; return (attribute2 + 5);}
   float method3(int, float);
}; // end ClassName

void ClassName::method1(int i) {
   i += 10; cout << i; }

float ClassName::method3(int i, float f) {
   f = i + attribute1 + attribute3;
   return(f); }
```

This piece of code is not a complete C++ program since it does not contain a `main()` section; however, it serves to introduce the basic format for defining classes. The name of the class is `ClassName` and three attributes are defined. The attributes, in this example, can be accessed only by the methods (1, 2, and 3) and not by another method defined outside `ClassName`. The keyword `private` can be left out. The "methods" are just procedures and functions that operate on the attributes. In simulation, these routines will often update the attributes periodically as time increases (because some of the attributes designate state variables). The semantics for each method can be defined either (1) in-line (inside the class definition as for `method2`), or (2) outside using a scope specifier. The scope is specified along with the method in the following form: `classname::method`. The in-line approach is similar to using macros in other languages: use in-line code when the procedure is small (in terms of object code size) and called frequently, especially in a part of code that is computationally intensive.

3.4.2 Methods and Message Passing

From the perspective of an arbitrary C++ program, once one has set up all the classes, it is necessary only to create a main program and other routines that will invoke certain methods in order to set and retrieve attributes. Attributes are accessed by the methods defined within the same class.[2] For instance, note that `method2` changes attribute1. We call `method2` by first creating an object of class ClassName, `ClassName object1;`, and then specifying the call `object1.method2();`. All methods that are defined within classes are invoked in this fashion: `object.method(args)` where *args* represents an argument list.

Now, from the perspective of conceptual models for simulation, we will start with a conceptual graph. Consider, for instance, part of that graph where we have two objects A and B related by a relation (or association) called m. "m" is also known as a message that A sends to B. Therefore, you can look at the interaction between A and B as either (1) A is related to B by m, or (2) A sends a message m to B; both metaphors will map to the same C++ code.

Consider an example: say that object A represents "John" and object B represents "base ball" while m is "hits." The subgraph $A \rightarrow B$ represents the sentence "John hits the base ball." The question is now: How do we code this relationship within C++? Here are some general rules to follow:

- For a conceptual graph, we can code the graph as a C++ graph data structure where objects point to other objects.
- Let each object have two attributes: `inputs` and `outputs`. Each of these attributes is a pointer to a linked list of object items where each item contains (1) an object reference, (2) an optional value, and (3) a pointer to the next object.
- To cause message passing to take place, have a `switch` statement in an object's primary "execution" method. The switch statement will be of the form

```
switch (message) {
message1: method_message1();
message2: method_message2();
. . . .
messageN: method_messageN();
}; // end switch
```

For our example, the "baseball" object would contain a method called "hit" or "throw" which would execute a method that handles "hitting." The hitting of the baseball involves

---

[2]This is not necessary if we were to make the attributes `public`; however, it is generally good programming practice to design for this sort of encapsulation of data.

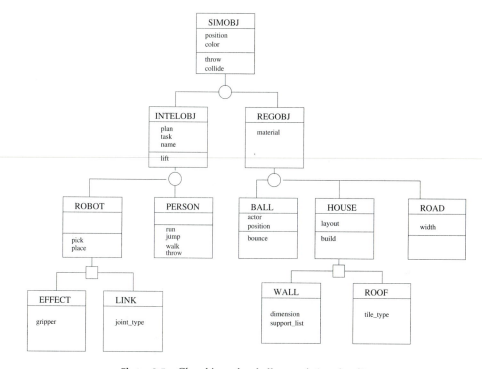

Figure 3.5   Class hierarchy: ball scenario(version 2).

semantics which update certain attributes of the ball, such as position and orientation in space. With many messages, there is a corresponding effect on the actor (i.e., object performing the action) as well as the object acted upon. In our throwing example, John may well move 1 or 2 feet forward in space while he performs an action. Figure 3.5 shows another possible class hierarchy for ball throwing. We now represent the class hierarchy in Fig. 3.5 in C++ code.

```
//
// C++ Class for the ball-throw scenario
//

#include <stream.h>
class Simobj {
   struct {float x, y;} position;
   enum {blue, red, brown, orange, green, white, black} color;
 public:
   void throw(void);
   void collide(Simobj*);
};
```

```
class Intelobj: public Simobj {
   char* name;
   Task current_task;
   Plan current_plan;
 public:
   void lift (void);
};

class Regobj: public Simobj {
   public:
   enum {concrete, plastic, rubber, cloth} material;
};
class Robot: public Intelobj {
   EffectorList effectors;
   LinkList links;
 public:
   Robot(void);
   ~Robot(void);
   void pick(void);
   void place(Simobj&);
};

class Person: public Intelobj {
   Regobj& object;
 public:
   Person(void);
   ~Person(void);
   void run(void);
   void jump(void);
   void walk(void);
   void throw(void);
};

Class Ball: public Regobj {
   Person& actor;
   struct {double x, y, z;} position;
 public:
   Ball(void);
   ~Ball(void);
   void bounce(void);
};

class House: public Regobj {
   Plan layout;
   int walls [4];
   Roof root;
 public:
   House(void);
   House(void);
   void build(void);
};
```

```
class Road: public Regobj {
   int width;
public:
   Road(void);
   ˜Road(void);
};
```

The highest-level object is `Simobj`, which stands for "simulation object." There are two types of objects in our simulation: an intelligent object (`Intelobj`) and a regular, unintelligent object (`Regobj`). To create an object, you use the classname in front of the object (i.e., `Person john;`). You will want to create an object from a class that is a "leaf node" in the generalization hierarchy. So you will want to create a `House` object, but not a `Regobj` object. The intermediate class nodes in the generalization hierarchy exist for purposes of organization and inheritance of class properties.

Inheritance of attributes and methods occurs where there are generalization hierarchies. So, for instance, `Person` will inherit attributes and methods from `Intelobj`, which, in turn, will inherit from `Simobj`. Attribute `position` in class `Ball` has the same name in class `Simobj`. In this case, C++ will use the correct version of `position`, depending on which object is created. If a ball is created, `ball's position` is used, whereas `Simobj's position` is used when a `House` or `Road` object is created.

There are two aggregation hierarchies in Fig. 3.5: one for `Robot` and the other for `House`. These hierarchies can be represented in C++ as data structures; they are not part of the class hierarchy that operates using inheritance. Therefore, if we create a `Puma` robot from class `Robot`, we need a pointer inside the object to point to all children of `Puma` which will specify the types, positions, and orientations of links, joints, and end effectors. Associations are coded as pointers as well. For instance, there is a relationship between the classes `Person` and `Ball`: that is, a person can have an effect on a ball. But, as we discussed earlier—in general—both actor and object can be affected, so we included a reference to the regular object in `Person` (`Regobj&`) as well as a reference to the actor in `Ball` (`Person&`). Note the difference, in C++, between a reference and a pointer. A reference denotes the object itself and can be thought of as the address of the object. A pointer creates a separate memory location that *contains* the address of the object.

3.5 EXERCISES

1. Conceptual models are nonexecutable and without a time base; however, this makes one wonder why we create such a model in the first place. Provide justification for conceptual modeling.

2. Build a conceptual model using objects and classes of someone who is going through the process of building a simulation using what you have learned in Ch. 1.

3. Text-to-object translation: translate the following pieces of text into class/object models by first performing a grammatical analysis and then by drawing the models. It is all right to specify knowledge that you have about the domain that is not in the text itself.
   (a) "The person waited in the queue for lunch. Lunch was served quickly."
   (b) "The population of fish in Lake Alice increased at an exponential rate."
   (c) "The reactor vessel rods became overheated after the coolant level dropped."
   (d) "The wheel turned the gear which turned the steel shaft."

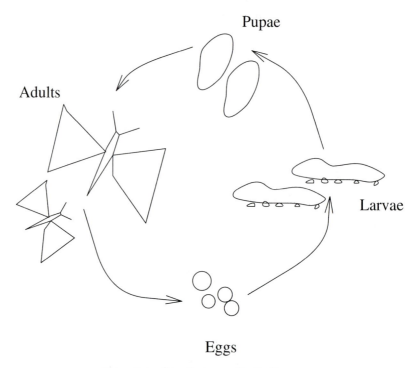

Figure 3.6   Growth stages of butterflies.

   (e) "The chess queen moved down a file to take the opponent's rook."
   (f) "The disk head was automatically retracted before a seek command was issued."

4. Create an object model for Fig. 3.6 and add new classes to form generalizations and aggregations.

5. Create class and aggregation hierarchies associated with tree growth models.

6. Given a scenario involving a population of ants on a patch of ground, provide an object-oriented definition, including C++ data structures.

3.6 PROJECTS

1. Build a natural language parsing tool that can parse a sentence and automatically produce the skeletons of objects—that is, their general frames with one or more slots (i.e., object name, methods, attributes) filled in, depending on how much information is in the sentence. For this project, focus on narrow domain and include a compiled database (or file) of objects whose embedded information can help in resolving ambiguities.

2. Pick an artist and create a report on how an object class hierarchy (with attributes and methods) could be constructed to capture the features of the artwork. Try to identify paintings that contain not only attributes (static and dynamic), but also methods in the form of perceived action.

3. Explore the area that lies between creating a conceptual or simple semantic model for a process and a more formal model that is executable and time-based. You should do this project after reading some of the upcoming chapters. How can you guide the user to going from the rough model to the more expressive model?

4. Use the *Exodus* database package (see the reference in the *Software* section) to build a conceptual model, stored as persistent objects, for any particular microworld. When you choose the world, you will need to (1) identify all objects, attributes, and methods, (2) create persistent objects in *Exodus*, and (3) build a simulation program which executes object methods in order to change their state attributes. This project is best when combined with one of the system modeling problems given in the remaining chapters.

3.7 FURTHER READING

Conceptual models have been discussed in a variety of technical communities. Sowa [209] provides an excellent introduction to the various approaches utilized to support this kind of modeling. A methodology for "soft" systems is specified within systems science by Checkland [38] and Flood and Carson [75]. Van Gigch [223] provides a broad systems-theoretic view of modeling. In the artificial intelligence literature many concept or "mental" models [87] are discussed under the area of knowledge representation [29]—specifically, in the discussion of semantic networks [59, 231] and causal networks [108]. In social science, causal methods have been used for some time in path analysis and probabilistic specification of causal links [23–25, 88].

Within the simulation community, there have been studies of the use of natural language as a language for model specification [7, 17, 18, 103] and the qualitative nature of models for simulation [35, 36, 69, 70, 72, 86, 182, 229]. Simula [22] was the pioneer of object-oriented languages for simulation. Since the advent of Simula, simulation has often been associated with Smalltalk [28, 142], which encourages a uniform view on using objects for all manipulation, and more recently the object-oriented system entity structure development [112, 192, 193, 235, 237, 238]. The DEVS approach has been implemented using an OO approach by Kim [122]. Kreutzer [130] provides a discussion on the use of object-oriented programming in simulation. Full-fledged model engineering environments (supporting model design from conceptual model to detailed model) [62, 63, 73] will be available in future simulation packages. Software engineering methodology has recently focused on real-time systems and systems modeling where object models are used first to define system knowledge in the abstract [26, 27, 195].

There are several good programming texts on C++ [46, 215], and there are, additionally, books that help move from C to C++ [171].

3.8 SOFTWARE

For developing conceptual models, it is often useful to use a good prototyping language interpreter with dynamic binding capabilities such as LISP or Prolog. For efficiency and portability reasons, C and C++ are good choices for implementing models that have gone beyond the conceptual stage. Gnu C and C++ are all that are necessary to use the SimPack and SimPack++ toolkits. Free versions of most computer languages exist over the Internet. For persistent object storage, you may want to go beyond C++ to storing objects and performing queries within an object-oriented database (OOD). Many OODs work directly with object-oriented languages so that the OOD works as an *extension* to the language. For instance, the *Exodus* persistent database software can be obtained from `ftp.cs.wisc.edu`.

Chapter **4**

Declarative Modeling

STORYBOARD: A large board on which are pinned sketches telling a story in comicstrip fashion.[1]

4.1 OVERVIEW

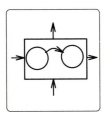

If you consider all of state space as a multidimensional jigsaw puzzle, declarative models tell you how to get from puzzle piece A to piece B by acting upon input or control from the outside. The declaratively modeled transition from one state to another within such a state space is similar to many common processes, such as building a storyboard for an animated movie. Declarative models contain two primary components: states and events. The declarative modeling approach suggests that we look at the world through a sequence of changes in state. The focus is on state rather than

---

[1] Bob Thomas, *Disney's Art of Animation* [220], p. 204.

on the function that transfers input to output. There are two classes of declarative model that concern us: (1) element mapping methods and (2) set mapping methods with respect to state space. The simplest forms of declarative model are where we specify points in a multidimensional state space with transitions between point pairs. A point in state space is an element of that space. Although this provides a very simple method of modeling, there remains the deficiency of having to denote explicitly every state space element and its associated transitions to other elements. A classic way around this difficulty is to allow a mapping of patterns rather than a mapping of point elements. A pattern will correspond to a subset of state space. By using patterns, we are in effect creating *anonymous* functions, which map one set onto another.

Declarative models are very good for modeling problem domains where the problem decomposes into either discrete *temporal phases* or *irregular spatial phases*. Temporal phases are identified as natural partitions over time where each phase corresponds to a cognitive or linguistic element. For instance, we can take any action and break it into subactions. The subactions are phases of the action. Phase transitions are accomplished through events which move a system through phase space. Declarative models are not the best model when the system is seen as a collection of objects, or when state changes are continuous in nature. In the first instance, a system with distinct objects is best modeled as a functional model where each function contains a declarative model specifying discrete state or phase transitions. In the second instance, even though declarative models are perfectly capable of modeling continuous systems, continuous systems are best modeled using functional or constraint models. Heuristics for employing declarative models are:

1. If the system has discrete states or events, specify them using a declarative model.
2. If there are *phases* of a process, use a declarative model to model phase transitions. Phase transitions can be *temporal* or *spatial* if the spatial regions are irregular.

All declarative models are construed to be *embedded* within an object. Therefore, a finite state machine representing phase changes in a four-cycle engine is embedded within the object called `engine`. Declarative models are suitable for representing the dynamics within an object that is part of a larger *functional model* to be discussed in Ch. 5. Figure 4.1 shows the basic form for a declarative model. The model is associated with discrete state changes inside the `object`. The state changes are performed by the method `trans()`, which encodes the state transition function (or table). `object` is generally coupled with other functions that accept input from `object`'s output, to form a functional block model. Let's consider the two inputs and two outputs of `object` since they will play a major role in linking declarative models together and helping us to form multimodels (Ch. 8). There is a single external input line into `object`. This is where inputs are accepted by the declarative model. The declarative model changes state when a condition on the relevant transition arc is satisfied. Time is incremented as follows: `object` will stay in state x for a period of time that comes directly from the time base associated with the input signal. That is, if the input signal is unchanged in value and none of the output arc conditions evaluate to true, the system will stay in that state until a condition is satisfied. `object` is displayed as being on abstraction level i. It receives input from some other object on

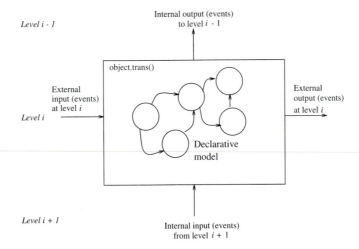

Figure 4.1   Form for a declarative model.

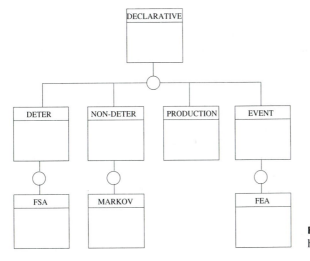

Figure 4.2   Declarative model class hierarchy.

the same level. The other type of input comes from an abstraction level $i + 1$ which is below level i. These two types of inputs (from level i and $i + 1$) are termed *external events* and *internal events*, respectively, since external events come from the *outside* and internal events come from *inside* the system but from a lower abstraction level. Internal events are inputs to the system from the next-lowest abstraction level, $i + 1$, and this type of event is particular to *multimodels*, where there exists more than a single abstraction level defining the system model. We have now covered the inputs. There are two outputs that can be described in much the same way as for the inputs. One output is on the same level i, and the other serves as input to an unspecified model on level $i - 1$.

Figure 4.2 displays the subclass hierarchy of Fig. 1.8 for declarative models. The basic types of declarative model are deterministic, nondeterministic, production rule/logic, and event.

4.2 STATE-BASED APPROACH

4.2.1 Deterministic Automata

Finite state automata (FSA) permit us to model systems with a minimum of components. Compared with the behavioral approach, FSAs permit more flexibility for modeling since they permit the integration of *state*, *input*, and *output*. Behavioral modeling allowed us to implement only *input* and *output* as our system description. An FSA is a system with *states* and *transitions*. A state represents the current condition or "snapshot" of a system for some length of time. Transitions enable the system to move from one state to another during the simulation while under the control of the system input (internal or external). The transition is labeled with a set of inputs. If the system, for instance, is in state A and there is a transition from state A to state B labeled with $i = 1$ (input equals 1), and an input equal to 1 arrives at time t, the system moves directly to state B.

A basic definition of system $\langle T, U, Y, Q, \Omega, \delta, \lambda \rangle$ provides the semantics for declarative models.

- T is the *time* set (or time base). $T = \mathcal{R}$ (reals), and for discrete time systems, $T = \mathcal{Z}$ (integers).
- U is the *input* set containing the possible values of the input to the system.
- Y is the *output* set.
- Q is the countable *state* set.
- Ω is the set of *admissible* (or *acceptable*) input functions. This contains a set of input functions for an automaton. Due to a discrete state space, input functions will map specific times to input values.
- δ is the transition function. It is defined as $\delta : Q \times \Omega \to Q$.
- λ is the output function, $\lambda : Q \to Y$.

Input for an FSA can be treated as *synchronous* or *asynchronous*. For an asynchronous treatment, whenever an input changes value, a state change can occur in the FSA. For synchronous operation, however, a single state change occurs as a result of an input event. Our treatment of FSA operation will revolve around synchronous operation.

The algorithm for simulating an FSA is:

Algorithm 4.1. FSA Simulator
Procedure Main
        Set current time to zero
        Set start state to any state in the automaton
        While not(end of input) do
                Output current state information

> *Update current time by* ΔT
> *Transition to next state using current input*
> *End While*
> *End Main*

A state describes a "snapshot" of a system at one point in time. A state may stay in effect for some interval of time, or state-to-state changes may be instantaneous. When input is not defined within an FSA, we simply have a state trajectory defining the model, in which case the FSA can alternatively be formally represented by a table that stores the invariant sequence of states. Input to a system is described either as a discrete string or a continuous trajectory. If the input is defined as a string, it may or may not have time specifically associated with it. For instance, consider a simple FSA with n states, s_1, s_2 through s_n, where $s_i \rightarrow s_{i+1}$ defines each transition. Furthermore, label each transition from s_i to s_{i+1} with a value of 1 and then make reflexive transitions from s_i to s_i with a label of 0. This means that when the input string value is 0, the system stays in the current state, and when it is 1, a transition to the next state occurs.

A four-stroke gasoline engine has four phases (see Fig. 4.3) which repeat until the engine is turned off. In the *compression* phase, injected fuel vapor is compressed. Then the fuel is ignited in the *ignition* phase. Ignition causes the piston head to move back in the cylinder in the *expansion* phase, and finally, the *exhaustion* phase pushes the resulting fumes to exit through the exhaust manifold and on through the tailpipe of the car. We will assume that the four phases continue in a cycle unless the ignition is turned off. At such a time, the system will move to an *off* state. The system is depicted in Fig. 4.4. There is one input signal, for our purposes, to the FSA in Fig. 4.4. This signal has the value 0 when the FSA is to remain in its current state. An input pulse[2] of 1 will drive the FSA to the subsequent state, and an input of 2 reflects that the ignition is turned off. A sample input trajectory is shown in Fig. 4.5, along with the corresponding state trajectory in Fig. 4.6. Let's more clearly define the input signal values by representing the input as a two-bit binary number. The lower-order bit is a control signal from the fuel injector logic encoded on an integrated circuit (IC). Thus, inputs of 00 (0) or 01 (1) come from the IC. An input of 10 (2) comes from the ignition when it is off. With the given input signal in Fig. 4.5, the engine is initially off (input is 2). At time 4, there is an input of 1, which causes a state change from off to compression. The system is in the state compression for 5 time units. A pulse causes the next change of state to ignition, which lasts 1 time unit. Finally, expansion and exhaustion occur. The amount of time spent in a state depends on its time-based driving input function, which controls the time duration. The definition of the engine FSA is:

1. $FSA = \langle T, U, Y, Q, \Omega, \lambda \rangle$
2. $T = \mathcal{Z}_0^+$

---

[2]The pulse causes a single-state change in a synchronous fashion since the state change results from an input *event* (and not an input *value*), where the event is a "leading edge" trigger.

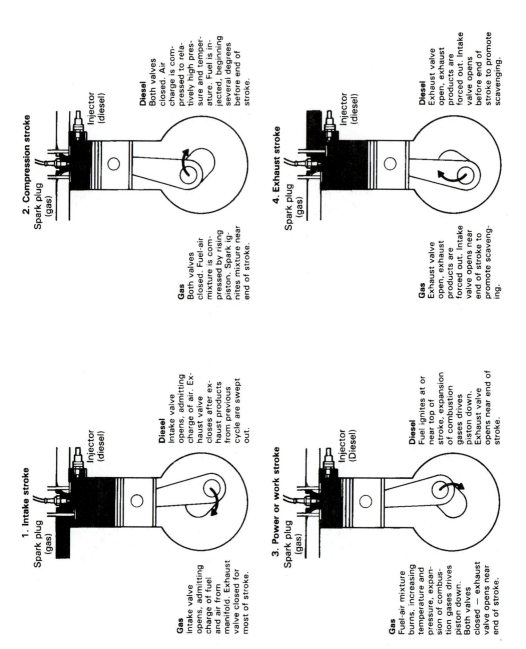

1. Intake stroke

Spark plug (gas)

Injector (diesel)

Gas
Intake valve opens, admitting charge of fuel and air from manifold. Exhaust valve closed for most of stroke.

Diesel
Intake valve opens, admitting charge of air. Exhaust valve closes after exhaust products from previous cycle are swept out.

2. Compression stroke

Spark plug (gas)

Injector (diesel)

Gas
Both valves closed. Fuel-air mixture is compressed by rising piston. Spark ignites mixture near end of stroke.

Diesel
Both valves closed. Air charge is compressed to relatively high pressure and temperature. Fuel is injected, beginning several degrees before end of stroke.

3. Power or work stroke

Spark plug (gas)

Injector (Diesel)

Gas
Fuel-air mixture burns, increasing temperature and pressure, expansion of combustion gases drives piston down. Both valves closed — exhaust valve opens near end of stroke.

Diesel
Fuel ignites at or near top of stroke, expansion of combustion gases drives piston down. Exhaust valve opens near end of stroke.

4. Exhaust stroke

Spark plug (gas)

Injector (diesel)

Gas
Exhaust valve open, exhaust products are forced out. Intake valve opens near end of stroke to promote scavenging.

Diesel
Exhaust valve open, exhaust products are forced out. Intake valve opens before end of stroke to promote scavenging.

Figure 4.3 Phases of a four stroke gasoline or diesel engine. (From *Mitchell Automechanics* © 1986, pg. 309. Reprinted with permission of Prentice Hall, Englewood Cliffs, NJ.)

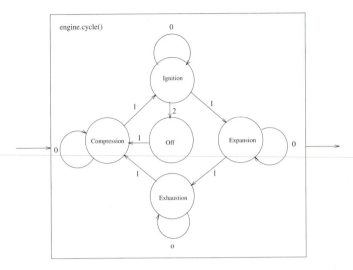

Figure 4.4   An FSA for a four stroke engine.

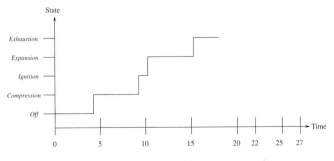

Figure 4.5   Engine input trajectory.

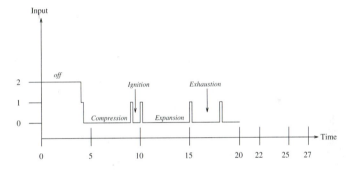

Figure 4.6   Engine state trajectory.

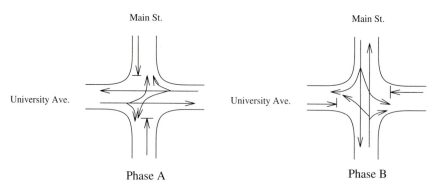

Figure 4.7   Two phase traffic control.

3. $U = \{0, 1, 2\}$

4. $Y = Q$

5. $Q = \{Compression, Ignition, Expansion, Exhaustion, Off\}$

6. Ω is defined in Fig. 4.5.

7. $\delta : Q \times \Omega \rightarrow Q$

8. $\lambda : Q \rightarrow Y$

Automobile traffic control modeling.   Control of vehicular traffic through intersections involves the use of a traffic control box located at one corner of the intersection,
or in a more central location if traffic flow uses remote control. During a phase, the traffic
flow directions do not change—for instance, cars may be moving straight ahead or turning
to the right during one phase. The traffic control box contains four sets of three lights, and
the number of states in the the control system depends on the type of switching that one
wants to implement in the control box. A simple two-phase control method is one where
there are two different flow patterns permitted (shown in Fig. 4.7).

Even though this standard control mechanism is termed "two-phase," there are actually
four states if we represent a state as a cross product of colors being displayed at a time. The
need for four states arises because there is always a transition from green to amber and then
from amber to red. One side of a traffic box is composed of three lights. From bottom to
top they are red (R), amber (A), and green (G). Since there are two directions for traffic, we
can represent a state as a 2-tuple. The traffic on University Avenue will be the first value in
the tuple, and the traffic on Main Street will be the second value. If, for instance, traffic is
moving east-west on University Ave., the state is (G, R). Consider the following situation.
Our intersection of University Avenue and Main Street (in Fig. 4.7) will now be controlled
whenever a car triggers a sensor located just before the intersection is reached. At first,
this kind of controller seems ideal; cars can continue on University Avenue until another
car needs to cross the intersection on Main Street; there is no need to wait if there are no
cars waiting to cross. The problem, though, is that if there is more than a little traffic on
both roads, the traffic lights will change erratically. The light switches whenever triggered.

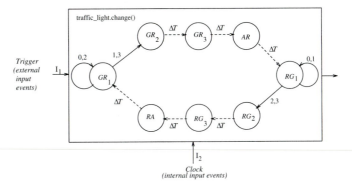

Figure 4.8   Segment of traffic FSA with triggering.

What is needed is a "minimum delay," where once the north/south light is green, it will stay green for *at least* a certain delay time regardless of east/west triggering.

We achieve the minimum delay with an internal event[†] representing a clock that runs within the states GR_2, GR_3, AR, RG_2, RG_3, and RA. The clock begins at zero when the state is entered and runs until the elapsed time is ΔT. The internal event is a binary-valued signal which is normally set to 0. When a state, with an internal event transition, is reached the internal clock pulses and sets the internal input to 1 after ΔT seconds have elapsed. This transition is denoted in Fig. 4.8 by labeling the transition with "ΔT". We create two triggers: one for east/west and the other for north/south. An input of zero on either sensor means "not-triggered" and an input of one means "triggered." Call the east/west trigger binary variable T_{EW} and the north/south trigger T_{NS}. Then we can encode the external input as a single number using a binary string created from the pair (T_{EW}, T_{NS}). For example, an input value of 2 [i.e., $(1, 0)$] means that the east/west trigger is on and the north/south trigger is off; and input of 3 means that both roads are triggered. Figure 4.8 displays an FSA set up to handle input triggering. Consider the case where the east-west traffic is flowing. This means that the trigger value can be either 0 (00) or 2 (10). As soon as a car passes the trigger point on the north-south road, the trigger value will be either 1 (01) or 3 (11). At this point, the FSA transitions from state GR_1 to GR_2. State GR_2 is active for ΔT seconds, as is state GR_3. The lights do not change, then for $2\Delta T$ seconds after the north-south trigger. At time $2\Delta T$ seconds after the trigger, the lights change to AR and stay in that state for ΔT seconds. Then the lights change to RG (when in state RG_1). Let us now define an FSA for the traffic control box. First, in order to convert the internal state events afforded by the clocks, we create an additional input (I_2) to the FSA from the bottom. This input will have a pulse of 1 when ΔT seconds have elapsed and be zero at all other times. The external input will be I_1. When specifying input, we use a (I_1, I_2) pair and let X represent a *don't care.*

   1. $FSA = \langle I, O, Q, \delta, \lambda \rangle$
   2. $I_1 = \{0, 1, 2, 3\}$

[†]The definition of "internal event" is defined more precisely in Ch. 8.

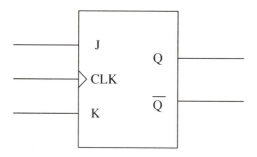

Figure 4.9   Schematic of a master slave JK flip flop.

3. $I_2 = \{0, 1\}$

4. $I = I_1 \times I_2$

5. $Q = \{GR_1, GR_2, GR_3, AR, RG_1, RG_2, RG_3, RA\}$

6. $O = Q$

7. $\delta : Q \times I \rightarrow Q$

8. $\lambda : Q \rightarrow O$

9. $\delta(GR_1, (0, X)) = GR_1, \quad \delta(GR_1, (1, X)) = GR_2, \quad \delta(GR_1, (2, X)) = GR_1,$

$\quad \delta(GR_1, (3, X)) = GR_2, \quad \delta(GR_2, (X, 1)) = GR_3, \quad \delta(GR_3, (X, 1)) = AR,$

$\quad \delta(AR, (X, 1)) \ = RG_1, \quad \delta(RG_1, (0, X)) = RG_1, \quad \delta(RG_1, (2, X)) = RG_2,$

$\quad \delta(RG_1, (1, X)) = RG_1, \quad \delta(RG_1, (3, X)) = RG_2, \quad \delta(RG_2, (X, 1)) = RG_3,$

$\quad \delta(RG_3, (X, 1)) = RA, \quad \delta(RA, (X, 1)) \ = GR_1$

Digital counters.   State-based models are widely used in digital logic design and simulation. Consider the basic operation of the master-slave JK flip-flop with schematic depicted in Fig. 4.9. Three of the JK flip-flops can be coupled together in sequence to create a three stage counter. The first JK flip-flop is denoted LSB for least significant bit, and the third flip-flop is denoted MSB for most significant bit. The JK flip-flop is used for memory and delay purposes. A flip-flop has the ability to remember its last state and hence can be modeled using an FSA. The relationship between the inputs and states of the JK flip-flop circuit is shown in Fig. 4.10.

Figure 4.10 demonstrates that this flip-flop is a system with more than one input; however, for the purpose of using a connection of flip-flops as a counter, we will use a unique input string. The behavior of the three-stage counter is presented in Fig. 4.11 and the FSA is shown in Fig. 4.12.

The definition of the binary counter FSA is:

1. $FSA = \langle I, O, Q, \delta, \lambda \rangle$

2. $I = \{1\}$

3. $Q = \{0, 1, 2, 3, 4, 5, 6, 7\}$

Inputs		Current state	Next state
J	K	Q_n	Q_{n+1}
0	0	0	0
0	0	1	1
0	1	0	0
0	1	1	0
1	0	0	1
1	0	1	1
1	1	0	1
1	1	1	0

(a)

$Q_n \rightarrow Q_{n+1}$		J	K
0	0	0	★
0	1	1	★
1	0	★	1
1	1	★	0

(b)

Figure 4.10   Behavior of the JK flip flop.   (a) Input/state transitions.   (b) State to state transitions. ★ means "don't care."

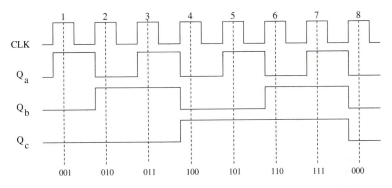

Figure 4.11   Behavior of three stage binary counter.

4. $O = Q$

5. $\delta : Q \times I \rightarrow Q$

6. $\lambda : Q \rightarrow O$

7. for $i \in \{0, \dots, 7\} : \delta(i, 1) = i + 1 \bmod 8$

Dining philosophers.   For the next two examples of graph-based models, we include "the dining philosopher's (DP)" example, which is based on the DP problem referenced in operating systems literature—it is a benchmark often used to illustrate a particular method for handling concurrency and resource allocation. Our discussion will revolve not around *solving* resource *problems*, but around DP as a process suitable for simulation purposes. In this sense, DP is treated simply as a physical process—which is no different than a running motor or a dynamically changing ecological system. To eat, each philosopher must pick up two chopsticks, ingest his food, and then place the chopsticks back on the table. Due to the

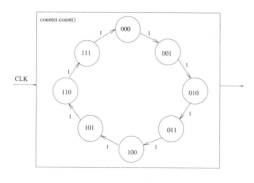

Figure 4.12 FSA for three stage binary counter.

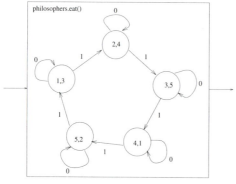

Figure 4.13 Finite state machine for DP.

contention for chopsticks (the resource), at most two philosophers (process components) may be eating simultaneously. Figure 4.13 displays an FSA that we will create to model the DP process. The DP model is specified below:

1. $FSA = \langle I, O, Q, \delta, \lambda \rangle$
2. $I = \{0, 1\}$
3. $Q = \{q_0, \ldots, q_4\} = \{(1, 3), (2, 4), (3, 5), (4, 1), (5, 2)\}$
4. $O = Q$
5. $\delta : Q \times I \rightarrow Q$
6. $\lambda : Q \rightarrow O$
7. for $i \in \{0, \ldots, 4\}$: $\delta(q_i, 1) = q_{(i+1) \bmod 5}$, and $\delta(q_i, 0) = q_i$
8. for $i \in \{0, \ldots, 4\}$: $\lambda(q_i) = ((i) \bmod 5 + 1, (i + 2) \bmod 5 + 1)$

I is the input set, Q is the state set, δ is the state transition function, and λ is the output function. The FSA models provide a completely deterministic description of DP. Pairs of philosophers simultaneously eat as the sequence progresses. We will presume a clockwise progression even though a counterclockwise progression would also be acceptable. (Note: The order present in FSA is independent of physical ordering of the philosophers.) The number of iterations is a function of $|X|$, the size of the initial input string ($X \in I^*$). The FSA models do not allow for any fluctuations in timing—for instance, the following partial sequence would be impossible: (1) philosophers 1 and 3 are eating, (2) philosopher

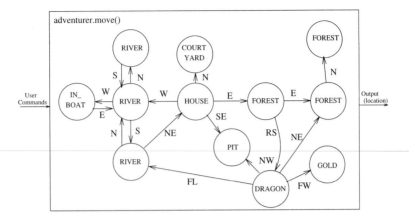

Figure 4.14   An adventure state automaton.

3 stops eating, (3) philosopher 4 starts eating. A Petri net model (Sec. 4.4.3) will remedy this problem.

Role-playing models.   There are some high-level languages for finite state machines. Normally, most adventure games treat time in a simplistic manner, by allocating a single time unit to every state transition (or "move"). For instance, it might take an entity (i.e., the player) one time unit for a move that can be made by specifying the direction of the move (north, east, south, west). Figure 4.14 displays an example FSA model that can serve as a basis for an adventure language. A simple adventure language can be considered as:

- An FSA with a state being an aggregate location (polygon) and a transition being a player move.
- An output for an adventure model would not be a simple short string (such as IN_BOAT), but instead, paragraphs describing the current location and other pertinent information for the player.
- The input control string, not specified in a static manner prior to the actual simulation, but as a series of asynchronous user prompts and interactive commands.
- The output connected to each state can be either a string pointer or an image pointer. String pointers would associate paragraphs of natural language text with one pointer, and image pointers would associate color images which visually depict the current simulation state.

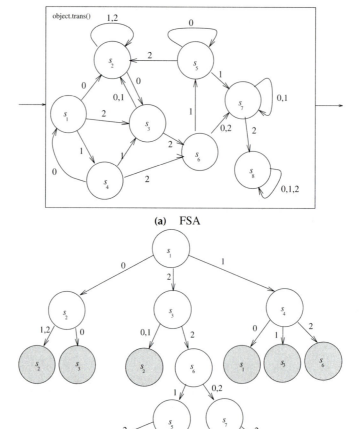

(a)   FSA

(b)   Reachability tree equivalent

Figure 4.15   Creating a reachability tree.

Reachability trees.   A reachability tree is an FSA with input drawn as a tree rather than a graph. This tree type of representation focuses on the number of states that can be assumed by the model; however, if drawn as a tree (rather than a graph), it does not specify if the individually drawn states are unique. Figure 4.15 displays an example FSA first drawn as a graph and then drawn as a reachability tree. The shaded states in Fig. 4.15(b) represent states that have already been reached, so it is not necessary to continue expansion from

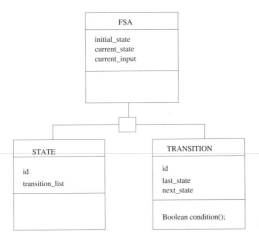

Figure 4.16   FSA model hierarchy.

that state. "Reaching" is accomplished by a state-to-state transition based on the current input value. For decision trees, sometimes the input is masked as a control value; however, control on a system is always affected by its external input signal (string). A "choice" at a given *decision point* plays the same role as an external input to the system.

Object implementation.   Figure 4.16 shows the subtree of FSA within the class hierarchy shown in Fig. 4.2. An FSA is composed of two types of objects: states and transitions. The state value is stored within the `id` attribute and the transition list is either an array or a linked list of all transition objects that leave that state. Transition objects also have an id attribute along with attributes that keep track of the previous state and next state since a transition has exactly one state at either end. The pointer to the `last_state` is useful when you need to backward chain from a state to determine a previous state. The methods `input()` and *output* are inherited from the `model` class shown in Fig. 1.8.

4.2.2 Nondeterministic Automata

Markov models are state machines like finite state automata; however, the major difference is that Markov machines are nondeterministic. That is, each arc in the Markov graph model is labeled with a probability of taking that arc when transitioning to another state.

A Markov model is defined as follows:

1. $MARKOV = \langle P, O, Q, \delta, \lambda \rangle$
2. $P = [0.0, 1.0]$
3. $\delta : Q \times P \rightarrow Q$
4. $\lambda : Q \rightarrow O$

P is the set of probabilities, Q is the state set, δ is the state transition function, and λ is the output function.

Another definition for Markov models is one where the model is defined as a sequence of random variables where each random variable can equal a state. Specifically, a stochastic process $X = \{X_n : n \in \mathcal{N}\}$ is called a *Markov chain* provided that

$$P\{X_{n+1} = j \mid X_0, \ldots, X_n\} = P\{X_{n+1} = j \mid X_n\}$$

for all $j \in Q$ (where Q is the state set) and $n \in \mathcal{N}$. This definition says that a state X_{n+1} of a system is independent of past states $X_1, \ldots, X_{n-1}$ as long as state X_n is known (i.e., state X_{n+1} depends solely on state X_n). It is common to represent a Markov model as a matrix P of joint probabilities where

$$P\{X_{n+1} = j \mid X_n = i\} = P(i, j)$$

$P(i, j)$ means "the probability of transitioning from state i to state j." The transition matrix looks like

$$
\begin{bmatrix}
P(0,0) & P(0,1) & P(0,2) & \ldots \\
P(1,0) & P(1,1) & P(1,2) & \ldots \\
P(2,0) & P(2,1) & P(2,2) & \ldots \\
\cdot & \cdot & \cdot & \ldots \\
\cdot & \cdot & \cdot & \ldots \\
\cdot & \cdot & \cdot & \ldots
\end{bmatrix}
$$

and can also be more conveniently represented as a graph (just as with the FSA).

The algorithm for simulating Markov models is:

Algorithm 4.2. Markov Chain Simulator
Procedure Main
        Set current time to zero
        Set start state to one state in model
        While not(end of simulation) do
                Output current state information
                Update current time by state time
                Transition to an adjacent state by sampling
                      a probability distribution
        End While
End Main

The first model will be a process described by the flowchart shown in Fig. 4.17(a). The corresponding Markov model is shown in Fig. 4.17(b). Note how flowchart assignment blocks (rectangles) and decision blocks (diamonds) are mapping directly to states. Also, the start and stop chart symbols are mapped to states. The percentages shown in Fig. 4.17(a) reflect the overall frequency with which certain decisions are made. The probability matrix

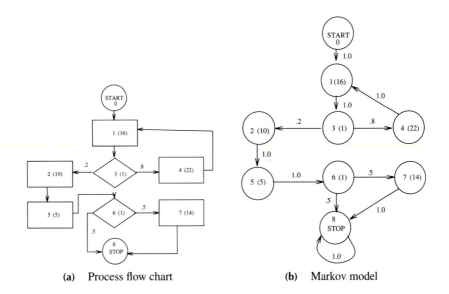

(a)   Process flow chart   (b)   Markov model

Figure 4.17   Markov model for a computer program.

for the Markov model in Fig. 4.17(b) is

$$
\begin{bmatrix}
0.0 & 1.0 & 0.0 & 0.0 & 0.0 & 0.0 & 0.0 & 0.0 & 0.0 \\
0.0 & 0.0 & 0.0 & 1.0 & 0.0 & 0.0 & 0.0 & 0.0 & 0.0 \\
0.0 & 0.0 & 0.0 & 0.0 & 0.0 & 1.0 & 0.0 & 0.0 & 0.0 \\
0.0 & 0.0 & 0.2 & 0.0 & 0.8 & 0.0 & 0.0 & 0.0 & 0.0 \\
0.0 & 1.0 & 0.0 & 0.0 & 0.0 & 0.0 & 0.0 & 0.0 & 0.0 \\
0.0 & 0.0 & 0.0 & 0.0 & 0.0 & 0.0 & 1.0 & 0.0 & 0.0 \\
0.0 & 0.0 & 0.0 & 0.0 & 0.0 & 0.0 & 0.0 & 0.5 & 0.5 \\
0.0 & 0.0 & 0.0 & 0.0 & 0.0 & 0.0 & 0.0 & 0.0 & 1.0 \\
0.0 & 0.0 & 0.0 & 0.0 & 0.0 & 0.0 & 0.0 & 0.0 & 1.0
\end{bmatrix}
$$

For the second example we will elaborate upon the algorithm necessary to simulate a Markov model. Let's reconsider the DP model, where we allow the following:

- Five states are defined to be the same as the FSA model shown in Fig. 4.13.
- When a particular pair of philosophers are eating, they will continue to eat with the following probabilities:

    —Pair $(1,3) = 0.1$
    —Pair $(2,4) = 0.4$
    —Pair $(3,5) = 0.1$
    —Pair $(4,1) = 0.2$
    —Pair $(5,2) = 0.2$

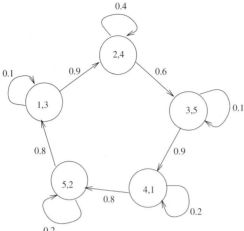

Figure 4.18   Markov model for **DP**.

We are supposing that philosophers 2 and 4 spend more time eating than do the other philosophers, hence the higher probabilities are eating again before handing over the chopsticks to the next pair. As in all simulation models, time is present. States and transitions can both have time tags attached to them. We will let each state take time 3 units, and each transition (from one state to another) take time 1 unit.

Figure 4.18 displays the Markov model. The Markov model shown in Fig. 4.18 is defined as:

1. $MARKOV = \langle P, O, Q, \delta, \lambda \rangle$
2. $P = [0, 1]$
3. $O = \{(1, 3), (2, 4), (3, 5), (4, 1), (5, 2)\}$
4. $Q = \{q_0, \dots, q_4\}$
5. $\delta : Q \times P \to Q$
6. $\lambda : Q \to O$

Markov models may also be defined with input. To provide for input as well as the influence of probability over a change in state, we specify a Markov model for each value that an input can assume. Figure 4.19 shows a two state/three input Markov model. The two states are $S1$ and $S2$ and the three inputs are 0, 1 and 2. This system is controlled by the input but is not determinate. Supposing our initial state is $S1$ and our input string is 0211. We have a 50% chance of staying in state $S1$ or continuing to $S2$—we'll assume that we stay in $S1$. Now, since the input changes from 0 to 2, we refer to the third model in Fig. 4.19. After sampling from a probability mass function (pmf) we move to $S2$ with a probability of 0.25. Then, we switch to the second Markov model since our last two input values are both 1. We now move back to state $S1$ with probability of 1.0. The FSA with input is a special case of the Markov model with input where the probabilities are either 0 or 1. For instance, in Fig. 4.19, if we replace all of the probabilities with those of 0 or 1, we now have a determinate system that is more appropriately drawn as a single FSA model.

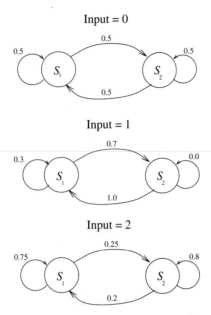

Figure 4.19   Markov model with input.

Scheduling in operating systems.   In a computer operating system, many processes are permitted to execute in a simultaneous fashion. Using a time slicing approach, one process may obtain access to the single central processing unit (CPU) for an interval of time. Then, the CPU process must yield the CPU to another process either because it has run out of time, or the process is waiting on an external resource such as a tape drive, disk drive, keyboard, or other peripheral device. We define that a process may be in one of five states:

1. *Running*: The process is running (i.e., has control of the CPU).
2. *ReadyA*: The process is ready to gain access to the CPU.
3. *ReadyS*: Same as *ReadyA* except that the process is suspended (as a result, for instance, of being swapped out to disk).
4. *BlockedA*: The process is blocked while requesting a peripheral device.
5. *BlockedS*: Same as *BlockedA* except that the process is suspended.

An FSA or Markov chain can adequately model the control process of scheduling jobs and processes for eventual (and iterative) execution. Figure 4.2 displays a Markov model assuming that statistics have been formulated based on data on the execution traces of a set of processes.

Object implementation.   Figure 4.21 shows the subtree of *MARKOV* within the class hierarchy shown in Fig. 4.2.

A Markov model is composed of two types of objects: states and transitions. The layout for this type of model is identical to the FSA hierarchy in Fig. 4.16 with the exceptions that (1) the default Markov model does not accept input, and (2) the transition is based on a value obtained from a random number generator method `rng()`.

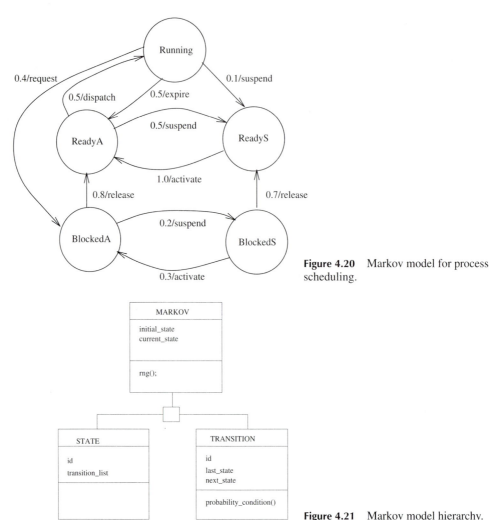

Figure 4.20 Markov model for process scheduling.

Figure 4.21 Markov model hierarchy.

4.2.3 Production-Based Models

Superstates. So far, we have treated a state vector as representing the entire state of the system. Diagrammatic representations of state-oriented models have used circles for states where a circle captured a set of distinct values for each state vector variable. For instance, the traffic light state was a two-tuple instantiation such as (R, G) or (R, A). We introduce some extensions to this basic idea by allowing for the concept of a *superstate*. A superstate is the state of a part of the system. Formally, a superstate is a simplified way of working with the homomorphism concept discussed in Ch. 8. For a traffic light, we might consider the east-west direction components of the traffic light box as having a state R, A, or G. These state values of the east-west components of the traffic box represent partial state information for the overall system: $(R, ?)$, where ? is a symbol that can match R, A, or G. The syntactical form $(R, ?)$ represents a morphism between three low-level states (R, R), (R, A), (R, G) and a lumped (i.e., super) state $(R, ?)$. A superstate is really just a subset of the system state space as discussed in Ch. 2; however, its power is to permit more complex declarative modeling constructs. Without being able to specify patterns, against which state vectors unify, declarative modeling would be limited to models delineating transitions from one unique state to another—each state in the model has to be denoted explicitly in all state-to-state transition specifications.

The superstate concept permits us to bridge the gap between the state space systems approach and advanced declarative modeling methods used in artificial intelligence, such as production systems and logic calculi (i.e., predicate, first order, second order). Superstates are nothing more than normal states; they are simply states for a higher-level process. Production system modeling can be seen as an application of superstates. The basic limitations of the earlier state-based modeling methods are seen in Fig. 4.22(a): one can specify a state trajectory (in the phase plane) by linking from one state to another. This creates our FSA or Markov model. However, for more power we will want to map a *group* of states to another *group* of states (refer to Fig. 4.22(b)). This, in effect, is how functions are defined, so such a group mapping method allows us to maintain a declarative approach to modeling while gleaning the power of functional mappings.

For example, consider the following set of mappings that are reflective of an FSA model:

$$(2,3) \longrightarrow (7,2)$$
$$(3,1) \longrightarrow (6,1)$$
$$(2,1) \longrightarrow (5,1)$$
$$(4,1) \longrightarrow (7,1)$$
$$(1,2)$$
$$(1,4) \longrightarrow (5,2)$$
$$(1,3)$$
$$(2,2) \longrightarrow (6,3)$$
$$(3,3)$$

One can use a production system (i.e., with the superstate definition) to create a mapping of groups of states which are equivalent to the previous set of individual mappings. We will number each production so that each can be cross-referenced when we draw the phase space.

1. $(2, 3) \rightarrow (7, 2)$
2. $(?x, 1), x \in \{2, 3, 4\} \rightarrow (x + 3, 1)$
3. $(1, ?x), x \in \{2, 3, 4\} \rightarrow (5, 2)$
4. $(?x, x), x \in \{2, 3\} \rightarrow (6, 3)$

In these productions, ? means "match any number" and $?x$ means match any number and store the number in the variable x. Figure 4.23 displays this mapping where the relevant production number is positioned within each mapped area.

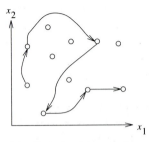

(a)   State-to-state mapping

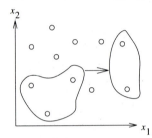

(b)   Mapping groups of states

Figure 4.22   Mapping within the phase plane.

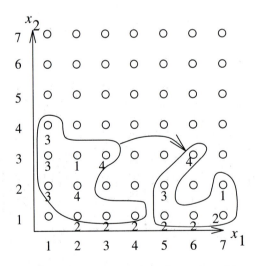

Figure 4.23   Phase space group mapping.

For production systems, inputs are treated as operators which modulate (i.e., control) which productions are to be used in cases where more than one left-hand side can be matched. If we had added the rule $(3, 1) \rightarrow (5, 2)$, this would conflict with the other rule $(?x, 1) \rightarrow (x + 3, 1)$, which defines $(3, 1) \rightarrow (6, 1)$. Which rule do we choose? In such a case, we create two operators—one for each of these two rules, so that an input signal or string will serve to guide the system through state space.

Production systems. The declaratively specified models encountered so far are fairly simple and the models associated with the methods lack the general power that we will find of functional models in Ch. 5. This is because, with declarative models, we must enumerate all state-to-state transitions specifically; there has been no general way of making it easier for the modeler to specify a large group of transitions using a mathematical shorthand of some sort. We will see, though, that the declarative approach need not be limited to simple state-to-state transitions. Note, for instance, that the declarative approach is based on state structure. That is, if a state matches a certain structure, then we delineate in our model the transition to the next state. We can, in fact, use pattern-matching capabilities so that state "pattern structures" are used instead of state "instantiations" with built-in constant values. The production system and logic approach utilize this form of calculation to afford declarative methods the capability of representing complex behaviors with a modicum of mathematical notation.

Methods of production systems and formal logic (either standard or temporal) may be used as a basis for simulation modeling. We need to define the concept of state, input, and time with respect to these models:

- State is defined as the current set of facts (truths) in a formal system. For predicate logic this equivalences to a set of predicates. For expert systems, the rule or "knowledge" base is the state of the system.

- Operators are treated as inputs. A sequence of parameterized calls to the operators serves as the input stream that controls the system. We will associate time durations with input events by saying that whenever an input event occurs (which causes a change in state), the ensuing state will last for some specified period of time.

- Time is assigned to each production or inference so that the process of forward chaining produces a temporal flow. We can make state variables vary continuously or discretely.

Water jug problem. We will now consider a problem often found in artificial intelligence search literature and rephrase it in systems terms by explicitly adding time into the model design and execution. First we note that in artificial intelligence the term *search space* with respect to dynamical system modeling is rephrased in systems terminology as a deterministic automaton. Consider a model with two water jugs (one 4-gallon capacity, and the other, 3-gallon). Jug A is the 3-gallon jug, and jug B is the 4-gallon jug. There is a water spigot and there are no markings on either jug. There is a goal to obtain exactly 2 gallons into the 4-gallon jug. The state of the water jug model will be identified by a set of predicates. Note that predicates and arguments are both in lowercase. The state is defined as (X, Y), where X is the amount of water (in gallons) in the 3-gallon jug, and Y is the amount of water in the 4-gallon jug. We define states to vary using discrete jumps so that filling an initially empty jug A, for instance, would cause a jump from state $(0, 0)$ to $(3, 0)$. We will assume an initial state of $(0, 0)$ (i.e., both jugs are empty). The goal state is $(X, 2)$. Time will be measured in minutes; therefore, rates are measured in terms of gallons per minute. Filling is somewhat slow and proceeds at a rate of 2 gallons per minute. All other operations take 10 gallons per minute. There are four operators. We will define them as follows:

- The operator and description.

- Conditions for the operator to be applied (i.e., fire).
- The time duration ΔT of the operator if it is applied. The duration is associated with the current state.

1. OPERATOR 1: *empty(J)*.

    (a) Empty jug $J \in \{A, B\}$.
    (b) For $J = A$, $(X, Y | X > 0) \rightarrow (0, Y)$ and $\Delta T = X/10$.
    (c) For $J = B$, $(X, Y | Y > 0) \rightarrow (X, 0)$ and $\Delta T = Y/10$.

2. OPERATOR 2: *fill(J)*.

    (a) Fill jug $J \in \{A, B\}$.
    (b) For $J = A$, $(X, Y | X < 3) \rightarrow (3, Y)$ and $\Delta T = (3 - X)/2$.
    (c) For $J = B$, $(X, Y | Y < 4) \rightarrow (X, 4)$ and $\Delta T = (4 - Y)/2$.

3. OPERATOR 3: *transfer_all(J1, J2)*.

    (a) Transfer all water from jug $J1$ to jug $J2$.
    (b) For $J1 = A$, $(X, Y | X + Y \leq 4 \wedge X > 0 \wedge Y < 4) \rightarrow (0, X + Y)$, and $\Delta T = X/10$.
    (c) For $J1 = B$, $(X, Y | X + Y \leq 3 \wedge Y > 0 \wedge X < 3) \rightarrow (X + Y, 0)$, and $\Delta T = Y/10$.

4. OPERATOR 4: *transfer_full(J1, J2)*.

    (a) Transfer enough water from jug $J1$ to fill $J2$.
    (b) For $J1 = A$, $(X, Y | X + Y \geq 4 \wedge X > 0 \wedge Y < 4) \rightarrow (X - (4 - Y), 4)$, and $\Delta T = (4 - Y)/10$.
    (c) For $J1 = B$, $(X, Y | X + Y \geq 3 \wedge Y > 0 \wedge X < 3) \rightarrow (3, Y - (3 - X))$, and $\Delta T = (3 - X)/10$.

The application of various operators will change the current state to new state. For instance, given the initial system state as being $(0, 0)$, we see that we can apply only operator *fill*. Specifically, we can do either *fill(a)* or *fill(b)*. Both operations take a certain amount of time ΔT. The ΔT is associated with the current state. Let's look at the time taken to fill jug A. The production rule associated with this operation is: "For $J = A$, $(X, Y | X < 3) \rightarrow (3, Y)$ and $\Delta T = (3 - X)/2$." To go from state $(0, 0)$ to $(3, 0)$, for instance, will take a total time of $T = (3 - 0)/2$ or $T = 1.5$. This is interpreted as: "If the system is in the state $(0, 0)$ and the operator *fill* is input to the system, the system will enter state $(3, 0)$ in 1.5 minutes." We do not specify any intermediate states in this model.

   If we consider each operator as representing an external, controlling input from outside the jug system, we can create an FSA that represents the production system. Figure 4.24 displays a partial FSA where the following acronyms are defined:

1. Ex: Empty jug x.
2. Fx: Fill jug x.
3. $TAxy$: Transfer all water from jug x to jug y. No overflowing of water permitted.
4. $TFxy$: Transfer all water from jug x to jug y *until* jug y is full.

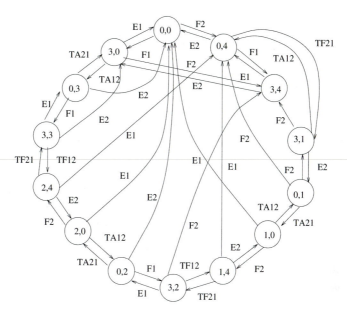

Figure 4.24 Partial FSA for the jug system.

Temporal logic models. The use of logic as a programming language provides the use of predicate logic as a basis for expressing general algorithms in a computational framework. Logic-based methods of programming provide a comprehensive method for expressing state-to-state transitions using the declarative modeling approach. Programming in logic is an extension to programming using production rules covered in the preceding section. We saw that with production rules, we could specify a multitude of state-to-state transitions by using patterns within rules. We utilize the same shorthand in temporal logic. Temporal logic is an extension of classical logic in that reasoning about actions over time is possible. With logic programming, in general, we utilize variables and unification to specify clusters of state to state behaviors.

In addition to the normal logical operators of classical predicate logic such as $\wedge$ (conjunction), $\vee$ (disjunction), $\supset$ (inference) and $\models$ (satisfaction), temporal logic includes temporal operators such as $\bigcirc$, $\Box$, and $\Diamond$. In general, there needs to be a way to augment traditional logic programming languages such as Prolog with the idea of "time advance." In logic, we can specify the state transition $s_1 \rightarrow s_2$ as: $\exists t_1, t_2 \colon ([t_1 < t_2] \wedge [X(t_1) = 0] \wedge [X(t_2) = 1])$. In Prolog, this would take the form

```
state(X,t2,1) :- state(X,t1,0), t1 < t2.
```

Our treatment, here, is to make the meaning of "logical inference" equivalent to "state transition function" for the purpose of using the mechanical features of logic programming languages.

Beetle population growth. The following first-order logic expressions represent a logistic growth rate of beetles (let p = population size, g = population growth, and t = current time).

$$t(t_1) \wedge p(small, t_1) \wedge g(slow, t_1) \supset t(t_2) \wedge (t_2 > t_1) \wedge p(medium, t_2) \wedge g(fast, t_2)$$
$$\supset t(t_3) \wedge (t_3 > t_2) \wedge p(steady, t_3) \wedge g(slow, t_4)$$

$$(4.1)$$

Traditionally, simulation is strictly a forward-chaining operation (i.e., time always moves forward). We can see how simulation can be seen, more generally, as a set of state trajectories upon which one can operate bidirectionally, and that it is possible to ask a question such as "Does the beetle growth level off?" and to have this question answered by proof that involves the states associated with well-formed formulas in logic. For instance, we could set as a goal: `?- p(steady,T), g(X,T), t(T)` (in the Prolog language) and have the system provide an answer in terms of previously occurring states and time constraints. If the proof cannot be performed due to incomplete information, various expert system techniques may be employed, such as (1) querying the user for missing information, or (2) using constraints to fill in empty frame slots.

Even though first-order logic provides one method of incorporating time into modeling, it is preferable to have a language that has a better integrated concept of time advance. The class of languages that specify time as an integral component are known as "temporal logic programming languages." We choose the language Tempura[3] to illustrate this approach. Tempura is based on the concept of truth within a time *interval* σ. σ is an interval with a discrete number of states $\langle \sigma_0, \sigma_1, \ldots, \sigma_{|\sigma|-1} \rangle$. From our simulation perspective, σ is a discrete-time state trajectory. In Prolog, the usual output is a list of variable values. For Tempura, the output is also a list of variable values with two distinctions when compared against the Prolog output: (1) the variables are termed "states," and (2) an interval of states—representing a temporal sequence of state value lists—is output (with a *display* function) rather than a single state value list. The following operators are relevant to time dependency:

- $\Box e$ (always). Expression e is true throughout the interval. This is interpreted as "for all time values."
- $\Diamond e$ (sometimes). Expression e is true at some time in the interval. This is interpreted as "for at least one time value."
- $\bigcirc e$ (next). Expression e is true at the next time step.
- e_1 *gets* e_2 (gets). Expression e_1 at the next time step equals e_2 at the current time step for all times (except the last) in the interval.
- $e_1 \leftarrow e_2$ (temporal assignment). Expression e_1 at the last time step equals e_2 at the first time step in the interval.
- $e_1; e_2$ (chop). There is an interval which can be subdivided into two sub-intervals so that e_1 holds in the first subinterval and e_2 holds in the second subinterval.

[3]See ref. [154].

Consider the following interval σ:

	X	Y
σ_0	4	5
σ_1	2	5
σ_2	3	1
σ_3	5	2
σ_4	8	3

This interval is defined as $\sigma = \langle \sigma_0, \sigma_1, \sigma_2, \sigma_3, \sigma_4 \rangle$ and is of length 5 (i.e., $|\sigma| = 5$). The state space is two-dimensional, say, $\mathcal{Z}^2$, where (X, Y) is a point in the state space. $\Diamond(X = 2 \wedge Y = 5)$ (i.e., at least one time in the interval, the state of the system will be (2,5)) is true since state σ_1 satisfies this expression. Also, while $\Box(Y = 5)$ isn't true on the above interval, it is true on the subinterval of length two: $\sigma' = \langle \sigma_0, \sigma_1 \rangle$. $\Diamond \bigcirc (X > 5)$ is true—it states that at some point [i.e., for some state) in the interval the next X component of the state will be greater than 5. This holds for σ_3 since at the subsequent state, $X = 8 > 5$. Here are some additional expressions that are true for interval σ:

1. $\Diamond(X = 2) \wedge \Diamond(Y = 1)$. True for X in σ_1 and Y in σ_2.
2. $\bigcirc \bigcirc \bigcirc(Y = 2)$. True for Y in σ_3.
3. $\Box(X > 1 \wedge X < 9)$. True for all states in σ.
4. $\Box(X < 5); \Box(Y < 4)$. Subinterval $\langle \sigma_0, \sigma_1 \rangle$ concatenated with $\langle \sigma_2, \sigma_3, \sigma_4 \rangle$ satisfies this compound expression. Another possibility is $\langle \sigma_0, \sigma_1, \sigma_2 \rangle$ concatenated with $\langle \sigma_3, \sigma_4 \rangle$.

Consider the production system for the water jug example. One of the operations was $fill(J)$. In temporal logic, filling jug A is expressed as: $I = \langle fill, A \rangle \wedge A = X \wedge B = Y \wedge \bigcirc(A = 3 \wedge B = Y)$, which is true for an interval where $\sigma_0 = (X, Y)$ and $\sigma_1 = (3, Y)$.

Many of the questions that we have asked in temporal logic are of the sort "Expression e is true for interval σ." That is, we provide an interval ahead of time, and then specify expressions that hold during that interval. While this helps us to understand temporal logic, it is important to see how the implementation Tempura operates with respect to interval specification. Specifically, Tempura interprets a program that *generates* an interval. To generate an interval successfully, we must give it many specifics. For instance, it would not suffice to specify $\Box((I = 1) \wedge (J = 2))$ and expect Tempura to provide an interval. Given this expression, we understand the set of intervals that would hold for this expression [namely, intervals where each state equals $(1, 2)$]; however, Tempura will need to have more information to produce output:

- How do "next states" depend upon "current states"?
- What variables are to be output?
- What is the length of the interval to be generated?

Here is a piece of Tempura code to give a list of 10 integers with the corresponding double value for the integer:

```
define temp() = exists I,J: {
len(9) and I=1 and J = 2
and box I gets I + 1
and box J gets 2 * (I+1)
and always format("%d %d \ n",I,J)
}.
```

4.3 EVENT-BASED APPROACH

Since events are the counterpart components of states—since states are nothing more than what occurs between events—we should be able to construct modeling methods that focus on events. As an example of events, consider the *calendar of events* [139] for September 1993 in Table 4.1.

　　We often see this type of calendar, and it is a natural way of modeling a process declaratively: simply note key events and the time when they occur. Even though there are no constraints on when events occur, they are usually defined when a system marks a change from one phase to another or from one invariance to another. Events such as when Mars passes 0.9° south of Jupiter are marked because this appears as a collision of planets from the projection of the sky afforded to us as a result of being based at a fixed position on Earth. Also, some events appear to take some time, so an argument could be made that they are not really events, but states. However, as discussed in Ch. 2, the definition of a system in terms of events presupposes a certain level of abstraction. After all, this is true of building any model: the model reflects only one abstraction level. In Ch. 8 we resolve the issue by allowing for multiple layers which hierarchically organize systems.

TABLE 4.1　ASTRONOMICAL EVENTS FOR SEPTEMBER 1993

Date	Event
3	The moon is at apogee (252, 353 miles from Earth), 1 pm EDT.
4	Ceres is stationary, midnight EDT.
6	Mars passes 0.9° south of Jupiter, 8 pm EDT.
9	Last Quarter Moon is at 2:26 am EDT.
13	The Moon passes 6° south of Venus, 11 pm EDT.
15	New Moon is at 11:10 pm EDT.
16	The Moon is at perigee (222, 082 miles from Earth), 11 am EDT.
21	Venus passes 0.4° north of Regulus, 2 am EDT.
22	First Quarter Moon is at 3:32 pm EDT; Autumnal equinox is at 8:22 pm EDT.
24	The Moon passes 3° north of Neptune at 3 am EDT; the Moon passes 4° north of Uranus at 3 am EDT.
27	Uranus is stationary, 11 pm EDT.
29	Neptune is stationary, 11 pm EDT.
30	Full Moon is at 2:54 pm EDT; the Moon is at apogee (252, 540 miles from Earth), 5 pm EDT.

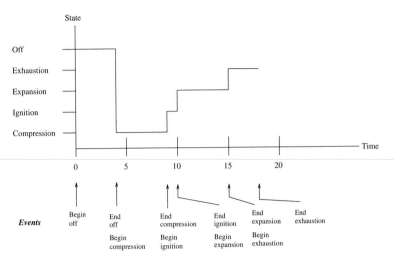

Figure 4.25   State trajectory with events for the engine model.

4.3.1 Finite Event Automata

Finite event automata (FEA) models represent the event analog to state automata presented in the preceding section. The formalism for an event graph is identical to that of an FSA except that (1) nodes in the graph represent events instead of states, and (2) transitions (not the states) are labeled with time durations used for scheduling future events.

Events change the state of a system. Reconsider the engine model shown in Fig. 4.4. The events in Fig. 4.4 are not shown *explicitly*. An FSA is a model type that emphasizes states instead of events. Figure 4.25 shows the events that are associated with the states in Fig. 4.5.

An event graph may be generated automatically from the state graph in Fig. 4.4 by:

1. Taking the times of the states, and assigning them to the respective event graph transitions.

2. Taking the events in Fig. 4.25 and using these instead of the state names.

Figure 4.26 displays an event graph for the engine.

We present a formal method for taking any FSA and converting it to an equivalent event graph EG. Given an FSA defined as $\langle I, O, Q, \delta, \lambda \rangle$ with n states $Q = \{q_1, q_2, \ldots, q_n\}$, we form an event graph (EG) defined as: $\langle I', O', Q', \delta', \lambda' \rangle$, where:

$Q' = \{q_1', q_1'', q_2', q_2'', \ldots, q_n', q_n''\}$ where $\forall i \in \{1, \ldots, n-1\} q_i'' = q_{i+1}'$.

$\delta': Q' \times I' \to Q'$

$\lambda': Q' \to O$

$I' = \mathcal{R}_0^+$

$O' = Q'$

The meaning of q_i' is "beginning of state q_i," and q_i'' means "end of state q_i." I' is the set

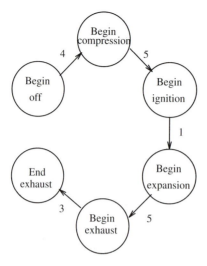

Figure 4.26 Four stroke engine event graph.

of all possible elapsed times when the system is in any $q_i' \in Q'$. δ' is defined as follows:

$$\delta'(q_1', \Delta t_1) = q_2'$$
$$\delta'(q_2', \Delta t_2) = q_3'$$
In general, $\delta'(q_i', \Delta t_i) = q_{i+1}'$ for $i \in \{1, \ldots, n-1\}$
$$\delta'(q_n', \Delta t_n) = q_n''$$

A similar method can be used to convert event graphs to state graphs. Specifically, we take events occurring sequentially and create states to keep constant the value of state variables between the events.

4.3.2 Keyframe Animation

In the field of computer graphics, we find certain state-to-state modeling methods as exemplified by the technique of *keyframe animation*. Keyframes are events (or instantaneous states) and keyframe animation involves the creation of in-between states from events which demarcate the event boundaries. Keyframe animation is used to specify the behaviors of objects over time using only state specifications at discrete points in time. An expert animator creates the *key* positions of all objects in a scene to be animated, and then these positions are handed to another less-qualified animator who fills in the in-between frames. One item used in computer animation is called a *track*. A track is a list of keyframe positions for a single object. For instance, a track might be associated with a human figure, or it may be associated with an object such as a ball. Each object associated with a track is undergoing change of some sort, and so keyframes are used to specify this motion. Consider the action of a bouncing ball. Figure 4.27 displays three keyframes and five instantiations of those frames. Each ball position shown reflects the state of the system at times 0, 10, 20, 30, and 40 (i.e., event times). When the keyframes are interpolated (to locate the "in-betweens"), the keyframe animation system will fill in ball positions for the other times in the system $(1 \rightarrow 9, 11 \rightarrow 19, 21 \rightarrow 29, 31 \rightarrow 39)$. In the case of the ball our states would be of

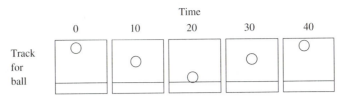

Figure 4.27   Animation track with five events for a ball in motion.

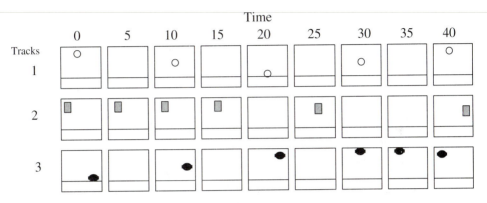

Figure 4.28   Multi-track with events for 3 moving objects.

the form $\mathbf{x} = \langle x_1, x_2 \rangle$, where x_1 represents the horizontal coordinate, and x_2 represents the vertical coordinate of the ball's centroid.

A multitrack keyframe animation system allows for multiple tracks so that we could have other actions occurring in parallel. Suppose that there are two other small objects being thrown near the path of the ball (but not intersecting with it). Figure 4.28 displays this scenario using three tracks. A multitrack animation is very similar to the concept of a "script." Scripts are low-level event tables that have been widely used in computer animation and artificial intelligence. The origin of scripts comes from the concept of script in the theater. That is, playwrights compose scripts for actors who must operate in parallel and interact with one another on stage. In a script, one has a set of objects or *actors* each of which undergoes a set of different operations. The beginning and end of such operations are termed events. The term "operation" is identical to our simulation term "activity." The special event "-END-" means that this is the end of the script for that particular object. Event-based methods as defined in Chapter 3 are useful since the objects are all operating in parallel and require coordination for simulation purposes. In our script objects are people who drive to a store to shop. Each person takes a different amount of time for each activity. Each event is defined by a set of words in parentheses. Since events are "timeless," we will assume, for compactness, that each defined event is automatically prefaced with "begin to." So "(get in car)" really means "(begin to get into car)."

It is most convenient to consider a script in the form of a table. The table form presents dynamics by juxtaposing all events next to a global time column. Consider a table

TABLE 4.2   SCRIPT FOR THREE OBJECTS

Time	Object 1	Object 2	Object 3
0.0	(get in car)	(get in car)	
0.4			(get in car)
1.5		(drive to store)	
2.0	(drive to store)		
5.5			(drive to store)
6.0		(shop)	(shop)
7.2	(shop)		
14.4	(drive home)		
16.0			(drive home)
17.3	END	(drive home)	
20.0		END	END

script composed of three objects (Table 4.2). This script can easily be converted to an event scheduling execution (Sec. 2.3.3) by forming a program that:

1. Assigns one entity per script object. We have, then, three entities (or tokens).
2. Schedules all events for each object at the beginning of the program.
3. Enters a loop that causes all events to be simulated.

An "object" can mean many different things in a script-based simulation. An object can be a car, a projectile, a person, light sources, or a camera (from which the visual simulation report can be generated). Large objects may have multiple parts each of which move in different directions—in this case, large objects are best broken into small subobjects where each subobject has a distinct behavior.

4.3.3 Augmented Event Graphs

Simple event graphs illustrate models with points in event space, where the points are connected with transitions from one event to another. While this provides a direct analog to FSAs for state-based declarative models, we are left with a modeling technique (FEA) that is not very powerful. What we would like is to use a modeling method where we partition event space in much the same way as we partitioned state space to use production rule and logic-based models.

With state space, we created partitions that resembled jigsaw puzzles. For event space, it is not as straightforward; if we break event space into partitions whose elements are contiguous, we realize that there is no cohesive *glue* that binds these elements together. It was easy to create partitions in state space where a partition possessed a natural cognitive mapping such as a natural language word or phrase. For event space, we can still create partitions, but the elements in a partition will be related, not by Euclidean distance but by an object associated with an event. Figures 4.29 and 4.30 show the generic relationship between state and event space partitioning. For the state-based approach in Fig. 4.29, the transition is made based on external input, and for the event-based approach in Fig. 4.30, a transition is based on scheduling; the symbolism $e_i \rightarrow s(\Delta t) \rightarrow e_{i+4}$ means that event e_i *schedules* event e_{i+4} to occur in time Δt from the present clock time. In the case of events

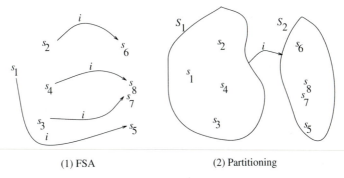

(1) FSA   (2) Partitioning

$$S: s_1 \xrightarrow{i} S_2$$
$$s_i \xrightarrow{i} s_{i+4}$$

(3) Invariant

Figure 4.29   State space partitioning.

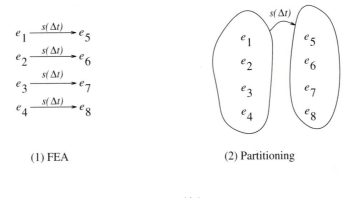

(1) FEA   (2) Partitioning

$$E: e_i \xrightarrow{s(\Delta t)} e_{i+4}$$

(3) Invariant

Figure 4.30   Event Space partitioning.

(Fig. 4.30), a partition such as $\Pi_1 = \{e_1, e_2, e_3, e_4\}$ might be called a "begin service" in a queuing application, whereas $\Pi_2 = \{e_5, e_6, e_7, e_8\}$ would be called an "end service." The *invariant* is an expression that conveniently defines the function that maps one partition to another. For a state-based approach, we used production rules and logic-based methods. For events, the invariant is often defined by an augmented event automaton (i.e., an augmented FEA).

Let us reconsider a previous example of a single-server queue with more detail on the events that occur. There are three *event types* for our system: (1) entity arrival (A_i), (2) entity begins service (B_i), and (3) entity ends service (E_i). Even though these are three

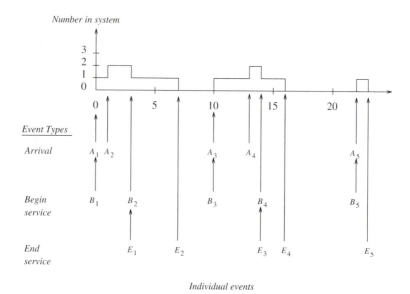

Figure 4.31   Event types for a single server queue.

different events, we can partition event space on the basis of the object (i.e., entity) which is common to a set of events. That is, the arrival of token i always precedes the beginning of service for token i, and the beginning of service for token i always precedes its end of service. The phrase "event types" is normally never used. Instead, an "arrival" is generally considered an event; however, it is a different event for each token. Figure 4.31 displays events that occur and the state of the system over time. For this example we will have the following components:

1. *Arrival event type*: $A = \Pi_1 = \{A_1, A_2, A_3, A_4, A_5\}$
2. *Begin service event type*: $B = \Pi_2 = \{B_1, B_2, B_3, B_4, B_5\}$
3. *End service event type*: $E = \Pi_3 = \{E_1, E_2, E_3, E_4, E_5\}$

4.4 HYBRID STATE-EVENT METHODS

4.4.1 State-Event Graphs

A state-event (or object-interaction) graph displays both events and states for a system. Some systems have a lot of interaction either in the form of communication or exchange of matter. A classic example of object interaction can be taken from space-time diagrams in physics. In Fig. 4.32 we have two objects: A and B. A is not moving, so its position over time remains the same and is represented by the vertical "axis" line drawn through the origin. B, however, moves to the right at a constant velocity that is less than the speed of light. At a time denoted by the horizontal axis, some time after B has begun to move, a light flash occurs which sends light in all directions. The two light rays make a light "cone" which just

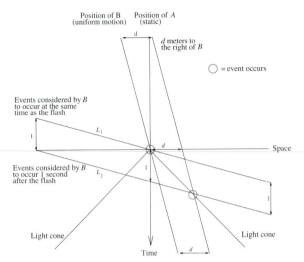

Figure 4.32 Space-time diagram with two objects: A and B.

specifies the position of the light photons over time.[4] Points of interaction (where position lines cross each other) are called "events" in physics. Even though *potential* events can occur anywhere in space-time, we are mainly interested in those reflecting object interaction (where one object is a light photon). In the theory of relativity, laws of physics appear the same to all observers within an inertial frame. Moreover, light is measured at the same rate of 300,000 km/s regardless of who is observing it (*A* or *B* in Fig. 4.32). According to *B*, all potential events along L_1 occur at the same time and all potential events along line L_2 also occur simultaneously (that is, 1 second after the light flash). Two key events are shown: (1) where *B* meets *A*, and (2) what occurs *d* units to the right of *B* 1 second later. *B*'s conception of time is different from *A*'s conception since *B* is moving.[5] *A* considers all potential events to occur at the same time if they intersect with horizontal lines parallel to the horizontal axis.

In computer science, the study of distributed systems also makes use of object inter-action models such as the one in Fig. 4.33. There are three processors that are distributed physically. They are all, perhaps, part of a local area network. Events occur much as they did in Fig. 4.32—through object interaction, although in this case objects "spawn" messages which interact with objects, whereas in physics, objects, themselves are in motion. Processor *A* first sends a message to processor *B*. This message transmission can take an arbitrary amount of time and it depends on network traffic characteristics such as load influenced by other processes and messages that are not being modeled here (but which have an effect). *B* responds with an acknowledgment (*ack*) of *A*'s request to initialize a protocol. *A* then sends the *data* to *B*, which *B* subsequently acknowledges. While this is

---

[4]In Fig. 4.32, space is one-dimensional. In a two-dimensional model $X \times Y \times T$ (where T is time), the movement of photons over time would look like a real three-dimensional cone.

[5]The rotation of *B*'s axes are a result of the Lorentz transformation, which affects all moving masses but whose effects are considerable as velocities approach the speed of light.

TABLE 4.3 THREE EVENT TYPES FOR A SERVER

i	A_i	σ_i	B_i	E_i
1	0	3	0	3
2	2	4	3	7
3	10	4	10	14
4	13	2	14	16
5	22	1	22	23

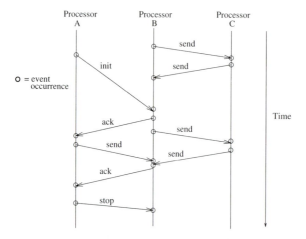

Figure 4.33 Object protocol trace with three processors.

happening, processor C is also communicating with B. B's protocol with C is fairly primitive, and possibly hazardous, since no acknowledgments are transmitted (only the sending of message packets back and forth). Finally, A's protocol with B is finished, and A sends a termination message to B. Within each processor it is common to create interrupt service routines which handle incoming messages or interrupts.[6]

Object interaction models can easily be mapped to finite event automata because the action of the system is most naturally defined in terms of events that occur when something is either sent or received (with respect to the protocol example in Fig. 4.33). A subset of the list of events is depicted in Fig. 4.34 starting with the event defined when A's first message is received by B.

4.4.2 Inventory Control

Let's use a modified augmented event graph to model an inventory problem involving a *mining* company that sells raw material to a *product* company. An example would be a company that makes ceramic plates as a product for the general public. The plates are made

---

[6]An alternative is for processors to *poll* for messages, however, this can be inefficient since messages can accumulate at channel buffers and overflow them.

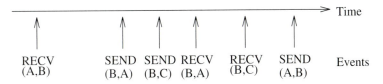

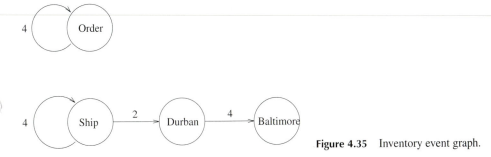

Figure 4.34   Subset of system-wide events for the communications protocol in Fig. 4.33.

Figure 4.35   Inventory event graph.

from a number of raw materials obtained from several locations. We will consider only one raw material. Shipments of a mineral are sent from Zimbabwe by train to Durban in South Africa. As the mineral is mined, it is placed into containers. When all containers are full (taking four weeks), a train is loaded with the containers as proceeds to Durban, which takes two weeks. From Durban, the containers are loaded onto a cargo ship for Baltimore, Maryland. The ocean trip takes 4 weeks. In Baltimore, there is a product company that takes the mineral and creates ceramic plates. The company has its own silo, where it can store the raw material. The mining company also has its own warehouse, where it stores material that it receives from Durban. This is the basic scenario. The mining company knows the size of the product company's silo and it is eager to send material to the product company; however, there are significant storage costs associated with storing the material at the Baltimore warehouse. How can the mining company reduce its warehouse costs to a minimum?

This is known as an *inventory problem*, and this problem, along with transportation flow problems, represents a key domain for which simulation is ideally suited. An initial event graph in Fig. 4.35 displays a model for the problem. The event graph is split into two separate graphs because there are two processes occurring: (1) the product company, which orders material from the mining company, and (2) the mining company, which sends material to the product company. The missing part of Fig. 4.35 is how both companies interact. The product company effectively removes material containers from the warehouse, whereas the mining company fills the warehouse. A slightly modified event graph could show this *storage* element, perhaps as in Fig. 4.36. Here are some key problem data:

- The arcs in Fig. 4.36 are labeled in the number of weeks it takes for the transportation.
- Containers carry tons of material.
- The cost of warehouse storage in Baltimore is $3 per ton per month. This is the same as $0.75 per ton per week.

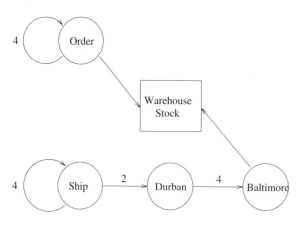

Figure 4.36 Hybrid inventory event graph with connecting storage element.

- Accumulated cost over time is calculated as cost = cost + 0.75 × time in warehouse × stock in warehouse.

This modified graph combines concepts of events and states (the warehouse content) to define the problem more completely. Now that we have a model, we need to perform an optimization to answer our original question of how to minimize the costs. To do optimization, we need to vary some controllable parameters in the simulation and simulate the system for each parameter setting. If we are the mining company, consider the only controllable under our control as being the number of tons to ship from Durban to Baltimore. We can ship, say, anywhere from 200 to 1000 tons in 10-ton increments. Taking the minimum over many simulation runs provides us with an answer in between 330 and 350 tons. We certainly must make sure that the warehouse never empties when the product company asks for material; otherwise, we would have to include the *cost* of the delay to future business. The following algorithm illustrates the optimization procedure.

Algorithm 4.3. Multiple Simulations to Calculate Minimum Cost
Procedure Main
      While more simulations to run while varying parameters
         Switch on event
            Case ORDER: calculate cost of warehouse storage
               remove material from warehouse
               schedule ORDER in 4 weeks from NOW
            Case SHIP: schedule DURBAN in 2 weeks from NOW
               schedule SHIP in 4 weeks from NOW
            Case DURBAN: schedule BALTIMORE 4 weeks from NOW
            Case BALTIMORE: add material to warehouse
         End Switch
         Keep track of minimum cost
      End While
   Output minimum cost scenario
End Main

This example is quite simple and there are a number of things that could be done to make it more realistic, such as:

- Vary other parameters besides the shipment amount. You can only vary those parameters over which you have control.
- Make some of the variables time dependent (such as the order amount) or stochastic.
- Vary more than one parameter and use a gradient method to search for the optimum, thereby avoiding a search of the complete space. The cost value over the search space is called the *response surface*.

4.4.3 Petri Networks

Definition. Petri nets are used primarily for studying the dynamic concurrent behavior of network-based systems where there is a discrete flow. A Petri network (net) is a 4-tuple defined as follows:

$$C = (P, T, I, O)$$

$P = \{p_1, p_2, \ldots, p_n\}$ is a finite set of places (also known as conditions). $T = \{t_1, t_2, \ldots, t_m\}$ is a finite set of transitions. I is the *input function* specifying a function from a single transition to a *bag* of places, $I : T \rightarrow P^r$ (where r = the number of places). Output is similarly defined: $O : T \rightarrow P^q$. A marking μ is an assignment of tokens ($\mu = \mu_1, \mu_2, \ldots, \mu_n$) to the places of a net.

Flexible manufacturing. Let's first consider the Petri net in Fig. 4.37, which represents an assembly operation in a flexible manufacturing system. The scenario is one where M1 and M2 represent machines for milling, cutting, or another operation. A1 represents assembling of parts that come from M1 and M2. Material flows through the Petri net from left to right. Note how each transition is drawn with a straight line, and each place is drawn with a circle. Places are also called "conditions" and transitions are called "events." Since Petri nets have two types of node elements (places and transitions) in the graph, they are called *bipartite graphs*. The small solid black circles (or dots) in each place are called *tokens* or *markers*. A marking for a net specifies the instantaneous state of that net for any moment in time. For instance, the state of the net in Fig. 4.37 is defined by $\mu = (1, 1, 0, 0, 0)$, where μ represents the current marking. Places are "place-holders" for the tokens, while transitions specify a resource or an event. Transitions t_1 and t_2 can fire immediately since they have tokens in their input places. When a transition fires, it does the following:

1. Removes one token from each input place.
2. Holds the token at that transition for a period of time. In normal Petri nets, each transition can be thought of as taking the same delta-time (such as 1 or zero). Timed Petri nets allow one to associate a different time with each transition, and stochastic Petri nets permit one to associate a probability distribution that can be sampled for the time.
3. Adds one token to each output place for that transition after the appropriate time lapse.

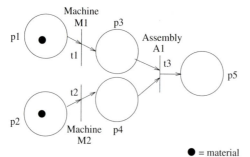

Figure 4.37   Assembly operation from M1 and M2 processed parts.

● = material

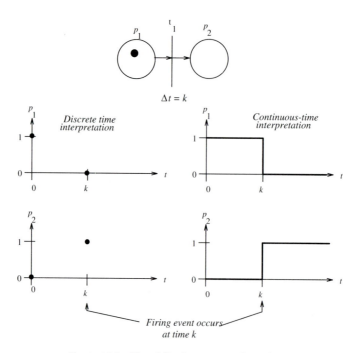

Figure 4.38   Timed Petri net state trajectories.

Extending the basic Petri net. The *generic* Petri net does not involve the passage of time, so we must add this if Petri nets are to be useful as models for simulation. Figure 4.38 displays a simple Petri net along with two different interpretations for the state trajectories. Since a Petri net state space refers to the cross product of places, an untimed Petri net for Fig. 4.38 would simply involve a state transition of the form $(1, 0) \rightarrow (0, 1)$. The two interpretations in Fig. 4.38 are shown for the execution of the net. On the left-hand side, we have two state discrete time trajectories for the net, whereas the right-hand side displays the continuous interpretation. While the discrete time interpretation is less assuming, we will choose the continuous time approach where the system will always have a known state for a continuum of time values. The system begins at time zero and t_1 fires after a delay

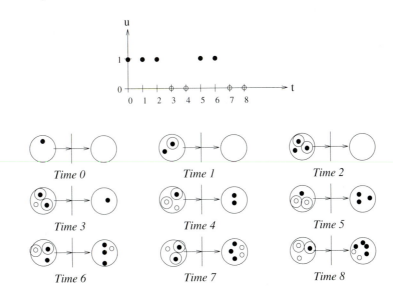

Figure 4.39   Petri net with input for times $0 \to 8$.

of $\Delta t = k$. The best way to interpret the action of a transition is through the scheduling concept presented in Ch. 2. That is, when a token appears in place p_1 at some time t, the token *schedules* t_1 to fire at time $t + \Delta t$. When a token is scheduled (at time t) to fire (at time $t + \Delta t$), that token is marked *reserved* for that transition. Tokens, in timed Petri nets, can be in one of two states: *reserved* or *free*. A *stochastic Petri net* is a timed Petri net where the time delay Δt can be sampled from a probability density function.

There are two other extensions to the basic Petri net which are of use. First, a *colored Petri net* is one where tokens have attributes or *color*. Moreover, a Petri net should be able to respond to external input if we are to use it effectively. We can combine the discussion of input with colored tokens if we consider our simple one-transition Petri net reacting to a discrete input *signal* where the signal values can be either 0 or 1. We will let 0 and 1 be two colors that a token can assume. Furthermore, reserved tokens are marked by enclosing them within a second circle. Figure 4.39 illustrates a simulation of the Petri net for the first nine time units where $\Delta t = 3$.

Industrial part manufacturing.   Terms such as "computer-integrated manufacturing" (CIM) and "flexible manufacturing" guide the development of more productive plant configurations for building products from raw material. We introduce the following categories and definitions:

1. *Material.* Plants are built to process material—often called raw material stock—and shape the material into a product. As raw material goes through its changes, it turns into a *part* to be processed.
2. *Machines.* Plants are composed of machines of all kinds which process material and parts. Some examples are ovens, lubricators, flame cutters, lathes, and robots.

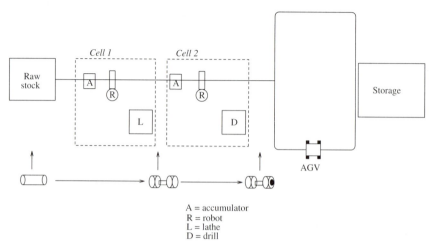

A = accumulator
R = robot
L = lathe
D = drill

Figure 4.40   Manufacturing line with two robots and two machines.

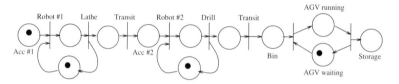

Figure 4.41   Petri net for manufacturing line.

3. *Transportation*. Material flows through a network of machines. The method of transport is effected by devices such as conveyors and automated guide vehicles (AGVs). During this transit, it encounters storage areas and accumulators which buffer parts until the machines can operate upon them.

Figure 4.40 shows a sample manufacturing system containing nine parts. The raw stock arrives from the left via a central conveyor. At this point, the material stock is a cylinder shape. The cylinder parts are loaded into a spiral accumulator (ACC) which holds parts for the pick-and-place robot until both it and the lathe are ready. Once both are ready to work with the part, the cylinder is turned into a barbell shape by the lathe and sent on toward a second spiral accumulator using a conveyor belt (CNV). A second robot also performs a pick-and-place operation and hands the barbell part to a drill machine which punches a longitudinal hole through the part. That is the final product part, which proceeds to a small storage bin taken by the AGV, which runs around a closed track while dropping the bin contents into longer-term storage. This type of application involves discrete parts flowing through a network of resources which proceed when constraints are met. This suggests the use of a Petri net to model the system as in Fig. 4.41.

Object-oriented Petri nets. To capture Petri nets within an object-oriented framework, we must begin, as usual, with objects and determine attributes and methods. Consider

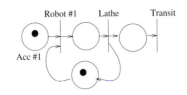

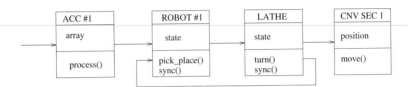

Figure 4.42   Object implementation for a piece of Petri net.

part of the Petri net in Fig. 4.41. This is shown with an object-based implementation in Fig. 4.42. Parts are messages that flow through each object as each object serves to process the incoming parts. We see that the ACC object releases its stock item to the robot when the lathe is ready for the robot to provide it.

Human task coordination.   When humans work together in a structured manner, they perform tasks. These tasks involve checklists that are processed by humans while they coordinate with one another. The coordination is important since the highest-level task is often not decomposable into disjoint parts. Instead, humans must work together to share resources and coordinate their activities. Three examples of task coordination are:

1. *Space Shuttle or Station Activity.* Operations in space are expensive and are performed in small, closed areas. For astronauts to achieve their mission—whether it involves deploying a satellite or performing scientific experiments in zero gravity—they will have detailed checklists to follow. These checklists will have times when activities must begin, end, and halt waiting on another activity.

2. *Nuclear Power Plant Control.* All power plants have operators and detailed control panels. Depending on the condition of plant components, the operators will perform tasks and work collectively to solve potentially hazardous situations from arising.

3. *Military Command and Control.* During military exercises, it is important to maintain the command and control hierarchy. If a company commander within a battalion does not follow the battalion leader's order, chaos will ensue and battle plans will have to change suddenly, perhaps altering the outcome. In general, a commander at some level (division, battalion, company, platoon, section) must follow orders and acknowledge actions by reporting back up the chain.

In general, humans synchronize when an activity is to be executed efficiently. We will study a microcosm for human coordination by looking at the *dining philosophers*, which is usually presented within the context of operating systems literature; however, we employ

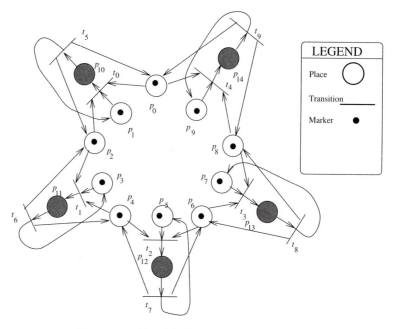

Figure 4.43   Level 1: Five synchronized figures.

it as a basis to do model design characteristic of human task coordination, and not to solve analytic problems in concurrency. Each philosopher requires two chopsticks to eat (no two adjacent philosophers, therefore, may eat simultaneously). The tokens within the Petri net represent the chopstick resource and general flow of control. Initially, all tokens (or "markers") start out in the center of the net in Fig. 4.43. Then two philosophers grab their respective chopstick pairs, then proceed to eat at some rate defined by the modeler. *PNET*, shown in Fig. 4.43, has places and transitions which are labeled counterclockwise (using two concentric passes) starting with p_0 and t_0, respectively. Figure 4.43 illustrates the model for *PNET*. Two Petri nets are shown in Figs. 4.43 and 4.44. Figure 4.43 displays the higher-level net, while Fig. 4.44 displays just the "eating" subnet that replaces the shaded place in Fig. 4.43. Figures 4.43 and 4.44 can be compressed into one net, although it is useful to represent two levels of aggregation as shown. We define *PNET* as follows:

1. $PNET = \langle P, T, I, O \rangle$
2. $P = \{p_0, \ldots, p_{14}\}$
3. $T = \{t_0, \ldots, t_9\}$
4. $I : T \rightarrow P^\infty$
5. $O : T \rightarrow P^\infty$
6. $\mu : P \rightarrow Z_0^+, \mu(p_i) = 1$ for $i \in \{0, \ldots, 9\}$ else $\mu(p_i) = 0$
7. $I(t_0) = \{p_0, p_1, p_2\}, I(t_1) = \{p_2, p_3, p_4\}$
8. $I(t_2) = \{p_4, p_5, p_6\}, I(t_3) = \{p_6, p_7, p_8\}$

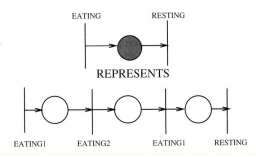

Figure 4.44   Level 2: The Eating Process Sub-Network

EATING   RESTING

REPRESENTS

EATING1   EATING2   EATING1   RESTING

9. $I(t_4) = \{p_8, p_9, p_0\}$, $I(t_5) = \{p_{10}\}$

10. $I(t_6) = \{p_{11}\}$, $I(t_7) = \{p_{12}\}$

11. $I(t_8) = \{p_{13}\}$, $I(t_9) = \{p_{14}\}$

12. $O(t_0) = \{p_{10}\}$, $O(t_1) = \{p_{11}\}$

13. $O(t_2) = \{p_{12}\}$, $O(t_3) = \{p_{13}\}$

14. $O(t_4) = \{p_{14}\}$, $O(t_5) = \{p_0, p_1, p_2\}$

15. $O(t_6) = \{p_2, p_3, p_4\}$, $O(t_7) = \{p_4, p_5, p_6\}$

16. $O(t_8) = \{p_6, p_7, p_8\}$, $O(t_9) = \{p_8, p_9, p_0\}$

To simulate Petri nets, we must be able to scan a set of transitions to determine if these can fire. There are two key aspects to this problem: (1) choosing the set of transitions to check, and (2) choosing a particular transition from this set. One can certainly scan **all** transitions in the network; however, this is wasteful since we need only check transitions in the outset of the current transition that has just fired. Consider the net in Fig. 4.37. The outset of $t1$ is $\{t3\}$; $t2$: $\{t3\}$; $t3$: $\{\}$. The outset of a transition t is defined as the set of transitions that can potentially be affected by t if t fires. An algorithm for outset determination (assuming that the topology of the net is stored as an array indexed by transition number) follows:

Algorithm 4.4. Calculating the Transition Outset

Procedure Outset
For transition T ($t1$)
    For each output place (P) of transition T
        For each transition T ($t2$)
            For each input place of $t2$
                Is this place equal to P?
                If so, place $t2$ in outset of $t1$
End Procedure Outset

This algorithm provides the necessary outset. We should be fair in our choice of the order in which the outset transitions are checked. To promote fairness we randomize the order of the outset array using the following algorithm:

Algorithm 4.5. Randomizing Elements of a Set

Procedure Randomize (*max, array*)
For all elements of the array ($i = 0 \ldots \max -1$):
        Pick a random index (*elt*) *between* 0 *and* $\max -i$
        Swap array[*elt*] *with array*[$max - i$]
end For
end Procedure

The complete algorithm for the timed Petri net is listed below.

Algorithm 4.6. Timed Petri Nets

Procedure Main
        Set SIMCLOCK to zero
        Set # of iterations
        For each transition do
               schedule BEGINFIRE at time zero
        While more iterations do
               ProcessNextEvent
End Main

Procedure ProcessNextEvent
        Get event from event list
        Update SIMCLOCK

        For event BEGINFIRE:
        Def: Begin to fire this transition if possible
        At least one token in each input place?
        If yes,
               (1) Transition can fire
               (2) Mark one token from each input place as reserved
               (3) Schedule ENDFIRE at SIMCLOCK + delta TIME
        For event ENDFIRE:
        Def: Complete transition firing and scan through
               all other transitions in random order to see if they can fire
               (1) Add 1 *token to each output place*
               (2) Randomize order of transitions
               (3) Schedule BEGINFIRE on all transitions in
                     this transition's outset
End ProcessNextEvent

     The DP network is closed (no input; no output), so our system-wide statistics must be ignored since there are no arriving or departing tokens—just the number of tokens placed inside the net at the beginning. As tokens come back to their respective places (representing the return of chopsticks), we often have the case where more than one transition can fire.

Figure 4.45   Keyframe # 2—Eating
Stage 1 (EATING1).

How do we pick which ones should fire? This is called a *conflict*, and to ensure fairness
our Petri net simulator should allow a different ordering of checking transitions each time a
"sweep" is made for that simulation time. In this way, a transition does not get locked out
since it happens to always be the one of last transitions to be checked for potential firing.

Event-controlled animation.   The total action of any single philosopher is a period
of eating followed by a period of rest. The *eating process* is shown in the first three
transitions and places of Fig. 4.44. The eating process is described by three actions:

1. EATING1: The right hand is halfway between the table and mouth. See Fig. 4.45.
2. EATING2: The right hand is at the mouth. See Fig. 4.46.
3. EATING1: See Fig. 4.45.

The *resting process* (RESTING) is represented by a single net transition. The resting
configuration is shown in Fig. 4.47. Our Petri net model is simulated for some period of
time; each transition can be set arbitrarily to some ΔT. The output from the simulation is
a single file consisting of a sequence of 3-tuples, as the following example illustrates:

```
[... tuples deleted ...]
  40   man3 RESTING
  45   man1 RESTING
  46   man3 EATING1
  65   man1 EATING1
  74   man3 EATING2
  82   man3 EATING1
  85   man1 EATING2
  89   man1 EATING1
  90   man3 RESTING
[... tuples deleted ...]
```

Figure 4.46   Keyframe # 3—Eating Stage 2 (EATING2).

Figure 4.47   Keyframe # 1—Resting (RESTING).

The meaning of the individual fields in a tuple is as follows:

1. The time when the respective event (node firing) is to take place. An integer time is used here, as it corresponds directly to the frame number in the animation.
2. The animation object associated with the event. In this example, the object is one of the five philosophers (`man1–man5`).
3. The event to take place. This event corresponds to a keyframe as it was defined previously in the animation program. For this particular animation, we used three keyframes to represent the eating sequence (EATING1 → EATING2 → EATING1) and one keyframe for the period of rest (RESTING).

We do not have to create keyframes for each of the five philosophers, which would be a total of 20 keyframes. Rather, we define the sequence in the *local* coordinate frame of one generic object. Then, during the simulation we translate the keyframe to the *global* world coordinate of the respective philosopher.

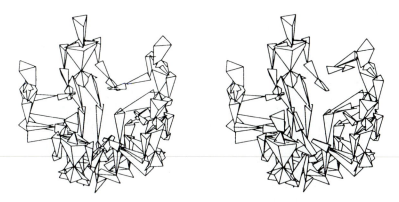

Figure 4.48   Frames 1 and 35. From Fishwick, *IEEE Systems, Man and Cybernetics* ©1988 with permission of the Institute of Electrical and Electronics Engineers.

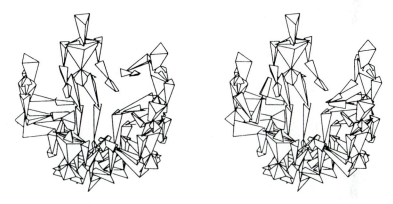

Figure 4.49   Frames 106 and 121. From Fishwick, *IEEE Systems, Man and Cybernetics* ©1988 with permission of the Institute of Electrical and Electronics Engineers.

Figures 4.48 through 4.52 show 10 frames from an animation of the Petri net DP model by assigning Petri net transitions to individual keyframes. The animation was prepared as follows:

1. Create two key frames (keyframes) for each philosopher body. Keyframe 1 positions the body in a sitting position with both hands resting on the table. Keyframe 2 positions the right arm so that the hand is located at the mouth.

2. Use an inverse kinematic method to determine the trajectory of each link so that the initial and final states are satisfied. This provides a smooth path for the end effector (hand) to follow on its path from the table to the mouth.

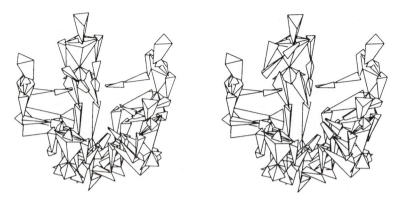

Figure 4.50   Frames 195 and 214. From Fishwick, *IEEE Systems, Man and Cybernetics* ©1988 with permission of the Institute of Electrical and Electronics Engineers.

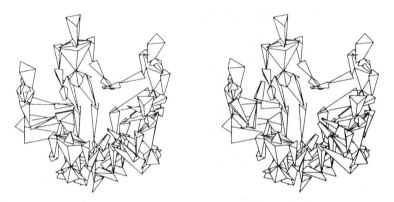

Figure 4.51   Frames 290 and 314. From Fishwick, *IEEE Systems, Man and Cybernetics* ©1988 with permission of the Institute of Electrical and Electronics Engineers.

3. Use the timing specified in each of the Petri net transitions to determine how many frames should be generated and at what time they need to be generated. This synchronizes the Petri net model with the frames.

4. Run the Petri net simulation and use the output to drive an animation.

A more accurate algorithm would include the dynamics associated with robotic linkages; however, in many instances interpolation approach can generate high-level simulations that serve the requirements for training.

The end result of a simulation as driven by the Petri net DP model is a set of more completely rendered video frames, such as the one shown in Fig. 4.53.

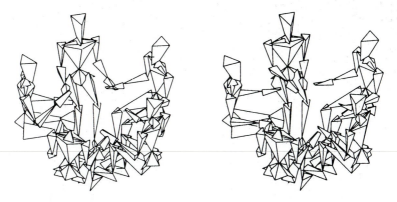

Figure 4.52   Frames 330 and 387. From Fishwick, *IEEE Systems, Man and Cybernetics* ©1988 with permission of the Institute of Electrical and Electronics Engineers.

Figure 4.53   Sample DP video frame.

4.5 EXERCISES

1. An object interaction graph is similar to a set of trajectories defining the system behavior of the components: *input*, *state*, and *output*. What are the differences between the two approaches?

2. In many instances, physical machines such as motors can be operated at multiple speeds. A motor is attached to a set of control buttons each of which represents the desired speed of the motor. Let's consider a cooling fan with three speeds: low, medium, and high. Much like an automobile gear-shifting procedure, the motor cannot be immediately placed in high speed; the buttons represent the *desired final speed*. Instead, the motor must undergo a more gradual speed transformation, always starting with the lowest speed and progressing to a faster rate until the desired speed

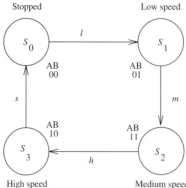

Stopped   Low speed

Figure 4.54   Fan motor FSA.

High speed   Medium speed

is reached. The phasing of motor control is modeled using states and the states can be connected into an FSA. We let S_0 be the "off" state of the fan, S_1 the low speed, S_2 the medium speed, and S_3 the high speed. Figure 4.54 displays the FSA for the cooling fan controller. Pressing the buttons for stop, low, medium, and high are represented by the inputs s, l, m, and h—known as the *primary* signals. The state is encoded using *secondary* signals A and B. Perform the following.

(a) Formally specify the FSA for the motor control.
(b) Derive an entirely different domain example that has the same FSA structure (with different labels).

3. Consider the first two tracks of the track animation model in Fig. 4.28. Model this composite behavior with a single FSA.

4. It is not possible to draw an "infinite" state automaton because there would be an infinite number of states and transitions. Is there another way that you might model such automata? Consider two cases.

(a) There are a countable number of states.
(b) There are an uncountable number of states.

5. [From Ashby [6] (12/10, p. 228).] Consider the "rat maze" containing nine cells shown in Fig. 4.55. G is the goal. The rat cannot obtain sensory clues in cells 1, 2, 3 and 6 (lightly shaded), so when in one of these cells, it moves at random to such other cells as the maze permits. Thus, if we put it repeatedly in cell 3, it goes with equal probability to 2 or to 6. In cells 4, 5, 7, 8, and G, however, the clues are available, and it moves directly from cell to cell toward G. Thus, if we put it repeatedly in cell 5 it always goes to 8 and then to G. Construct a matrix of probabilities and draw a Markov graph for this scenario.

6. Consider a process where two dimes are tossed repeatedly, simultaneously, on a flat surface. Draw a Markov model (with labels) for this process. Also, specify this model formally (using the "tuple" notation).

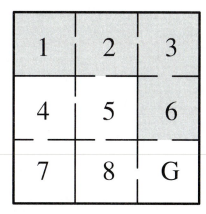

Figure 4.55   Rat maze.

7. Consider a physical system defined by one bucket with two water spigots. The first spigot fills the bucket at a constant rate of 1 quart per second, and the second spigot fills the bucket at a constant rate of 3 quarts per second. The only other possibility is that either spigot might be turned off at any time. Formally define a production system model for this system. Show a partial reachability tree from the initial state where the bucket is empty.

8. Under what circumstances is a Markov model equivalent to an FSA?

9. Temporal logic:

   (a) What is the difference between a previously specified formula $\Box(X < 5); \Box(Y < 4)$ and $(X < 5); (Y < 4)$? Does one statement imply the other?

   (b) Consider the following interval σ:

Time	X	Y
0	0	1
1	1	1
2	0	1
3	1	0
4	0	0

   Specify a temporal logic formula for each of the following.

   (i) Does X ever equal zero?

   (ii) Is it true that $Y = 1$ at three consecutive times?

   (iii) Are X and Y always less than 2?

   (iv) Are there two subintervals where $Y = 1$ in the first and $Y = 0$ in the second?

10. Specify a declarative model for the following systems:

   (a) Driving a car along the turnpike

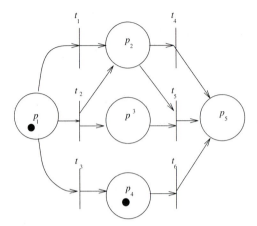

Figure 4.56   Petri net.

  (b) A school of fish swimming in a stream
  (c) Building a house

11. Consider the Petri net in Fig. 4.56. Answer the following questions.

  (a) Formally specify the Petri net using the $\langle P, T, I, O \rangle$ notation.
  (b) Draw a reachability tree using three levels.

12. Analyze the following two views and provide justification.

  (a) FSAs are easily encoded as Petri nets, so Petri nets, subsume FSAs. Therefore, it is easier to include all *sequential model semantics* in Petri nets without using FSA models at all.
  (b) FSAs are models that emphasize sequential behavior and are therefore useful in certain cases even if Petri nets are more inclusive and expressive.

4.6 PROJECTS

  1. Everything has a dynamical aspect to it. Even the buildings that you see, over generations, are constructed, lived in, and demolished. Most architectural structures of historical significance, such as castles and manor houses, can be modeled over time. Take castles as an example. Most castles had their humble beginnings as small fortifications. For instance, the castle of Warwick in England has gone through several major phases since its construction in 1068, two years after the Norman conquest. Over time, the castles went through many phases of construction where phases are often demarcated by events noted by fire, flood, or war. Choose a historical building, discuss its historical significance, and specify a declarative model for it. Specify all inputs, outputs, states, and events. Specify at least two levels of detail (a coarse level and a fine-grained level) for your model.

  2. Build a visual hierarchical automaton simulator that handles both deterministic and nondeterministic models. As the system changes state, you should highlight the

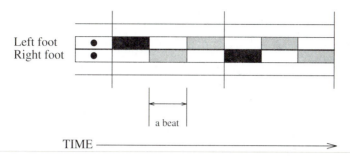

Figure 4.57   Waltz steps in Labanotation.

current state in the automaton, and show the state trajectory over time in a separate window.

3. Music and Labanotation use two similar notations for expressing declarative model semantics. With music, a state is represented by the current notes and chords being played, and with Labanotation the state represents the position of body parts during a dance. Labanotation—developed by Rudolph Laban—is a formal graphical method for recording dance. Dance positions are represented using graphical icons. Using Labanotation, we can specify the current state (i.e., position) of human body parts over time. Body parts can move in parallel or in sequence with respect to one another. Both music and Labanotation use the same basic method for representing time—the score is written onto a *staff* and the individual time increments of the staff are specified by the length of the note being played (in music) or by the number of the beat written vertically on a Labanotation diagram. The basic rhythm of a waltz is represented in Labanotation within Fig. 4.57. In Fig. 4.57, we see that time is marked off by beats where a beat can be either a quarter note, half note, etc. or a specified number of beats per minute as marked with a metronome. The motion of the feet over time is specified with rectangular icons that are filled with different colors:

   - If filled with a single dot in the center of the rectangle, the feet are together in a normal standing position.
   - If filled with a shade of gray, the body is raised by lifting the foot.
   - If filled with black, the body is lowered by flexing the knees.

   The waltz, therefore is a sequence of steps where the body goes low once, then high twice with a repetition of those basic steps.

   For a project, select a basic subset of Labanotation moves and build a simulator that takes as input an encoded or pictorial Labanotation script and produces a keyframe animation of a dance sequence. This project can be made less complex with the following restrictions: (1) the input can be in an encoded form rather than giving the user a graphical way of constructing measures, and (2) the output can be a simple stick figure. Is Labanotation state- or event-based (or both)?

4. Markov models have convenient representations in the form of a transition matrix. When a Markov model executes, it changes state by using one of the matrix elements. It is possible, then, to build a visual simulation of a Markov model by highlighting

the matrix element used in any given time. If the matrix element is represented as a small square on the screen, the square will move around. Write a program to illustrate Markov model dynamics by building this "matrix animator."

5. Declarative models are most often used within functional model blocks to be discussed in Ch. 5. Build a block model simulator whose blocks contain automata to define state change. This represents a simple form of multimodel (called an *FSA-controlled multimodel*), where a component of a functional model is *refined* into a declarative model. Add this capability as an option to any of the projects in Ch. 5.

6. Find three game simulations. Examples of game simulations are adventure games, multiuser dungeons, and space combat simulations. Determine the underlying declarative model(s) used in the game and prepare a report on the game models. Most game models are *hidden* from the user. This is done to give the user a challenge when working with the simulation. Unfortunately, it also obscures the model if you are doing research into making your own game.

4.7 FURTHER READING

Early in computer science research, there was a lot of activity in the general area of "theory of computation." Within this study the finite state automaton [106] has received much attention since it represents a model used to accept regular languages as specified in Chomsky's language hierarchy. From a systems theory perspective [165], declarative models are *local transition functions*, which specify how a system moves through discrete state changes under the influence of input. State trajectories are discussed by Klir [123] and Zeigler [237]. Klir refers to this kind of model data system. Zeigler developed the DEVS [237] formalism, which further refines the concept of *local* state change (called internal transitions) within the context of a function that responds to inputs from outside the system (external transitions). From our modeling perspective, this is achieved through multimodeling (Ch. 8).

Animation [41, 220] serves as a rich analogy for declarative models where keyframes are events and frames are low-level states. Badler et al. [9–11] and Zeltzer [239] serve as a good introduction to the problems of animating human and other articulated figures. Magnenat-Thalmann and Thalmann [140, 141] demonstrate a wide range of tools necessary for simulating and animating human figures. The images rendered in Figs. 4.46 through 4.48 and Fig. 4.54 were created using Porr's XKEY and animation software tools [173]. Labanotation [113] is one of two main forms of dance notation (the other being Eshkol-Wachman). Fletcher [74] provides a comprehensive discussion of digital circuit theory and counters, along with specification sheets for various integrated circuits. For the control of devices, traffic control methods are discussed in [11] and digital control approaches are found in [213]. In computer science, operating systems use a variety of digital control methods; high-level scheduling of processes is accomplished with FSAs. A discussion of space-time and the use of space diagrams can be found in [185] and [200]. The dynamics of adventure simulations and approaches to building simulators are covered in [133, 147]; generally, an FSA is used to drive the state changes; however, there is no inherent reason why more complex modeling types cannot be employed.

Formal programming languages and theories for logic [55] (including production systems [42], propositional, predicate, and modal logics) can be seen as the natural extension for comprehensive declarative modeling since not only are state transitions supported (in the form of an inference), but also pattern matching (i.e., unification) and recursion are supported in languages such as Prolog [129]. Temporal logic languages (such as Tempura [153]) explicitly support reasoning about time-stamped pieces of knowledge. Robertson et al. [190] have constructed a sizable simulation system in standard Prolog, primarily for studying ecological models. Narain and Rothenberg [157] and Radiya and Sargent [183] suggest ways that logic can be used as a simulation specification.

Event graphs developed by Schruben [199] are perhaps the best known simulation modeling technique which stresses events during the modeling process (rather than states) and how events are related (by causality or scheduling). The inventory problem discussed in Sec. 4.4.2 can be augmented with statistical approaches for measuring confidence intervals and gradient estimation [132]. Petri nets began with the work of Carl Petri [169, 170] but have been greatly expanded since to accommodate many new additions and variations. They have good coverage for modeling all sorts of systems from computer systems to manufacturing activities [167, 168]. Petri nets have a familiar role in manufacturing [44, 216, 222], where discrete entities flow through networks containing shared resources.

4.8 SOFTWARE

Tempura is an interpreter for the language Tempura specified by Ben Moskowski [154]. The interpreter was written by Roger Hale at Cambridge University and can be obtained by sending email to him at `rwsh@computer-lab.cambridge.ac.uk`. Simpack includes the following tools for declarative modeling:

- fsa: finite state automata with timed states
- markov: Markov model with timed states
- petri: timed Petri net simulator

GSPN is a general stochastic Petri net simulator available under `ftp.cis.ufl.edu:/pub/simdigest/tools`. Also, a Petri net simulator with a graphical front end, called *Olympus*, is available by sending mail to `nutt@boulder.colorado.edu`.

## Chapter 5

Functional Modeling

Note that the transformation is defined not by any reference to what it "really" is, nor by reference to any physical cause of the change, but by the giving of a set of operands and a statement of what *happens, not* why *it happens.*[1]

5.1 OVERVIEW

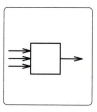

Functional models are graphs containing two primary components: functions and variables. Two approaches to functional modeling are identified by having either the functions or the variables as graph nodes. Recall that objects are composed of methods and attributes. Methods correspond to functions, and attributes to variables. If the problem suggests that you configure the system in terms of objects—each having a definite function—then choose the *function-based approach*; otherwise,

---

[1]W. R. Ashby, *An Introduction to Cybernetics* [6], p. 11.

choose the *variable-based approach*, where the focus is on the attributes, with functions playing a secondary role.[2]

When should you use functional models over other model types? Let's consider some heuristics:

1. If the problem is given in terms of distinct physical objects which are connected in a directed order, use a functional model.

   (a) If the objects are primarily functional in nature, use the *function-based approach*.
   (b) If the objects basically represent capacitance or storage, use the *variable-based approach*.

2. If the problem involves a material flow throughout the system, use a functional model.

Examples of domains for which functional models are appropriate are:

- Queuing models (where a system involves a flow of entities throughout a system of objects).
- Control engineering models where directionality comes from the need to *control* the system. Objects in a controlled environment have *one-to-one* mappings with physically engineered components.
- Compartmental models used in biology and medicine, where individual organs map to objects.

Functional models are not appropriate when the objects are *tightly interconnected* in a non-directional way, as with analog electrical networks. Digital logic networks are not a problem since directionality is built into the design; integrated circuits and other chip-based devices update their output given the current input signal when a clock signal triggers the device. If the problem domain is best viewed as a connected set of phases or maps, a declarative model would be better. The main keywords suggesting functional model use are *distinct physical components* and *directionality*.

Functional models map very easily into the OO paradigm since a function is represented as a method located within an object. Figure 5.1 displays the subclass hierarchy of Fig. 1.8 for functional models. The basic types of functional model are function-based (FUNC-BASED) and variable-based (VAR-BASED). Function-based models contain a transfer function which *relates* input to output. Values are passed on transitions where each transition may have an optional variable named attached. This variable is usually a state variable but may also be an auxiliary variable such as a constant or parameter. Variable-based models focus on variables which have values. In addition to values, the variable (level) must have pointers to the input transitions and output transitions. This is necessary to formulate the equations for execution. Transitions for variable-based functional models

---

[2]In many variable-based modeling approaches (to be discussed in Sec. 5.3) the functions are implicit—linear, for instance. Somewhere, you must specify the nature of f when deciding on the semantics of a graph arc from node X to Y, where $Y = f(X)$. If f is assumed to be of a certain type such as linear, all that remains is to label graph arcs with parameter values.

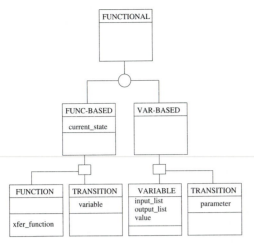

Figure 5.1   Functional model hierarchy.

carry a parameter value but there is no explicit function. Functions are implicitly defined to represent a linear algebraic combination in general.

5.2 FUNCTION-BASED APPROACH

The first type of functional model is one in which we focus on the function that transfers input to output,[3] possibly including a change in internal state (as with integrator blocks). We begin our study with modeling methods that focus on function rather than variables affected by functions. Functional (or *procedural*)[4] modeling relies on functional elements as the building blocks upon which to construct a dynamical model. While declarative models are useful for digital control and the modeling of discrete event systems with relatively small state spaces, we often use functional modeling for continuous systems and discrete event systems composed of coupled functional blocks. Functions, along with inputs and outputs, are often depicted in a "block" form, especially when a block is the iconic representation of physical device being modeled. A block model primitive is shown in Fig. 5.2(a), which contains a mathematical function that operates on inputs and produces outputs. There can be an arbitrary number of inputs and outputs and blocks may be coupled together to form block networks. An example block network with four inputs and one output is shown in Fig. 5.2(b). A functional block model such as the one represented in Fig. 5.2(b) is represented as an object-oriented model by realizing that each block contains a function and each function represents an object's method. A complete functional block model is composed of interconnected object methods where there are no restrictions on the number of objects used. Therefore, a block model might incorporate block *methods* which are all part of the same object, or multiple object methods can be employed. Figure 5.3 is the

---

[3] Such functions are often called *transfer functions*.

[4] The term *functional* is preferred over *procedural* since a function has an output, whereas a procedure need not have an output.

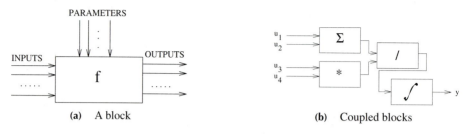

(a)   A block   (b)   Coupled blocks

Figure 5.2   Block primitive and network.

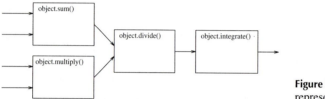

Figure 5.3   Object oriented
representation of Fig. 5.2(b).

object-oriented representation (assuming that one object contains all represented methods) of the functional block model in Fig. 5.2(b).

5.2.1 Block-Models

An algorithm for a functional block model simulator using the time-slicing method is represented as:

Algorithm 5.1. Time-Slicing Block Simulator (Version 1)
Procedure Main
　　　　Read in block network structure
　　　　For all blocks in the system:
　　　　　　　If the block is an integrator then initialize
　　　　　　　　　output according to initial condition
　　　　　　　Otherwise, set block output to zero
　　　　End For
　　　　While not(end of simulation) do
　　　　　　　Update_Blocks
　　　　　　　Output time and state variable(s)
　　　　End While
End Main

Procedure Update_Blocks
　　　　For each block in network do
　　　　　　　Given current block input(s)
　　　　　　　Calculate block output(s)
　　　　End For
　　　　Update current time by a fixed amount
End Update_Blocks

TABLE 5.1   BLOCK DEFINITIONS

Type	Symbol	Description	Input 1	Input 2	Param. 1	Param. 2
int	$\int$	integrator	input	N/A	Initial cond.	N/A
mult	$\times$	multiplier	input1	input2	N/A	N/A
div	$/$	divider	input1	input2	N/A	N/A
add	$\sum$	summer/adder	input1	input2	N/A	N/A
sub	$-$	subtracter	input1	input2	N/A	N/A
const	K	constant	value	N/A	N/A	N/A

To implement a basic time slice block diagram simulator, we need only set up a simple data structure for each node in the graph. A graph node is composed of a set of inputs, a transfer function, and a single output. When individual nodes are linked to one another in the form of a graph, we simulate the model by computing outputs from current inputs for each node, then repeating this process for a specified time duration. The input for such an interpreter would be as follows:

```
#blocks #outputblock time
block-num block-type input1 input2 .. param1 param2 ..
block-num block-type input1 input2 .. param1 param2 ..
. . .
```

The possible block types along with their inputs and parameter specifications are listed in Table 5.1.

Let's take the differential equation $x'' = -ax'$. We can represent this equation by the block diagram in Fig. 5.4. There are four blocks: two integrators (`int`), one multiplier (`mult`) and one constant (`const`). We will simulate behavior for 50 time units. The initial conditions for the two integrators will be 1.0, and parameter $a = 1$. Our block simulator will use the following input data to describe this block diagram:

```
4 3 50.0
0 const -1.0
1 mult 0 2
2 int 1 1.0
3 int 2 1.0
```

The output of the block model in Fig. 5.4 is shown in graph form in Fig. 5.5. Below we simulate a few iterations of the block simulator to get a better feel for its operation. The time slice is 0.01 and we will simulate the system for five steps (or, until the simulation time equals 0.05) and output block 3 (x):

```
0.010000 1.0099
0.020000 1.0197
0.030000 1.0294
0.040000 1.0390
0.050000 1.0485
```

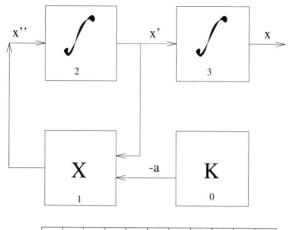

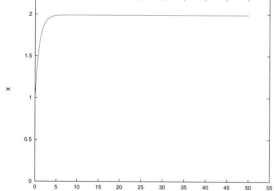

Figure 5.4 Block diagram for $x'' = -ax'$.

Figure 5.5 Plot of block 3 output (x).

The first column of output is the time (t), and the second column is the value of $x(t)$. If you refer to the time slice algorithm, you note that there is a sweep through all blocks while updating each block output during each sweep. Initial conditions give the values for $x'(0)$ and $x(0)$; however, other block output values may be incorrect until the correct values propagate through the block network. Initially, the simulator sets all outputs to an arbitrary value 0.0—with the sole exception of integrator blocks. Since integration assumes a knowledge of the present state, the output of each integrator block is initially set to its appropriate initial condition. Let's go through the algorithm during startup. The output of blocks 2 and 3 are set to 1.0 since those are the initial conditions specified in the input file. All other block outputs are set to 0.0. During the first block sweep, block 0 executes, providing a constant -1.0 for all time. Next, block 1 executes by multiplying the initial condition of block 2 (1.0) with -1.0 yielding -1.0. Euler integration is now performed: $1.0 + -1.0 * 0.01 = 0.99$. The output of block 2 feeds into block 3 with another integration operation: $1.0 + 0.99 * 0.01 = 1.0099$. We have completed the first full sweep of the blocks. This procedure continues throughout the simulation. As a second example of block model simulation, we simulate the third-order equation $x'' - 0.8x' + x = 2$ shown in Fig. 5.6. The

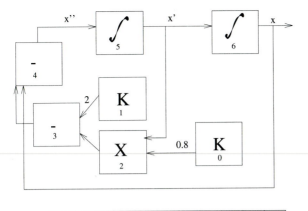

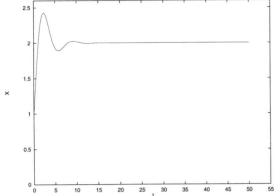

Figure 5.6   Block diagram for $x'' - 0.8x' + x = 2$.

Figure 5.7   Plot of block 6 output (x).

output from block 6 is in Fig. 5.7. The input to the block simulator is

```
7 6 50.0
0 const 0.8
1 const 2.0
2 mult 0 5
3 sub 1 2
4 sub 3 6
5 int 4 1.0
6 int 5 1.0
```

The algorithm specified for the time slicing block simulator is reasonable; however, there can be a problem with propagation of values. First, it is unnatural to set all block outputs to zero artificially (unless the blocks are integrators, in which case the outputs are set to initial conditions). Consider the ordering of blocks given in Fig. 5.4. This order causes no problems in propagation and no reason to preset all outputs to zero. However, we would prefer to have a more general method than this—we would like to have a simulation algorithm where the order can be determined by the program and not by the user. The network in Fig. 5.8

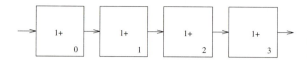

Figure 5.8   Partial block model illustrating the propagation problem.

displays the problem of ordering. Each of these blocks has the same function denoted by
1+, which means "take the input and add one to it for the output." If we choose the order 0,
1, 2, 3 for execution of the four blocks and the initial input line entering block 0 is, say, 12
then the propagation will occur correctly. That is, block 0 will yield 13, block 1 will yield
14, and so on. The issue is that functional composition should not *take any time* in a block
network. If we were to execute the blocks in the reverse order (starting with block 3, and
with output lines all set to zero at the beginning), we have a propagation problem; the input
of 12 does not propagate through block 3 until 3 time units later in the simulation.

The way around the propagation problem is first to notice that there are no cycles
in the block graph if the integrator blocks are removed. This is because the integrator
blocks represent pieces of differential equation(s) and without them there can be no cycles
(assuming that no other "delay" or "memory" blocks are permitted). Therefore, if we first
propagate all values through the rest of the network (i.e., the network minus the integrator
blocks), and then execute the integrator blocks, we will have an algorithm that does not
require the user to enter block ordering. This revised algorithm is listed below.

> **Algorithm 5.2. Time-Slicing Block Simulator (Version 2)**
> *Procedure Main*
> > *Read in block network structure*
> > *Create an acyclic graph for all nonintegrator blocks*
> > *Identify the leaf nodes (with no inputs)*
> > *For all blocks in the system:*
> > > *If the block is an integrator then initialize*
> > > > *output according to initial condition*
> > *End For*
> > *While not(end of simulation) do*
> > > *Update_Blocks*
> > > *Output time and state variable(s)*
> > *End While*
> *End Main*
>
> *Procedure Update_Blocks*
> > *Propagate values from the leaf nodes while calculating*
> > > *block output(s) from block inputs*
> > *For each integrator block do*
> > > *Given current block input*
> > > *Calculate block output*
> > > *Let output equal state*
> > *End For*
> > *Update current time by a fixed amount*
> *End Update_Blocks*

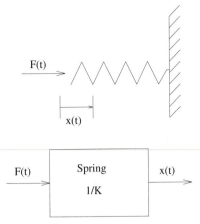

F = Kx

Figure 5.9   Schematic and block model for a linear spring.

The propagation problem is specific to block models where model components separate functions so that the model better reflects reality where functions map to physical components. In equationally based constraint models, the problem does not usually arise since the models are completely oriented around changes in state space variables; variables that are *intermediate* in nature are not made explicit in the equations, so model execution simply involves sweeps where the state variables are incrementally updated.

We have discussed some example block networks; however, we still need to ground our discussion by providing real-world examples. Most functional network blocks map directly to real-world objects. For instance, the spring, damper, and mass shown in Figs. 5.9 through 5.11 display devices whose block functions are defined using the known physical relationships between an input signal (such as force) and an output signal (such as position).

Even though these elements (spring, damper, and mass) demonstrate functional relationships between force and position, there is a difference in the function used: the spring function is derived from Hooke's law, which states that displacement of a spring is proportional to the applied force.[5] On the other hand, the force applied to the damper creates a viscous drag which is proportional to the first derivative of position (i.e., velocity of the piston), and the force on the mass is proportional to its linear acceleration.

These mathematical relationships can be verified experimentally; however, we should note that they represent *ideal* elements; real-world springs and dampers may be nonlinear in their force/position relationships. Most systems modeled with functional blocks are of the form shown Fig. 5.12, which serves as a *template* for simple control block models. The blocks in Fig. 5.12 are defined below.

- *Reference Input.* This signal specifies the goal or the setting for which the controller is constructed. The system will attempt to reach this goal in spite of the lags present in the controlled process.

___

[5]We assume the spring to be linear, so that we multiply the displacement by a spring constant K to determine the force.

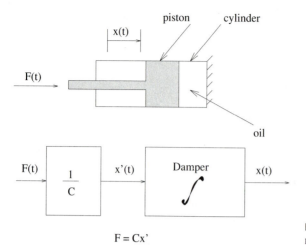

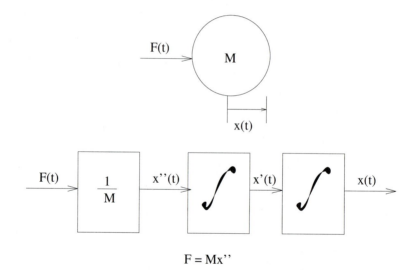

$$F = Cx'$$

Figure 5.10   Schematic and block model for an hydraulic damper.

$$F = Mx''$$

Figure 5.11   Schematic and block model for a moving mass.

- *Comparator.* This element compares (by summing the inputs and their associated signed weights) the reference signal with the feedback obtained from the transducer.
- *Error Signal.* This is the difference between the reference input and the feedback signal.
- *Controller.* The controller (or compensator) takes the error signal and "speeds it up," thereby reducing overall time lags.
- *Actuator.* The error signal, while representing a difference to reduce, is usually weak and must be amplified.

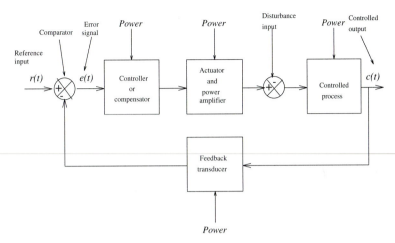

Figure 5.12   Generic block model.

- *Controlled Process.* This is the system (i.e., process) which is being controlled, and the dynamics of the process are located in this block.
- *Feedback Transducer.* The transducer is a measuring device that measures the current output of the signal. It then feeds this value to the comparator, which compares the output (what the system is "really doing") to the reference input (what we want the system to do).

5.2.2 Automatic Vehicle Control

Many ordinary day-to-day activities can be modeled with block networks—such as driving an automobile. We are controlled by our central nervous system, which has its control in the brain. We set a desired speed in our minds and then activate one foot to press on the accelerator pedal. The pedal controls flow of gasoline to the engine, which serves as the "amplifier" for our signal. Along with external disturbances such as the road gradient, this causes the car to move forward. The inertia of the car mass serves to characterize the transfer function for the automobile. The speedometer is the transducer that determines the current speed and reports this back to our eyes in the form of a digital or analog display. Figure 5.13 displays the "man in the loop" control for an automobile. This provides a generic look at manned automobile control where the human (and the central nervous system) serves to control the automobile. Now, let's adapt this to a fully automated vehicle control model. Automatic vehicle control will be central to future highway systems as cars are fitted with sensors and control systems so that collisions can be avoided when the driver does not react quickly. Automatic vehicle control, in general, is applicable not only to automobiles but also to monorails and airport subway shuttles, both of which are unmanned. Consider a subsystem of two cars (or cabs) on a monorail system: A and B, where B is following A as in Fig. 5.14. Our discussion will present control block models based on an equational presentation and on the example given in [211], where three different control algorithms (models) are studied. Each car will contain sensors for velocity and position for the car

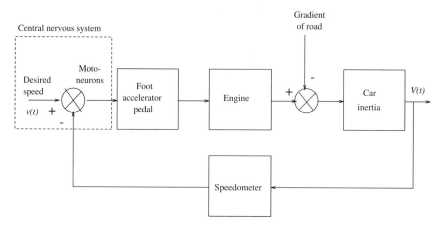

Figure 5.13   Block model for automobile control.

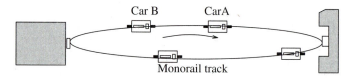

Figure 5.14   Monorail with cars: *B* following *A*.

ahead of it. Our discussion will center around the control system for car *B*, which has
sensors for detecting its own speed and the speed of *A*. We define:

1. *x* is the position of *A* (initially, 50 m).
2. *y* is the position of *B* (initially, 0 m).
3. *u* is the speed of *A* (constant: 15 m/s).
4. *v* is the speed of *B* (initially, 15 m/s).

x, y, u, and v are all functions of time (t). We need to come up with a control algorithm
which takes these four values as input and produces an acceleration for car *B*. Let D equal
the minimum safe distance ($x - y$) between the cars. We'll pick two different approaches:
(1) control based on distance between cars *A* and *B*, and (2) control based on relative velocity
of cars *A* and *B*. Based on these two approaches we design two algorithms for approach
1 and one algorithm for approach 2. The following values were used for parameters:
$a_{max} = 20.0$, $d_{max} = 20.0$, $D = 30.0$, and $K = k = 1.0$.

1. *Algorithm 1:* if $D > x - y$, then $a = -d_{max}$; if $D = x - y$, then $a = 0.0$; if
    $D < x - y$, then $a = a_{max}$.
2. *Algorithm 2:* Let $a = k[(x - y) - D]$. If $a > a_{max}$, then $a = a_{max}$. If $a < -d_{max}$,
    then $a = -d_{max}$.
3. *Algorithm 3:* Let $a = K(u - v)$. If $a > a_{max}$, then $a = a_{max}$. If $a < -d_{max}$, then
    $a = -d_{max}$.

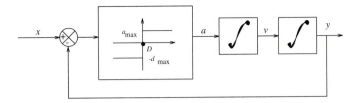

Figure 5.15   Block model for algorithm 1.

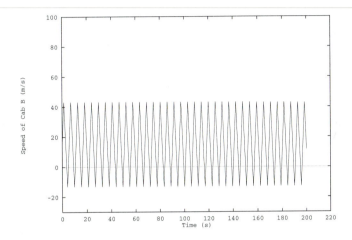

Figure 5.16   Speed of car B versus time for Fig. 5.15.

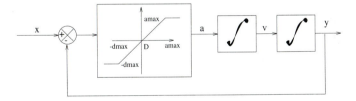

Figure 5.17   Block model for algorithm 2.

The three algorithms can be used to design block models as shown in Figs. 5.15, 5.17, and 5.19. The models are followed immediately by a graph of the velocity of car B vs. time for each model (Figs. 5.16, 5.18, and 5.20). The model (Fig. 5.15) based on algorithm 1 is very simplistic and has a *bang-bang* controller. This type of control is similar to the on-off control for an air conditioner or heater in which there is not a smooth gradient response. The speed of car B can be seen in Fig. 5.16 to oscillate with a kind of jerky behavior. Note how the car must reverse its direction just to maintain the correct distance. The second model has a smoother *ramp* type of transfer function, which takes the current distance error and creates a more gradual control to achieve a minimum distance error between the two cars. Unfortunately, even though the speed vs. time plot in Fig. 5.18 demonstrates a

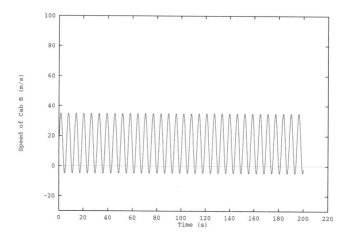

Figure 5.18   Speed of car *B* versus time for Fig. 5.19.

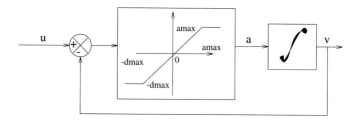

Figure 5.19   Block model for algorithm 3.

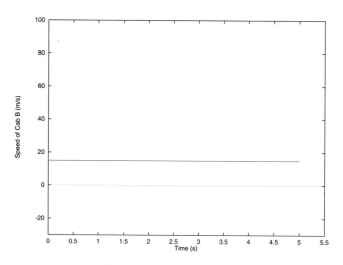

Figure 5.20   Speed of car *B* versus time for Fig. 5.19.

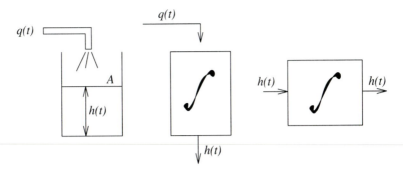

Figure 5.21   Single tank with input.

smoother ride and it has less severe backup requirements, it is still quite abrupt because the speed oscillates. The first two models exhibit harmonic oscillation like that of a moving mass spring system. The best controller is one where the control is based, not on distance measurement, but on relative velocity as shown in Fig. 5.20. This results in a first-order asymptotic behavior without oscillation.

5.2.3 Coupled Water Tanks

A single water tank with an input valve demonstrates a common "block" behavior: that of the integrator. An "integrator" is a device that "collects" or stores the input—just as a tank stores the water that flows into it at a rate $q(t)$. Figure 5.21 displays three subfigures: (1) a schematic of the tank, (2) the direct mapping from the topology of the design to block model format, and (3) the block model as normally shown. The schematic displays our first model of the device—a pictorial concept model. Then we consider the topology (or connectivity) in the design and abstract out a block model with an input flow and an output height. Note that the "output height" simply reflects our observation of the system state and does not imply that water leaks out of the tank. To make a block network where blocks line up in rows and columns, we standardize this second representation to form the third model in Fig. 5.21. The relationship between the flow rate and the height of the water in the tank is

$$A\frac{dh(t)}{dt} = q(t)$$

where A is the cross-sectional area of the cylindrical tank. By multiplying by A we ensure that the dimensions on both sides of the equation refer to volume.

Now, let's create another tank so that the flow from the first tank flows into the second one. This creates the schematic and block diagram shown in Fig. 5.22. Figure 5.22 shows no feedback from the second tank to the first tank even though there is feedback from the first tank to itself since $h_1(t)$ decreases over time in the absence of input (due to gravity). Tank 1 empties into tank 2 while simultaneously being filled by the tap with a time-dependent flow rate $q(t)$. The notation $h_1(t) \sim h_2'(t)$ means that $h_1(t)$ is linearly proportional to $h_2'(t)$. Figure 5.23 is a *closed network* with feedback, since not only do we have an input flow into tank 1, but there is also potential flow back and forth as long as the heights in each tank

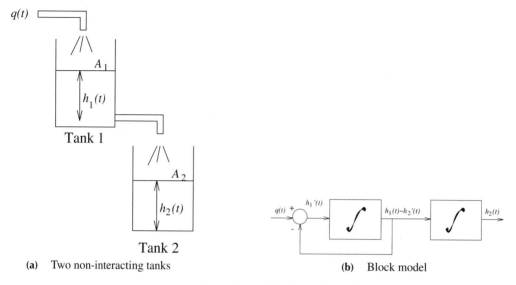

(a)   Two non-interacting tanks   **(b)**   Block model

Figure 5.22   Non-interacting tanks.

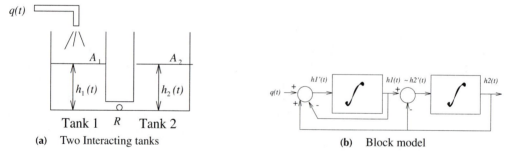

(a)   Two Interacting tanks   **(b)**   Block model

Figure 5.23   Interacting tanks.

$h_1(t)$, $h_2(t)$ differ. Below are the equations representing the dynamics for the Fig. 5.23 scenario:

$$A_1 \frac{d}{dt} h_1(t) = q(t) - \frac{1}{R}(h_1(t) - h_2(t))$$

$$A_2 \frac{d}{dt} h_2(t) = \frac{1}{R}(h_1(t) - h_2(t))$$

In Fig. 5.22(b) there is a fairly good resemblance to the schematic in Fig. 5.22(a); however, the same cannot be said for Fig. 5.23. Here, the schematic depicts a case where there are two storage tanks (or capacitances) connected to one another. Somehow, we should be able to represent this same connectivity in the model, yet the block model must specify the feedback loops so that the functional relationships are preserved. As a system tends toward a collection of highly interactive components, block models become a less appropriate modeling approach. If we recall our initial heuristic of *directionality* for block

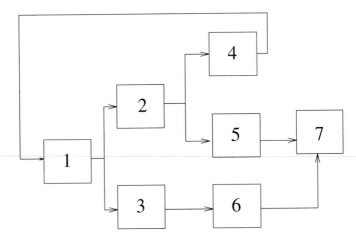

Figure 5.24   Sample functional network.

models, we see how the two tanks in Fig. 5.23(a) are connected in a nondirectional fashion. We will find that the constraint-based modeling approach in Ch. 6 yields the best models for such systems.

5.2.4 Encoding Functional Network Structure

Overview.   Many of the models in this chapter are built in the form of a directed graph (or network). There are two fundamentally different ways of translating such structure into code: *encapsulation in code* and *encapsulation in data*. Consider the network depicted in Fig. 5.24. This network can be either a network of functions, delay elements, or queues. Each block, ultimately, processes inputs and produces an output which can be routed to one of several other blocks. To encapsulate Fig. 5.24 in code, we create an algorithm whose structure is *isomorphic* to the network structure. That is, blocks will be represented by function calls or blocks of code, while network arcs are represented by transfers from the current code address to another address using a language feature such as a "go to" or "case/switch statement." The structure in Fig. 5.24 is encoded using the following code template:

Algorithm 5.3. Encapsulating Network Structure in Code
BeginSwitch (block-id)
    case 1: process(2); process(3); break;
    case 2: process(4); process(5); break;
    case 3: process(6); break;
    case 4: process(1); break;
    case 5: process(7); break;
    case 6: process(7); break;
    case 7: wrap-up; break;
EndSwitch

For time slicing applications, `block-id` represents the identification number of the block. For event scheduling and queuing, `block-id` will be referred to as a "token" or "entity." Also, the procedure `process()` will involve a call to the routine `schedule()`. In the case of time slicing with no delay, `process()` refers to the functional semantics of the specific block. Sometimes, there will be conditions attached to certain network arcs, and these can be handled easily using control functions of the programming language (such as `if..then..else`).

Writing simulation applications by encoding network structure directly within the code is fairly straightforward and somewhat less complicated than encoding network structure in data; however, the code can become unwieldy, especially when there are lots of network arcs—both feedforward and feedback. Also, whenever the structure changes, the code needs to be revised. In any event, simple networks are encoded easily in this manner.

For our second method, we consider encoding network structure in data. Here, we read an input file that encodes the data structure (for the network topology). This provides a very compact code representation where we are interested in encoding the action on a single block after reading the network structure from a file or database. Here is a code template for this method:

Algorithm 5.4. Encapsulating Network Structure in Data

BeginSwitch (block_type)

   Case TYPE1: apply-function(1); break;

   Case TYPE2: apply-function(2); break;

          . . . .

EndSwitch

For all nodes in the out set of this block:

   (1) Save the inputs of the out set node

   (2) Schedule out set nodes to execute

End For

In the second method, we must create a data structure to encode the network structure. The network is "read into" the program and processing begins. The processing of any single block occurs within the primary switch statement. Each block has a beginning (`BEGIN`) where inputs enter a block, and an end (`END`) where output comes out of the block.

Most sophisticated simulators use the second method of encapsulating network structure in data because then the user can easily modify the structure by changing a single data value within a file. Simulators that have graphical user interfaces overlaying them make this even easier by allowing the modification of values with an input device. The previous "code method" is still useful for small uncomplicated networks, although the "data method" is more convenient for the user of the simulator.

A third method, involving translation of an intermediate language, combines the advantages of both methods since the network is described using the language (data) and the translation can involve compilation (code), which produces executable code. This method was used in some of the SimPack tools such as Mini-GPSS (Sec. 10.5.15) and Xsimcode (Sec. 10.5.16).

Summary. Considering the "encapsulation of network in code" method, we can delineate certain advantages and disadvantages with respect to the "data" approach:

Advantages of Code Method

1. The code method is useful for simple networks that have a fairly linear structure. With nonlinear structures such as trees, the encapsulation of the model in the code entails keeping track of branching.
2. The code method operates in a "macro expansion" fashion in that adding an extra node to the network entails adding a chunk of code to take care of that node's functional semantics. This can produce code that is faster than an interpretive method.
3. In code, it is very straightforward to add special features that may not be available in a general-purpose network simulator that accepts a specific input format. You may, for instance, wish to perform all branching on a *fork* node based on the number of tokens in a previously visited server node. With the code method, this enhancement means adding one or two lines of code. With the data method, one would have to reorganize the way that the data is interpreted—a new data field would need to be designed.

Disadvantages of Code Method

1. The code method can be unwieldy when the model is complex with possible feedback loops; it is much easier to encode by translating the graph topology directly into a data file.
2. The size of the code becomes quite complex and grows linearly with the size of the model (in terms of node number). With a data approach, it is necessary to modify the relatively compact data file without touching the code.

5.2.5 Digital Logic Circuits

Let's study the method of event scheduling by considering the digital logic circuit presented in Fig. 5.25. If we model all logic gates as having zero propagation delay, the Fig. 5.25 circuit has the behavior shown in Table 5.2.

The behavior in Table 5.2 reflects a zero-propagation delay, for the two digital components. Time-based behavior would depend on how the inputs u_1, u_2, and u_3 change over time. Let's provide input trajectories that alternate every four time units. Then the inputs and single output are seen in Fig. 5.26. We see that the behavior is such that $\forall i\,(Y = U_i)$. However, if we assign a propagation delay of 1 time unit to the OR gate and 2 to the AND gate, the behavior changes. First, let's discuss the tools necessary to facilitate programming using event scheduling.

The SimPack routine `trace_visual(mode)` is the best method of seeing the step-by-step process used by the discrete event simulator. This routine takes one argument [either `INTERACTIVE` or `BATCH` (the `mode` argument)], and is discussed in more detail in the next section. For now, we can hand simulate the digital logic circuit in Fig. 5.25 to arrive at the following procedure. The future event list (FEL) is represented as a linked list of items. Each item has a time when the event is to occur (in parentheses) and the name of the event.

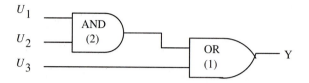

	AND		OR
0 0	0	0 0	0
0 1	0	0 1	1
1 0	0	1 0	1
1 1	1	1 1	1

Figure 5.25   Digital logic circuit.

TABLE 5.2
BEHAVIOR OF
FIG. 5.25

Inputs			Output
u_1	u_2	u_3	y
0	0	0	0
0	0	1	1
0	1	0	0
0	1	1	1
1	0	0	0
1	0	1	1
1	1	0	1
1	1	1	1

1. Schedule BEGIN_FUNC for three GEN blocks, providing the three four-time-cycle signal sources. All outputs are initialized to zero and the simulation clock is set to zero. The FEL is now

$$(0)GEN1 \rightarrow (0)GEN2 \rightarrow (0)GEN3$$

2. The first event is "caused" via next_event, and event routine END_FUNC is executed for the GEN events. The code associated with GEN schedules two new events to occur to change the cycle of the signal source (i.e., to enable oscillation and to initiate an AND event). The FEL is now

$$(0)GEN2 \rightarrow (0)GEN3 \rightarrow (2)AND \rightarrow (4)GEN1$$

3. The blocks in the out set of the remaining two GEN blocks are now processed. This means that all blocks directly connected to the GEN blocks are scheduled since their respective outputs may change. GEN1 and GEN2 cause the AND gate to be scheduled, and GEN3 causes the OR gate to be scheduled. The FEL becomes

$$(1)OR \rightarrow (2)AND \rightarrow (4)GEN1 \rightarrow (4)GEN2 \rightarrow (4)GEN3$$

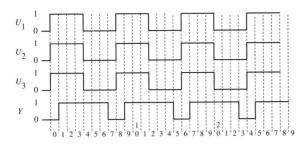

TIME

Figure 5.26   Behavior of Fig. 5.25 circuit with zero propagation delay.

4. The routine `Update` terminates and the `next_event` is processed. The simulation time is set to 1. The FEL is now

$$(2)AND \rightarrow (4)GEN1 \rightarrow (4)GEN2 \rightarrow (4)GEN3$$

5. The OR event code is executed, and the new OR block output value is determined. Then all blocks in the out set of OR are scheduled. Since there are no blocks, we go to the next step.

6. The routine `Update` terminates and the `next_event` is processed. The simulation time is set to 2. The FEL is now

$$(4)GEN1 \rightarrow (4)GEN2 \rightarrow (4)GEN3$$

7. The AND event code is executed and the new AND block output value is determined. Then the OR block is scheduled to be executed at the current simulation time (2) plus the OR block delay (1): $2 + 1 = 3$.

This general procedure is executed repeatedly until the simulation time is exhausted. The algorithm for the block simulator with event scheduling is shown below.

Algorithm 5.5. Block Simulator Using Event Scheduling
Procedure Main
        Set SIMCLOCK to zero
        Read in block network structure
        Initialize all output block values
        For each Generator Block, schedule GEN
        While more simulation time do
                Get next event
                Update
        End While
End Main

Procedure Update
        Update SIMCLOCK
        For event GEN:
                Generate an oscillating square wave
                Schedule GEN at SIMCLOCK + delay_for_wave

> *For all other events:*
> > *Given inputs saved, execute the function*
> > > *associated with this block*
> > *For all events:*
> > > *Save the inputs of all blocks in the out set of*
> > > > *this block*
> > > *Schedule out set blocks to occur using their delay times and stored inputs*
> *End Update*

The input data for a general digital logic network is

```
#blocks #outputblock time
block-num block-type delay input1 input2 ... #outputs output1 ... param1 param2
block-num block-type delay input1 input2 ... #outputs output1 ... param1 param2
...
```

This format is very similar to that of the time slicing simulator with the exception that, here, we specify the out set of each block (Table 5.3).

First, running the event scheduling simulator with zero propagation delays for the *and* and *or* gates with input

```
5 4 18.0
0 gen 0.0 1 3
1 gen 0.0 1 3
2 gen 0.0 1 4
3 and 0.0 0 1 1 4
4 or 0.0 3 2 1 5
```

yields the same behavior as in Fig. 5.25. If we let an and gate take 2 time units delay, and an or gate take 1 time unit delay, we can employ the event scheduling method as follows. Using the code for the simulator, and inputting the data

```
5 4 18.0
0 gen 0.0 1 3
1 gen 0.0 1 3
2 gen 0.0 1 4
3 and 2.0 0 1 1 4
4 or 1.0 3 2 1 5
```

we obtain Fig. 5.27.

The digital logic simulator provides a tool for basic logic simulation. For inputs to the same block, where event times are equal, *duplicate events* (equal times and block id) arise. In such a case, the last such event will contain the accurate input information for the block to execute. The future event list can be trimmed when duplicate events arise.

TABLE 5.3 BLOCK DEFINITIONS

Type	Symbol	Description	Input 1	Input 2	Param. 1	Param. 2
gen	GEN	generator	N/A	N/A	N/A	N/A
and	AND	and-gate	input1	input2	N/A	N/A
nand	NAND	nand-gate	input1	input2	N/A	N/A
or	OR	or-gate	input1	input2	N/A	N/A
nor	NOR	nor-gate	input1	input2	N/A	N/A
inv	INV	invertor	input1	N/A	N/A	N/A

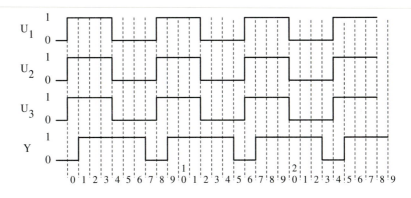

TIME

Figure 5.27 Behavior of Fig. 5.25 logic circuit with nominal delays.

5.2.6 Queuing Models

Queues occur in many physical systems, from lines at fast-food restaurants to waiting lines of computer jobs awaiting service from a processor. To model queuing systems, we will require the basic event scheduling method of the preceding section, but in addition we will need special tools for handling lines that occur at multiservice stations. Primitive queuing scenarios may be modeled using only the tools that we presented in the preceding section; however, when modeling queuing systems we also want to be concerned with special features such as (1) jockeying from one queue to another and (2) allowing queue entities to be assigned specific attributes and priorities. First, let's discuss some key statistics that will be of concern to us when reviewing the simulation output of a queuing system.

- *Givens.* Let T = total time for simulation (600 for our example), A = number of arrivals (6 customers into the bank), and C = number of completions. We will assume that $A = C$. The two will converge given a long simulation and relatively short queueing times (flow balance assumption).
- *Arrival Rate.* $\lambda = A/T$.
- *Throughput Rate.* $X = C/T$. If we let $A = C$, then $\lambda = X$. An important note about λ and X: they are relative measures. When you say that the throughput of an operating system, for instance, is 5 jobs per hour, you cannot accurately gauge

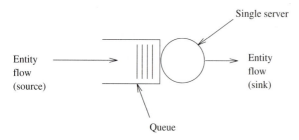

Single server

Entity
flow
(source)

Entity
flow
(sink)

Queue

Figure 5.28   Checkout line

whether this is good or a bad performance. You certainly can compare throughputs to one another; however, you must also know about how large and complex a job is; for large jobs, 5 jobs per hour may be very good or very poor.

- B = the total time that a server is busy, I = the total time that a server is idle. Note that these are mutually exclusive, so $B = T - I$.
- *Server Utilization.* $U = B/T$. The percentage of time that the server is busy. The server is idle I/T percent of the time.
- *Mean Service Time per Customer.* $T_s = B/C$.
- *Utilization Law.* $U = XT_s = \dfrac{C}{T} \times \dfrac{B}{C} = \dfrac{B}{T} = U$.
- *Little's Law.* Let w_i = time spent in system by the ith customer. Then let $W = \sum w_i/C$. W is the average time one customer spends in the system. Note that "being in the system" means being in one of two places: in the queue itself, or being served. Little's law states that $L = XW$, where L is the average number of customers in the system. If we separate people being served from people in the queue, we can define the following. First, rename T_s as W_s (mean service time per customer) and let U be L_s (the average number of customers receiving service). Let W_q = the average time a customer spends in the queue, and W_s = the average time a customer spends at the server. Then $W = W_q + W_s$. Likewise, let L_q = the average number of customers in the queue. Then $L = L_q + L_s$, where L is the average number of customers in the system. Little's law applies to the individual cases as well (i.e., $L_q = XW_q$).

Hand Simulations

Grocery Store Checkout 1.   Figure 5.28 shows a graphical model that represents a grocery store checkout. Customers buy food and then wait in line to check out their merchandise (i.e., pay for the food). Let us imagine that we have 5 customers in the store and that each of the customers arrive at the checkout register at different times. We will specify two critical times in this simulation: (1) the time it takes between "arrivals," and (2) the time it takes for the cashier to check out the customer. The first item is termed the "interarrival time," and the second, the "service time." Let's specify a constant interarrival time of 2 minutes, and a variety of service times ranging from 1 to 3 minutes. Consider Table 5.4. The columns are defined as follows: C# is the number of the customer, Arrives is the absolute arrival time in minutes (a constant 2-minute value). There are four possible times for the customer to leave the cashier. The first three are based on different service

TABLE 5.4   CUSTOMER ARRIVAL AND DEPARTURE TIMES

C#	Arrives	Leaves (1)	Leaves (2)	Leaves (3)	Leaves (C#)
1	0	1	2	3	1
2	2	3	4	6	4
3	4	5	6	9	7
4	6	7	8	12	11
5	8	9	10	15	16

TABLE 5.5   GROCERY STORE TIMES FOR 8 CUSTOMERS

C#	TSLA	ACT	ST	SB	SE	CW	SI
1	–	0	1	0	1	1	0
2	3	3	4	3	7	4	2
3	7	10	4	10	14	4	3
4	3	13	2	14	16	3	0
5	9	22	1	22	23	1	6
6	10	32	5	32	37	5	9
7	6	38	10	38	48	10	1
8	8	46	6	48	54	8	0
Totals						36	21

times (1, 2, and 3 minutes). The fourth time is equal to the customer number (i.e., the third customer takes 3 minutes to be serviced).

The problem with this model is that customers rarely arrive at precise time intervals such as 2 minutes. It would be more realistic to use a probability distribution that was obtained by studying an actual grocery store. Also, most grocery stores have more than one checkout line, slightly complicating the simulation—with multiple lines one must consider effects such as jockeying (moving among different queues before being served).

Grocery Store Checkout 2.   Let us consider a longer simulation of a checkout counter, where we specify the following column headers: C# for customer number, TSLA for time since last arrival, ACT for arrival clock time, ST for service time, SB for service begins, SE for service ends, CW for customer wait time, and finally, SI for server (i.e., cashier) idle time. Note that CW includes both the time waiting in the queue plus time waiting for service. This more detailed simulation is shown in Table 5.5.

Consider all absolute times to be on an integral scale so that when $SB = 14$ and $SE = 16$ this means that service begins at time 14 and service ends at time 16; simply subtract SB from SE to determine the length of service (i.e., 2). Let's plot the number of individuals in the queue over time in Fig. 5.29. The numbers above the horizontal bar represent the customer # currently being served. At time 13, for instance, there are two customers (3 and 4), which means that customer 4 is waiting in the queue for 1 time unit while customer 3 is being served. We see that for this process, a queue forms for a total of 3 time units; customer 4 is queued for 1 time unit and customer 8 is queued for 2 time units

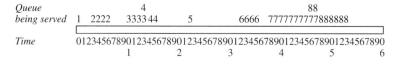

Figure 5.29 Picture of server and queue over time.

prior to obtaining service. We can summarize the simulation by noting some statistics associated with the behavior of this system. The average customer waiting time (w_i) is equal to the total customer waiting time (under CW) divided by the total number of customers (8) = 4.5 minutes ($\sum_i w_i / C$). The percent of time that the server is idle is equal to the amount of idle time (21) divided by the total time of the simulation (54; see last SE value) = 39%.

Values in columns are related to one another. Let each variable be indexed by the customer number. For instance, $TSLA(3) = 7$ and $ACT(1) = 0$. Note that C#, $TSLA$, and ST are all given at the beginning of the simulation, while the rest of the variables are computed as follows for $i \in \{1, \ldots, 8\}$.

1. Given: $TSLA$ and ST, $TSLA(1) = 0$.
2. $ACT(i+1) = ACT(i) + TSLA(i+1)$, $ACT(1) = 0$.
3. $SB(i+1) = \max(SE(i), ACT(i+1))$.
4. $SE(i) = SB(i) + ST(i)$.
5. $CW(i) = SE(i) - ACT(i)$.
6. $SI(i+1) = SB(i+1) - SE(i)$, $SI(1) = 0$.

To simplify the *transfer function* for a server, if we let D be the last departure from block B and A be the current arrival to block B, where Δt is the service time, we note that $D \leftarrow \max(D, A) + \Delta t$, where $D = 0$ initially.

This model is slightly more realistic than the previous model in that a *variety* of interarrival times are used; however, we have some of the same limitations otherwise. The main difference between this example and the previous one relates to the display of more variable values at each point in the simulation—a "point" in the simulation being a time when a new customer arrives.

Simple queuing simulations. All queuing model programs based on event scheduling will use a general *template* that is shown below.

Algorithm 5.6. Event-Driven Simulation Template
Initialization procedure
Set up all facilities
While not end of simulation
    Cause event to occur
    BeginSwitch (event)
        Case 1: event 1 code
        Case 2: event 2 code
        . . . .
        Case N: event N code

End Switch
End While
Print Simulation Report

The physical structure of the simulation program will follow the topological structure of the simulation graph that you create at the beginning.

Blocking in queuing networks. Our queuing networks so far have dealt with servers whose buffers could be of infinite capacity. But supposing that a buffer can be no larger than four or zero for example? There needs to be a method for making it so that a token is blocked from trying to request service from a server—to which it is connected—whose buffer is full. Consider the queuing network in Fig. 5.30. On the left-hand side of the figure, we have a generator of parts that generates 100 raw parts. The raw parts are fed into a spiral accumulator which stores the parts until they can be fed to a human lathe operator. The lathe operator takes only one part at a time, so has no buffer space. After the lathe operation is finished, the operator signals the accumulator to send the next raw part, and he sends the finished part to the part inspector, who can afford a buffer size of three while inspecting a part from the lathe. We define the following times:

- A raw part takes 2 seconds to move through the accumulator.
- The lathe operator takes an average of 2 minutes per part, using a normal distribution with $\sigma = 3.0$.
- The part inspector takes an average of 15 seconds per part with $\sigma = 1.0$.

Without blocking, parts would simply flow from left to right while holding in server queues until the server was ready. However, with blocking, we need to perform explicit *synchronization* between servers. Each of the three servers (accumulator, lathe operator, part inspector) is modeled by a facility in SimPack. Suppose that a part is being serviced by the accumulator facility. After it is finished being serviced, it wants to request the lathe facility but it cannot until it first checks to see whether the lathe buffer is full. Since the lathe buffer is set to zero, this is equivalent to checking whether the lathe operator is idle. If the operator is idle, the part in the accumulator requests the lathe and gets it. If the lathe operator is busy, the part stays in the accumulator server waiting for someone to release it. The accumulator part is released only when the lathe part is released. After the accumulator part is released, the first part in the accumulator queue begins to be processed, as usual.

This type of synchronization is shown by providing two algorithms. First, we present the switch statement assuming that we have unlimited (infinite) buffer sizes for the lathe and inspector:

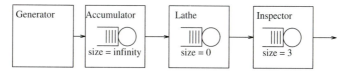

Figure 5.30   Lathe operation with blocking.

Algorithm 5.7. Lathe Operation without Blocking

For event GENERATE_PARTS:

Schedule REQUEST_ACC for 100 different times

For event REQUEST_ACC:

If the accumulator is free then

schedule RELEASE_ACC using 2 seconds

For event RELEASE_ACC:

Release the accumulator

Schedule REQUEST_LATHE now

For event REQUEST_LATHE:

If the lathe is free then

schedule RELEASE_LATHE using Normal(120,3) seconds

For event RELEASE_LATHE:

Release the lathe

Schedule REQUEST_INSPECTOR now

For event REQUEST_INSPECTOR:

If the inspector is free then

schedule RELEASE_INSPECTOR using Normal(15,1)

seconds

For event RELEASE_INSPECTOR:

Release the inspector

Next, with finite buffer sizes for the lathe operator and part inspector, we have to refine the algorithm to synchronize the release of the two servers.

Algorithm 5.8. Lathe Operation with Blocking

For event GENERATE_PARTS:

Schedule REQUEST_ACC for 100 different times

For event REQUEST_ACC:

If the accumulator is free then

schedule BLOCK_ACC using 2 seconds

For event BLOCK_ACC:

Set state to 'accumulator waiting'

If the lathe is free then

Set state to 'accumulator not waiting'

Schedule RELEASE_ACC now

For event RELEASE_ACC:

Release the accumulator

Schedule REQUEST_LATHE now

For event REQUEST_LATHE:

If the lathe is free then

schedule BLOCK_LATHE using Normal(120,3) seconds

For event BLOCK_LATHE:

Set state to 'lathe waiting'

If entities waiting for inspector < 3 then
  Set state to 'lathe not waiting'
  Schedule RELEASE_LATHE now
For event RELEASE_LATHE:
  Release the lathe
  If 'accumulator waiting' Schedule RELEASE_ACC now
  Schedule REQUEST_INSPECTOR now
For event REQUEST_INSPECTOR:
  If the Inspector is free then
       schedule RELEASE_INSPECTOR using Normal(15,1)
       seconds
For event RELEASE_INSPECTOR:
  Release the inspector
  If 'lathe waiting' Schedule RELEASE_LATHE now

General-purpose queuing networks.   As we have seen previously, we may encode a network model within code or data. So far, we have specified queuing networks within the code structure. This method is straightforward for small queuing networks; however, we should also consider defining queuing networks within data so that the topology and key queue characteristics are specified within a data file or from an interactive user interface that writes the data file. Our algorithm is specified below.

Algorithm 5.9. Queuing Network Simulator
Procedure Main
     Set SIMCLOCK to zero
     Read in block network structure
     For each Generator Block, schedule GEN
     For each Request Block, create a facility
     While more simulation time do
          Get next event
          Process-block
     End While
End Main

Procedure Process-Block
     Update SIMCLOCK
     For event GEN:
          Update arrivals
          Schedule GEN at SIMCLOCK + interarrival_time
          Schedule all nodes in the outset of the GEN block
               to occur now
     For event REQUEST:
          If queue is not busy, schedule RELEASE using service time
     For event RELEASE:

Release queue
Schedule all nodes in the outset of the RELEASE block
to occur now
For event FORK:
Pick a single fork output using a cumulative distribution
generated from the probability attached to each branch
For event JOIN:
Schedule all nodes in the outset of the JOIN block
to occur now
For event SINK:
Update completions
End Process-Block

Communication networks. We have already discussed unidirectional networks in the examples of the single-server queue model for the grocery store checkout and of the CPU/Disk network. These are sufficient to understand the basic concepts of the unidirectional approach. For a study of the bidirectional approach, we will consider a network of nodes where nodes serve as control points in a communications network. Each network will contain the following:

- A set of nodes
- A set of transceivers within each node
- A set of links between transceivers

To each link, there are two transceivers (one transceiver attached to a node), so that messages may be passed in either direction. We will model each transceiver as a facility resource. One will place messages into the network. A message is a string of integers where each integer represents a node number. In this way, our approach is one implementing static routing—that is, we specify the precise routing path for each message as opposed, say, to permitting the path to be determined in real time (dynamic routing). The pseudocode and the C program for the network simulation is shown below.

Algorithm 5.10. Communication Network Simulator
Procedure Main
Set SIMCLOCK to zero
Read in topology of network and message to be routed
Specify all transceivers as facilities
Each message is represented by a token
For each message, schedule LINK
While more messages do
Process-next-event
End Main

Procedure ProcessNextEvent
Get event from event list
Update SIMCLOCK

For event LINK:
Def: Message is crossing a link
  Determine which transceiver is being requested
  Schedule REQUEST for transceiver
    using link_time

For event REQUEST:
Def: Message is requesting a facility
  Schedule RELEASE of transceiver
    using node_time

For event RELEASE:
Def: Release transceiver
Update message pointer for this token
  (to go to next link in message)
End ProcessNextEvent

5.3 VARIABLE-BASED APPROACH

Functional modeling can also focus on the variables to be affected by the functions rather than the functions themselves. In the following modeling approaches, the variables (or levels) serve as the focal points of the modeling. The functional aspect is deemphasized to the point that a particular functional structure is often presupposed. For instance, a well-known structure such as a linear system is assumed and the modeler specifies only the variable dependencies and parameters.

5.3.1 Signal Flow Graphs

If we take the block model of two interacting tanks from Fig. 5.23, we can swap functions with variables and produce a *signal flow graph* where the emphasis is on the signals with functions labeled on the graph arcs. We introduce the following new variables: $a(t) = h1'(t)$ and $b(t) = h2'(t)$. A flow graph without specific parameters for the block model in Fig. 5.31(a) is shown in Fig. 5.31(b).

5.3.2 Kinetic Graphs

In chemistry, material A is transformed into material B through a reaction that takes place on A, usually with a catalyst of some sort. Also, one can combine two or more chemical constituents to form new chemicals. Consider the following simple sequence of changes from material A into material C:

$$A \rightarrow B \rightarrow C$$

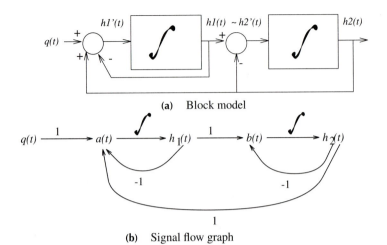

(a)   Block model

(b)   Signal flow graph

Figure 5.31   Block model and equivalent signal flow graph.

In this reaction, material A is transformed into material B, which is further transformed into material C. The following set of linear equations represents this reaction:

$$\frac{dA}{dt} = -k_1 A \tag{5.1}$$

$$\frac{dB}{dt} = k_1 A - k_2 B \tag{5.2}$$

$$\frac{dC}{dt} = k_2 B \tag{5.3}$$

These equations intuitively map to the model; material is transformed at a first-order exponential rate. Another model involves two more chemicals:

$$A \rightarrow B \rightarrow D$$
$$\updownarrow$$
$$C$$

This is represented by the following set of equations:

$$\frac{dA}{dt} = -k_1 A \tag{5.4}$$

$$\frac{dB}{dt} = k_1 A + k_2 C - k_3 B - k_4 B \tag{5.5}$$

$$\frac{dC}{dt} = k_3 B - k_2 C \tag{5.6}$$

$$\frac{dD}{dt} = k_4 B \tag{5.7}$$

An algorithm for taking a kinetics graph and representing it in equational form follows.

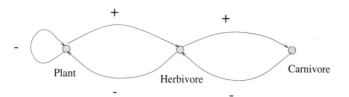

Figure 5.32   Pulse process of an ecosystem.

Algorithm 5.11. Kinetics Algorithm

Program Main
  We are given a kinetic graph with nodes and arcs
  The nodes require labeling
    : state variable for amount of material
  For each node (Y) with an input arc from X and an
  output arc to Z write:
$$dY/dt = k1 * X - k2 * Y$$
    where k1 and k2 are rate constants
  End For
End Main

Z is not part of the equation for node Y since we code only those variables which have an effect on the level of node Y (namely, its own decay and increase from whatever is input to Y).

5.3.3 Pulse Processes

Pulse processes are another type of graph model that can be used to determine influences on different nodes based on arcs that are signed. Consider the diagram in Fig. 5.32. This represents a simple ecosystem with three component populations: plant, herbivore, and carnivore. There are two vectors that are associated with pulse processes: the pulse vector and the value vector. The pulse vector represents a type of entity flow through the network; a pulse is "applied" to the network to activate it. The value vector represents the state vector. Value variables are state variables which keep track of accumulated changes affected by the moving pulses. Therefore, the value vector, for the process in Fig. 5.32, has the total populations of the plants, herbivores, and carnivores. We can start out with an initial population of 10 members for each of these groups, and then apply a pulse to the plant node. Note that pulses within a pulse process "travel" around the network—there is no capacitive effect with the pulses alone. The capacitive (or storage) effect is achieved by updating the values with the total sum pulse effect on a vertex. A value stored at each vertex does not affect pulse values in any way, so it is often instructive to hand simulate a pulse process simply by precomputing all pulse vector values at each time step. Once this is done, say for 10 time steps, one can then go back and start with some initial value vector. The value vector is updated by adding the pulse effect for each vertex. Let $V(t)$ and $P(t)$ be value and pulse vectors (at time t), respectively. $V(0)$ is equal to [10, 10, 10]; $V(1)$ is obtained by adding together [10, 10, 10] to $P(1)$ to get [9, 11, 10]. In general,

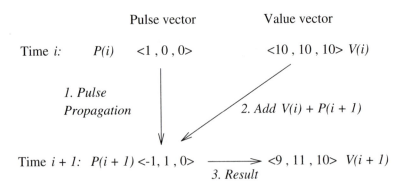

Figure 5.33   Simulation method for one time unit.

Figure 5.34   Model engineering within the System Dynamics method.

$V(t) = V(t-1) + P(t)$ for each vertex. There are two basic steps in simulating a pulse process for one time unit:

1. At time i, propagate the pulses defined by the pulse vector. This provides new pulse vector at time $i + 1$.
2. Add the value vector at time i to the pulse vector at time $i + 1$ to get the new value vector for time $i + 1$.

This procedure is shown graphically for the above pulse process starting at time zero (Fig. 5.33).

5.3.4 System Dynamics

System dynamics is a specific methodology for engineering simulation models. The modeler begins with a rough design for a system and then proceeds eventually to a DYNAMO program. The general flow of the modeling process is as shown in Fig. 5.34.

   Some simple examples of causal diagrams are diagrams that have two nodes and one arc:

- Food_Intake $\Rightarrow$ Weight
- Fire $\Rightarrow$ Smoke
- Fire $\Rightarrow$ Heat

The first graph says that "the amount of food taken has an effect on the body weight." The rule does not give specifics as to the type of effect. The other rules are abstractions for dynamical concepts such as "fire causes smoke" and "fire causes heat." The three graphs above can be thought of as informally specified simulation models. Just as software

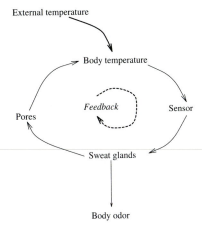

Figure 5.35 Heating control for humans.

engineering is concerned with how programs (in general) are developed incrementally starting with basic requirements, system dynamics is concerned with how simulation models are incrementally designed. Let's consider a more comprehensive graph with more than two nodes. Figure 5.35 shows a model of the heating control for the human body. So far we have not added any numerical or sign information. Let's first add sign information to a basic causal graph. There are two primary types of feedback: positive and negative. Feedback is shown as a cycle in a system dynamics graph—this is in accordance with our intuitive concept of feedback also. If a cycle contains an even number of "−" (minus) signs, the cycle represents positive feedback,[6] whereas an odd number of minus signs represents negative feedback. Figure 5.36 is an example of positive and negative feedback. Each graph has two extra features not previously discussed: a feedback sign and arc signs. The feedback sign is a circular arc with a sign in the middle, and it denotes the feedback and type (positive is +, negative is −). The arc signs provide information about the type of causality. An example of positive feedback is shown first: as someones salary increases, we (ideally) expect their performance on the job to increase. Look at the negative feedback in Fig. 5.36. In a grocery store, with no resupply of grocery items, the number of customers increases depending on the number of groceries; however, the number of groceries decreases with more customers. Since there is an odd number of minus signs in this feedback loop, the loop is negative. Consider now Fig. 5.37. We note the following trends that are indicated by this model:

- The biosphere is self-regulating. This means, essentially, that there exists enough negative feedback to produce stable conditions.
- When a feedback loop is within another loop, one loop must dominate in terms of overall behavior. Stable conditions will exist when negative loops dominate positive loops.

For an example of loop domination, consider the negative loop connecting evaporation and water amount. Suppose that this negative loop was weak compared with the positive loop surrounding it (evaporation → clouds → rain → amount of water → evaporation). We

---

[6]The case with no minus signs represents *zero*, which is an even number.

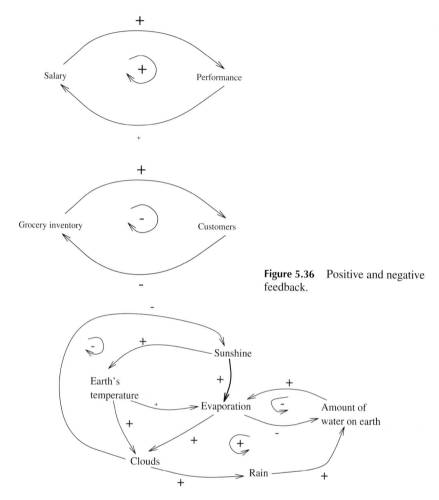

Figure 5.36 Positive and negative feedback.

Figure 5.37 Biosphere model with nested loops [191].

can the imagine a scenario where water evaporates but the process of evaporation would cause new water to be generated. This is counterintuitive but it agrees with our temporary assumption of positive feedback dominance. In a world such as this, the water level would continue to rise with continued rainfall. Luckily, we do not encounter this situation in earth's biosphere since evaporation causes water to be removed from the water surface, and this process occurs in conjunction with the process of positive feedback where clouds are formed and rain eventually occurs. In general, our water level will oscillate back and forth as a result of the positive and negative feedback, hopefully maintaining some sort of steady state. If steady state is not observed by executing the model, an imbalance will occur, resulting in a drought or flood.

The semantics of a graph arc are not always causal. For instance, we can say that an increase in evaporation causes a decrease in the amount of water; however, it sounds unusual if we say that a decrease in evaporation causes an increase in the amount of water.

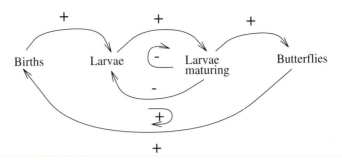

Figure 5.38   Two phase causal graph birth model.

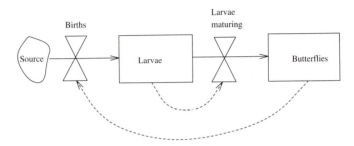

Figure 5.39   Flow graph birth model.

The arc semantics in the latter case should be something like "a decrease in evaporation is *associated with* an increase in the amount of water."

So far, we have seen a progression of model creation in the system dynamics methodology—we create a basic graph model and then we assign arc signs and feedback signs. Next, we will take this graph and convert it into a system dynamics flow graph. Let's consider the causal graph shown in Fig. 5.38. We see that this model provides us with information about how a change in the number of births has an effect on the number of larvae and butterflies which represent matured larvae in a second phase of biological development. This model is shown for illustrative purposes only; the model does not incorporate the concept of death, and therefore the population of adults can only increase. Figure 5.39 shows the flow graph for this model: Note the symbols in this flow graph. The symbols that look like water valves are examples of rates: birth rate and larvae maturation rate. The rectangles are examples of levels: number of larvae and number of butterflies. The system dynamics flow graph symbols provide us with a hydraulic (or fluidic) metaphor for systems problem solving. Water represents individual entities passing through the system (number of larvae or butterflies in this case). The water valve (or rate) can be adjusted just like a water tap, and the rectangles can be seen as containers (the levels) for the water flow. Figure 5.40 displays the valid symbols in a system dynamics flow graph.

After specifying a flow graph, the next step is to translate a graph into a DYNAMO program. First, we need to note that time in DYNAMO is represented in Fig. 5.41. This time scale can be used to represent differential equations (where dt is small) and difference equations (where dt is an integer).

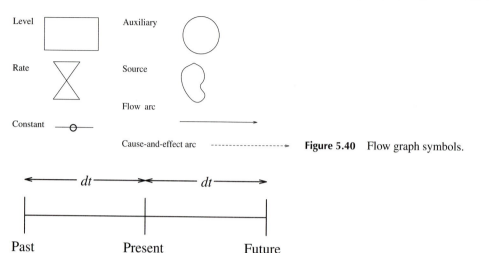

Figure 5.40   Flow graph symbols.

Figure 5.41   Time in system dynamics.

Consider a simple two-node causal graph which has positive feedback and two positive arcs between nodes GROWTH and BACTERIA. In this model an increase in growth will cause an increase in number of bacteria. This model defines growth for bacterial cells. Figure 5.42 displays two system dynamics components of this problem: the causal graph and the flow graph.

The following is the representative DYNAMO program, which—when executed—will simulate the growth of bacteria:

```
*      Bacterial Growth
L      BACT.K = BACT.J + DT*GROWTH.JK
N      BACT = 10
R      GROWTH.KL = (BACT.K)(PAR)
C      PAR = 0.10
SPEC DT = 1.0
```

The DYNAMO program can be translated into a set of difference equations which can then be simulated according to the rules for difference equations described previously. In many circumstances, though, the DYNAMO program is written so that it should be represented better as a set of differential equations. In the bacterial growth example, BACT.K is clearly updated using the *Forward Euler* method of integration. We see, then, that the DYNAMO program is really an implementation to model the following equation and initial conditions:

$$x(0) = 10$$
$$k = 0.10$$
$$\dot{x} = kx$$

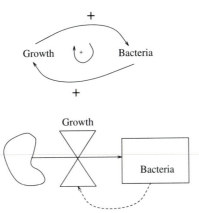

Figure 5.42   Bacterial growth model.

Here is the general algorithm for taking a causal graph and producing a set of equations which can be simulated:

Algorithm 5.12. System Dynamics Algorithm
Program Main
   We are given a concept graph with nodes and arcs
     The arcs require sign $(+, -)$ labeling
     The nodes require labeling:
        source,rate,level,constant,auxiliary
   For each level node (L) with an input rate node (R1)
     and an output rate node (R2) write:
     $dL/dt = k1 * R1 - k2 * R2$
     where $k1$ and $k2$ are rate constants
  End For
  For all other nodes (N) write:
     $N(t) = a$ *linear function of all inset members of this node*
  End For
End Main

To illustrate the algorithm consider the three basic delay system dynamics flow graphs in Figs. 5.43 through 5.45.

As a final system dynamics model, we will consider a simulation of building construction discussed in [191]. In this scenario, there is a fixed area of available land for construction. New buildings are constructed while old buildings are demolished. Our primary state variable will be the total number of buildings (in the fixed area) over time. We take the following approach to modeling this situation:

1. Create a causal graph with concepts at the graph nodes and causality, influence, or flow at the arcs (Fig. 5.46).
2. Add signs to each arc, and for each loop specify the sign of the loop: negative or positive feedback (Fig. 5.47).

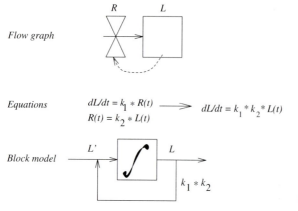

Figure 5.43   Simple growth model.

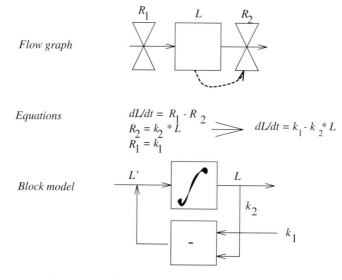

Figure 5.44   First order delay with growth and decay.

3. Determine the type of each node. A node will be one of the following: Rate (R), Level (L), Auxiliary (A), or Constant (C) Fig. 5.48.

4. Translate the existing augmented causal graph into a flow graph using the system dynamics icons (Fig. 5.49).

5. At this point, one can translate the graph into DYNAMO code; however, we will construct a set of equations directly from the flow graph.

We will first abbreviate the graph nodes before translating to equational form: B_l = industrial buildings; C_r = construction rate; D_r = demolition rate; CF = construction fraction; FLO = fraction of land occupied; AA = average area of land; LA = land available for building construction; and AL = average lifetime of buildings.

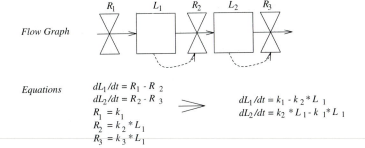

Flow Graph

Equations

$$dL_1/dt = R_1 - R_2$$
$$dL_2/dt = R_2 - R_3$$
$$R_1 = k_1$$
$$R_2 = k_2 * L_1$$
$$R_3 = k_3 * L_1$$

$$\Longrightarrow$$

$$dL_1/dt = k_1 - k_2 * L_1$$
$$dL_2/dt = k_2 * L_1 - k_1 * L_1$$

Block Model

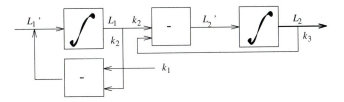

Figure 5.45   Second order delay "pipeline" model.

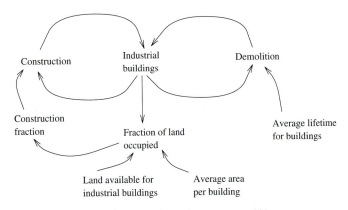

Figure 5.46   Step 1: Causal graph [189].

The generic equations are:

$$\frac{dB_l}{dt} = C_r - D_r \tag{5.8}$$

$$C_r = f_1(CF, B_l) \tag{5.9}$$

$$D_r = f_2(AL, B_l) \tag{5.10}$$

$$CF = f_3(FLO) \tag{5.11}$$

$$FLO = f_4(LA, AA, B_l) \tag{5.12}$$

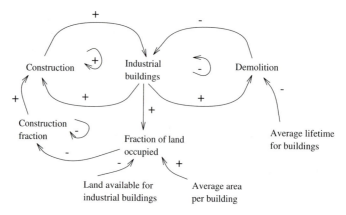

Figure 5.47   Step 2: Causal graph with signed arcs and loops [189].

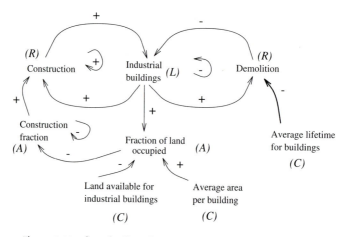

Figure 5.48   Step 3: Causal graph with added node types [189].

The f_i functions can be any function, although usually the default is for f_i to be a linear algebraic function. That is, one can substitute $C_r = k1 * CF * B_l$ for the above equation $C_r = f_1(CF, B_l)$. The causal graph does not usually have any information in it that would differentiate the equations $C_r = CF + B_l$, $C_r = CF * B_l$, or $C_r = CF - B_1$, so this knowledge must come from the user while the equations are being formulated.

5.3.5 Compartmental Modeling

Overview.   The coupled tank example shown in Fig. 5.22 permits us to model multiply connected tanks each with a specific volume, and with connections to other tanks that are modeled with a resistance dictated by either a valve or the cross-sectional area of the adjoining pipe. With the system dynamics approach to modeling, we modeled such tanks as *levels*. Compartmental modeling is very similar to system dynamics in that the physical phenomenon to be studied is represented as a connected network of *compartments*. Each

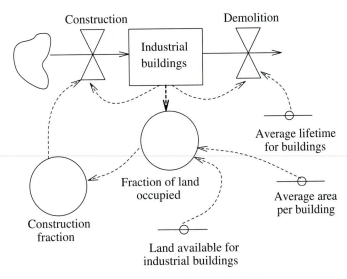

Figure 5.49   Step 4: Flow graph [189].

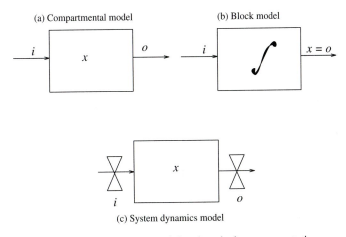

Figure 5.50   Compartment model and equivalent representations.

compartment has a specific volume and is functionally equivalent to the system dynamics *level*. As a matter of convenience, flow rates for the liquid flowing through compartments are labeled on the arrows representing the connecting pipes. Figure 5.50 shows a compartment with a single input and output along with equivalent block and system dynamics representations. *i* and *o* are the input and output flow rates, and *x* is the state variable reflecting the amount of material flowing through the compartment.

Blood circulation.   Blood flows around the body in a loop containing the heart, lungs, and the circulatory system composed of arteries and veins. Let's begin with the oxygen poor blood that comes back from the main system (body). Blood comes from

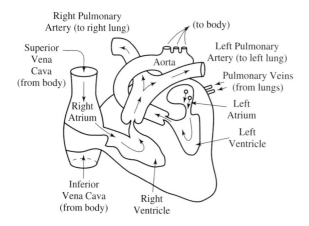

Figure 5.51 Human heart with its subcomponents.

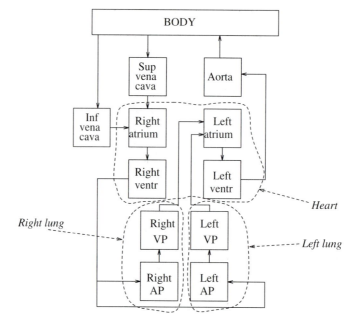

Figure 5.52 Compartmental model for blood flow.

the body via the superior and inferior vena cava and into the right atrium down to the right ventricle. The blood then proceeds to both lungs so that the blood is oxygenated. This blood is moved back into the heart via the left atrium and down into the left ventricle. From there, the blood exits the heart and back into the body through the aorta. The relative positions of the heart and lungs are shown in Fig. 5.51. Using a combination of the textual description of blood flow and Fig. 5.51, we arrive at a compartmental model shown in Fig. 5.52. We see in Fig. 5.52 that compartments are labeled with certain acronyms which are defined below.

- *Vena Cava.* Vena cava VC is divided into two flows: the superior VC and the inferior VC.

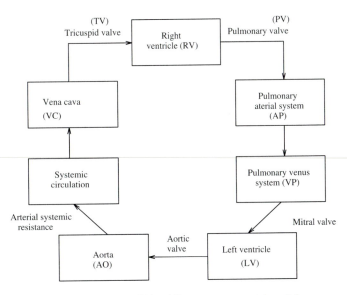

Figure 5.53   Lumped blood flow compartmental model.

- *Heart.* The four compartments are the right atrium, right ventricle, left atrium, and left ventricle.
- *Lungs.* AP refers to pulmonary arterial system and VP means pulmonary venus system.

We take some of the compartments in Fig. 5.52 together to provide the more aggregate, lumped compartmental model shown in Fig. 5.53. There are a set of equations for each compartment. The basic equation in each set is the one relating the derivative of the state variable to the input and output flows; however, there are also auxiliary variables and parameters. The equations for the right ventricle are

$$\frac{d(V_{RV})}{dt} = F_{TV} - F_{PV} \tag{5.13}$$

$$V_{RV} = 150.0 \tag{5.14}$$

$$P_{RV} = \frac{V_{RV}}{C_{RV}} \tag{5.15}$$

$$F_{TV} = 78 \max(P_{VC} - P_{RV}, 0) \tag{5.16}$$

Other equations may be found in the original PHYSBE paper [148] or in [127].

Epidemic model of HIV infection.   The model and discussion in this section follows the description by Murray [[155], pp. 624–630] of the transmission dynamics for AIDS. HIV (the human immunodeficiency virus) leads to the acquired immunodeficiency syndrome (AIDS). AIDS is a disease that could affect the entire human population and it is spreading as greater numbers of people become infected. Evidence to date demonstrates

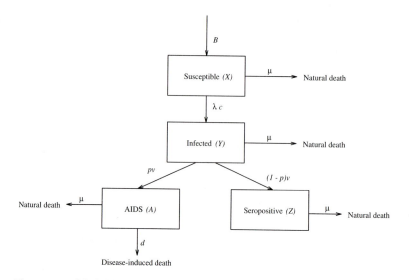

Figure 5.54 Model for AIDS transmission. (From Murray, *Mathematical Biology* ©1990, pg. 627. Reprinted with permission of Springer Verlag, New York, NY.)

that the HIV virus is transmitted through intimate contact with an infected individual. While sexual behavior is the most common vehicle for disease transmission, blood transfusions performed prior to adequate screening have resulted in a large number of people being infected, such as hemophiliacs.

Assume a constant influx of susceptible individuals into a population of size $N(t)$. Let $X(t)$, $Y(t)$, $A(t)$, and $Z(t)$ represent the number of susceptible individuals, infected individuals, individuals who have AIDS, and seropositive individuals who have not contracted AIDS. Each population is affected by input and output rates corresponding to influx and death rates. Figure 5.54 displays a compartmental model for AIDS transmission. The set of first-order equations generated from this model is

$$\frac{dX}{dt} = B - \mu X - \lambda X, \lambda = \frac{\beta Y}{N} \tag{5.17}$$

$$\frac{dY}{dt} = \lambda c X - (v + \mu)Y \tag{5.18}$$

$$\frac{dA}{dt} = pvY - (d + \mu)A \tag{5.19}$$

$$\frac{dZ}{dt} = (1 - p)vY - \mu Z \tag{5.20}$$

$$N(t) = X(t) + Y(t) + Z(t) + A(t) \tag{5.21}$$

5.4 EXERCISES

1. The two noninteracting tanks in Fig. 5.22 exhibit a single feedback loop since the height of rate of flow from tank one into tank two, $\frac{d}{dt}h_2(t)$ depends on $h_1(t)$. Consider a two-block model without *any* feedback loops. Now draw a schematic for such a system.

2. Consider a single-server queue. There are four customers who have interarrival times 10, 20, 15, 20 and server times 5, 28, 20, 12. The first customer's interarrival time is relative to 0. Hand simulate this system starting at time zero until all customers have been served. Produce a graph of *number in system* vs. *time* and compute the following metrics: throughput, server utilization, mean wait time per customer, and mean response time. Define response time to be wait time plus service time.

3. The automatic car example presented in Sec. 5.2.2 shows different algorithms for controlling the speed of a car as it senses the car directly ahead of it. How does this type of control approach differ from what a human does in the same type of situation? Describe a more sophisticated control mechanism based on human senses.

4. Consider the following scenario, and then model it using the SimPack queuing library. On a trip to an amusement park, you decide to try out the "3D Theater," where a 3D film is shown. First, you need to enter the park and wait in line to purchase a ticket. After that, you pass through ticket collectors who tear your ticket in half and give you the other half. Then, you proceed to the *3D Theater* to watch a 30-minute movie. Unfortunately, there is a queue forming in front of the theater. Once inside, you find that you are not actually in the main room of the theater but in a kind of "holding area" where you watch a short 10-minute 3D movie. After the 10-minute movie there are four possible doors leading into the main theater room where you sit down and watch the main movie. After the movie, there are two exits from the theater. Do the following.

    (a) Draw a class hierarchy model for this scenario.
    (b) Create and display the functional model.
    (c) How many queues and facilities did you use to model this situation?
    (d) What would be the effects on the system if the 10-minute movie was not shown?
    (e) If you change the number of internal theater doors and exits, what significant effect does this have on the results?

5. Show that the *time slicing* and *event scheduling* algorithms can be used interchangeably. For the first case, take a block model containing simple functions (such as add, subtract, multiply, and integrate). Show how *event scheduling* can be used to execute the model. For the second case, take a queuing network and show how time slicing can be used. Why are these algorithms mismatched in terms of computational efficiency?

6. Queuing models have discrete flows, whereas block models have continuous flows.

    (a) Describe a queuing situation that might be modeled as a block model with continuous data flow. Do such situations occur in the physical world?

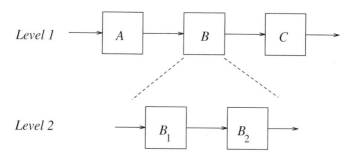

Figure 5.55   Two level functional block model.

(b) Consider a system containing moving balls that may collide. Under what circumstances can such a system be reasonably defined in terms of a functional model?

(c) Why is the server "function" more complex than functions such as *integral*, *multiply*, and *add*? In what ways does the server function different from other, simpler functions?

7. Figure 5.55 shows two levels for a functional block model. The highest level (level 1) shows a block model with three blocks in series (A, B, C). Function B, however, is further refined into blocks B1 and B2 in series. If we are interested in modeling both of these levels, derive a switch statement with event routines (cases) that would execute this model. You can assume that each function has some type of built-in delay, thereby necessitating an event scheduling approach. (*Note:* This is an example of homogeneous model refinement, which is discussed in more detail in Ch. 8).

8. The basic system dynamics graph has level and rate nodes. Consider the class of graphs where there are no feedback arcs from a level to a rate. Feedforward arcs from a level to a rate that provides *drainage* of the level are permitted. How is this type of graph different from the less restricted graph with respect to the equations that are generated?

9. Consider two populations: lizards and flies. Both lizards and flies are born and then die; however, flies have a higher rate for both birth and death. Moreover, lizards eat flies. Create a realistic system dynamics model for this system by specifying the following types of models: (1) a conceptual model (Ch. 3), (2) a causal graph, (3) a flow graph, and (4) a set of equations generated using the system dynamics algorithm.

10. Take Exercise 11 and specify how to make it object oriented given that functions are just object methods and variables are object attributes.

5.5 PROJECTS

1. The block simulator can be enhanced by adding features commonly found in commercial block-oriented CSSLs (continuous system simulation languages). Pick an off-the-shelf CSSL such as CSMP, ACSL, or TUTSIM and add more functionality to the given block simulator so that it has more features.

2. Build a window-based front end to the block simulator so that users can place blocks on a grid and connect them together. Permit the user to click on a functional block and be able to view and change the transfer function.

3. The models represented in Figs. 5.15 and 5.17 cause the car to back up. This is unacceptable, so you should revise the model so that backing up is not allowed. Rerun the simulation and describe the difference between the newly observed behavior and the behaviors in Figs. 5.16 and 5.18.

4. We have seen that the queuing modeling and block approaches are similar. The main difference is with respect to the *server transfer function* in the queuing model, which is significantly more complex to program than the functions in the block model. Combine the block modeler and the queuing simulator to allow mixed-mode simulation, where server blocks and algebraic blocks can be coupled in the same model.

5. Queues occur in golf courses when golfers must wait for golfers playing ahead of them. The general rule is that you must wait to hit until all players ahead of you are out of range for whatever club you pick from your bag. Choose a golf course and obtain the layout of the course from the score card. Now model each hole using a queuing model. For a par 3 hole, for instance, there is a server for the tee and a server for the green. For par 4 or 5 holes, you will need to construct queuing models with more servers. Do an analysis of the course given various settings for arrival rate and service rate. Write up your report and obtain feedback from golfers (or the club professional) who have played the course. Are the results realistic? What can be done to make a more detailed simulation of the golf course?

6. Build a system dynamics simulator complete with a window-based interface.

7. Take any of the functional modeling approaches and construct a hypertext document for part of this book. For instance, consider Sec. 5.2.1 on block models. Take one or two successive pages that contain text with at least one figure. To build a hypertext document for these two pages, you will need to have access to a hypertext toolkit. The best approach is to use the World Wide Web (WWW) on the Internet. WWW documents are specified in Hypertext Markup Language (HTML), which is a hypertext analog of the Standard Graphics Markup Language (SGML). The purpose of this project is to construct an HTML document that looks very similar to the part of this book that you will structure. When someone clicks on the figure, a simulator should come up and allow the user to run a simulation and change variables and parameters. When they click on a keyword, a window should come up which provides a *thread* of context that stems from the keyword. There are several available viewing tools that permit you to view HTML documents. XMosaic is one such tool. Add the following types of multimedia capabilities to your document: plain text, raster (pixel) image, a scanned photograph of something relating to a keyword or a region of text, sound, and a movie.

8. Build a visual pulse process simulation tool to show the user the dynamics of pulse process components over time using movement and color change. For instance, pulses can be shown as moving particles, and the variable values can be mapped to colors using a color lookup table.

5.6 FURTHER READING

1. *Function-Based Functional Models.* Although block models are used in areas as diverse as social science and software engineering, they are most often found in abundance within books on control engineering such as [161, 49]. This is because the blocks map directly to human-engineered objects such as pumps, motors, and transformers. In cases where naturally engineered objects have distinct components with inflow and outflow [151], the block model approach is useful here as well. Queuing networks are used to model systems at a fairly high abstraction level—without resorting to mechanics. My approach to modeling facilities using optional priority queuing was influenced by the works of Schwetman [201] (CSIM) and MacDougall [136] (SMPL). Both CSIM and SMPL are C-based simulation tools similar to SimPack. The SimPack C queuing routines are modeled after the syntax used in SMPL. Discussions of scheduling can be found in [166] and [205]. Simulation of logic circuits is discussed in [150, 184].

2. *Variable-Based Functional Models.* Pulse processes are discussed by Roberts [188] as a method for modeling discrete processes. Roberts et al. [189] provide a good text full of *System Dynamics* examples. Figure 5.37 and Fig. 5.46 through 5.49 are borrowed from this text. Richardson also provides a thorough treatment of the system dynamics approach [186] developed originally by Forrester [79, 81, 80]. Compartmental models have wide coverage in the medical community [115], especially when modeling bodily fluid flow and tracer kinetics.

5.7 SOFTWARE

There are a large number of functional model types supported by SimPack and described in Ch. 10. Also, there are other packages supporting models using event scheduling such as CSIM [available by contacting Herb Schwetman at MCC (hds@mcc.com)]. CSIM is a *process-oriented* compiler. SIM++ also supports a process view and can be obtained at ftp.telematik.informatik.uni-karlsruhe.de. SMPL is available under ftp.cis.ufl.edu:/pub/simdigest/tools. The SimPack queuing library was based on the original SMPL calling convention. SMPL is *event oriented*, but note that event *orientation* is a facet of code translation; functional models are designed in the same way regardless of how the model is translated into code. NEST is a general network simulator available under directory ftp.cs.columbia.edu:nest. REAL, which used NEST as a basis for its initial design, supports flow and congestion control protocols in data networks and can be obtained at research.att.com:/dist/qos/real.tar.Z. A general-purpose simulation language for the Linux operating system, Simula, can be obtained at sunsite.unc.edu: /pub/Linux/development/ linux.cim.tar.Z.

Chapter **6**

Constraint Modeling

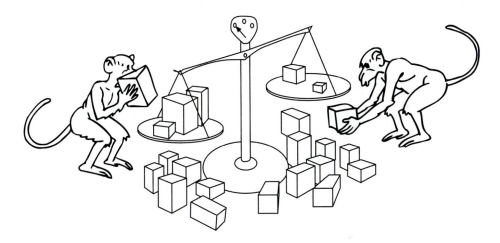

... A conservation law means that there is a number which you can calculate at one moment, then as nature undergoes its multitude of changes, if you calculate this quantity again at a later time it will be the same as it was before, the number does not change.[1]

6.1 OVERVIEW

Declarative and functional models always have a kind of direction associated with them. Declarative models have a linear ordering in state space; connect one piece of state space to another until it results in a connected graph. Functional models have explicit directionality with *flow* from one function into another by coupling the output of one function to the input of another. Constraint models are quite different. One builds a constraint model by thinking about balance, and any causality that appears is placed there for purposes of fostering explanation, since causality is not part of the model itself.

[1]R. Feynman, *The Character of Physical Law* [58], p. 59.

Constraint models are most useful to represent the laws of nature—which are all *symmetrical* and amenable to being specified in terms of a balance of certain things. A balance may involve one of several different types of quantities—usually force, mass, momentum, or energy. In the broadest and most concise sense, balance of various energies within a system captures all other quantity types since equations of force, mass, and momentum are special cases of energy balance.

Constraints are most naturally defined in terms of *equations* even though they can also be represented in graph or other forms. Difference and differential equations are the most applicable to simulation since they explicitly involve time as an independent variable. The basic principles underlying equational models are *invariance* and *balance*. Consider the following formulas with terms expressed in natural language:

- *General.* Net Rate = Input − Output.
- *Biology.* Change in Population = Growth − Death + Reaction.
- *Thermodynamics.* Heat Generated Internally = Heat of Products Entering − Heat of Products Leaving ± Heat Produced or Absorbed by Process.
- *Mechanics.* All Forces Acting on an Object Added Together = 0, or more generally we have Hamilton's Principle, which states that the Sum of Variations of Work (derived from potential and kinetic energy) = 0.
- *Electricity.* All Currents at a Node Added Together = 0, and All Voltages Around a Closed Circuit = 0.

Constraint models should be used when the components making up the model are not connected in a unidirectional way. Even though constraint models are general in that they automatically cover functional models as a subset, they are best to use when the modeler thinks in terms of a balance of terms or a conserved quantity at various points in a network.

Figure 6.1 displays the subclass hierarchy of Fig. 1.8 for constraint models. The basic types of constraint model are equation- and graph-based. We will focus on the manipulation of equations for the purpose of simulating them; however, let's talk about modeling connected with equations. First, one can model a system directly by considering the physical phenomenon and then immediately writing down equations of balance. Using this approach, we identify the terms as objects where a term in a population model would include objects such as *birth*, *death*, or *interaction*. While this is reasonable, it does not directly correspond to the OO approach. The best method in building a model that is composed of equations is first to consider the objects in the physical system. If the population model involves bees and their mating behaviors, we should create an object for the population of bees and one for each bee that we use. On a detailed level, individual bee objects interact with one another, and at the higher population level, we are concerned only with the aggregate bee population count and not in individual bee attributes. Some equational models may involve only one physical object, so it might seem burdensome to create programming objects when all of the model execution *semantics* have to be located in the interaction between the object's methods and its attribute state variables. Still, the OO approach is appropriate since it corresponds to a uniform way of viewing systems and mapping the systems to software.

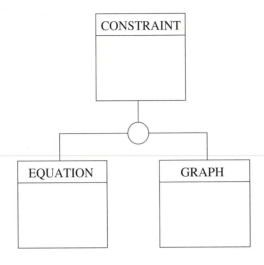

Figure 6.1   Constraint model hierarchy.

Let's begin the discussion of equations by introducing computational approaches to executing constraint models when they are specified in difference or differential form. Difference equations encode discrete changes in time and state space, while differential equations encode continuous changes.

6.2 EQUATION-BASED APPROACH

6.2.1 Discrete Delays in C

Difference equations represent state changes over discrete time intervals such as seconds, months, or years. In Ch. 5, we saw examples of difference equations such as the operation of a logic gate with a propagation delay. At that time, we used the discrete event simulation approach to executing functional models. The discrete event approach has the important property that it can be used to store past state information about a system. For discrete delays, however, we can derive simpler methods that do not involve the data structures necessary for using event scheduling methods.

The *order* of a model is determined by subtracting the lowest-value expression involving t (time) from the highest value. For instance, if an equation set included both $x(t)$ and $x(t-4)$ [where $x(t)$ contains the largest value t, and $x(t-4)$ contains the smallest value $t-4$], the order of this model would be $t-(t-4)=4$. Consider the following two-dimensional system:

$$x_1(t) = x_2(t-1) - 3x_1(t-1) + 4$$
$$x_2(t) = 8x_1(t-1) \tag{6.1}$$

We might consider using the following code: `x1 = x2 - 3*x1 + 4; x2 = 8*x1` to simulate (6.1); however, the value of `x1` in the second equation `x2 = 8*x1` would actually be $x_1(t)$ instead of the correct $x_1(t-1)$. Therefore, we must *hold* the value for a new $x(t)$ before changing the current value.

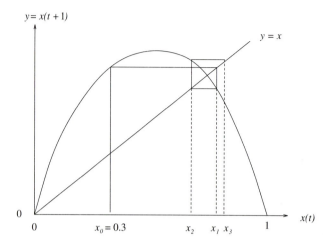

Figure 6.2   Cobweb cycles for difference models.

The following block of C code will correctly simulate the difference equations:

```
new_x1 = x2 - 3*x1 + 4;
new_x2 = 8*x1;
x1 = new_x1;
x2 = new_x2;
```

Note the use of the temporary variables to store the updated version [i.e., $x_i(t)$] so that it is not used immediately for subsequent equations.

For first-order difference equations, this previously defined method is sufficient; the "memory" of unit length is stored in the form of a variable. However, for higher-order difference models, we may choose first to convert the system into a set of first-order equations. This will lead to a uniform template for solving both difference and differential equations.

Let's consider a single first-order system called the *logistic equation*: (or recurrence)

$$x \leftarrow ax(1 - x) \tag{6.2}$$

With (6.2), we will introduce the graphical cobweb method of solving difference equations. This method is most popular when one is solving a difference equation by hand, although, even as an automated process, studying cobweb cycles is most instructive when considering issues of convergence and divergence. Consider Fig. 6.2. The method of solving for x in this case is to start x at some value such as 0.3, move a vertical line to the function $ax(1 - x)$, move a horizontal line to $y = x$, and continue (1) moving toward the function, and (2) moving toward the line $y = x$ until a termination criterion has been reached (say, for 200 iterations of the equation):

1. Let $x_0 = 0.3$.
2. Calculate $y_0 = ax_0(1 - x_0)$.
3. Let $x_1 = y_0$.

4. Calculate $y_1 = ax_1(1 - x_1)$.

5. Let $x_2 = y_1$.

6. Calculate $y_2 = ax_2(1 - x_2)$.

7. Let $x_3 = y_2$.

8. Repeat steps: $y_i = ax_i(1 - x_i)$ and $x_{i+1} = y_i$ for some period of time.

6.2.2 Fibonacci Growth

The Fibonacci series was originally devised to model natural growth of populations. For instance, consider the single second-order difference equation for the Fibonacci series:

$$x_t = x_{t-1} + x_{t-2} \tag{6.3}$$

where x_0 equals the initial size of the population at time zero, and x represents the state variable for population size. We will consider rabbit population growth where x_t is the population of rabbit *pairs* (each pair is one male and one female) at time t (measured in months). Let us justify (6.3) by considering the rabbit population given simplifying assumptions such as (1) Rabbits never die; (2) Rabbits can give birth only after a maturation period of 2 months; and (3) A rabbit pair, once mature, will give birth to exactly 1 pair of rabbits each month. Since t will represent month number, we can assume that there is initially one pair of baby rabbits in month 0 ($x_0 = 1$). Also, in month 1 ($x_1 = 1$), there will be still one pair—this is because the original pair of rabbits is not yet mature. In month 2, there will be the original pair plus the new pair that is born since the original rabbit pair is now 2 months old (for a total of 2 rabbits in the population). Therefore, $x_2 = 2$. If we let the number of births at time t be $B(t)$ and the total number of rabbits at time t be $R(t)$, our problem definition states that $B(t) = R(t - 2)$, and that $R(t) = R(t - 1) + B(t)$, which, when combined, yield the Fibonacci relation. The 2-month maturation period can be thought of as a memory of length 2 time units. This is why this is a second-order delay B which lags R by exactly 2 time units (months).

6.2.3 Difference Equation Coding with Circular Queues

As time is incremented for a difference equation, we need to update current values for $x(t)$, $x(t - 1), \ldots, x(t - n)$ for a difference equation of order n. Consider difference equation Eq. (6.4) of order 3:

$$x(t) = x(t - 1) + x(t - 2) - x(t - 3) \tag{6.4}$$

Figure 6.3 displays two ways of storing each of the values for x in (6.4). The first method is a linear linked list with dynamically allocated storage and the second is an array. Each method has its advantages. The advantage of the linked list is that it is well-structured and facilitates efficient memory allocation and deallocation for elements in the linked list. The array method, while not the most efficient approach, is very fast since array elements are usually stored in contiguous memory locations, and pointer dereferencing is simplified. We see from Fig. 6.3 that after $x(t)$ is calculated, there must be a shifting of the current pointer for $x(t)$. $x(t)$ will always replace the old $x(t - n)$ for an nth-order system with one variable. Figure 6.4 displays a sequence of six time-value calculations for Eq. (6.4).

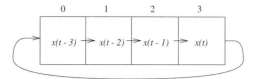

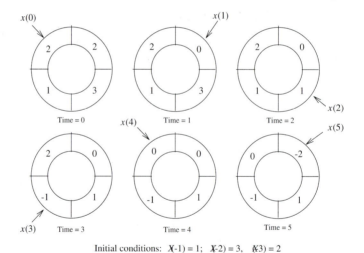

Figure 6.3   Circular queue storage and retrieval.

Initial conditions: $x(-1) = 1$; $x(-2) = 3$, $x(-3) = 2$

Figure 6.4   Queue entries for first 6 time values.

6.2.4 Canonical Form Difference Equations

It is often useful to view all difference equations by restructuring them into first-order form. Even though this method is more cumbersome than the previous one, the first-order form will provide a smooth transition when we discuss differential equations and how they can be coded. Assume that we have a single, multiple-order difference equation with variable x:

$$x(t) = F(x(t - n), \ x(t - n + 1), \ldots, x(t - 1)) \tag{6.5}$$

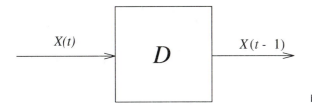

Figure 6.5 Delay block.

This represents a difference equation with an arbitrary order n. Initial conditions for (6.5) are (where $t = 0$ is common)

$$x(-n),\ x(-n+1),\ldots,x(-1)$$

We must convert this equation to a set of first-order difference equations by creating new variables:

$$x_1(t) = x(t-n+1),\ x_2(t) = x(t-n+2),\ldots,x_n(t) = x(t)$$

D is defined as a "delay" operator which operates $D(x(t-1)) = x(t-2)$, $D(x(t)) = x(t-1)$. In general,

$$D(x(t-k)) = x(t-k-1)$$

The function of D is shown in Fig. 6.5.

Now, we generate the following first-order set of equations using our D operator.

$$
\begin{aligned}
x_1 &= D(x_2)\\
x_2 &= D(x_3)\\
&\ \vdots\\
x_{n-1} &= D(x_n)\\
x_n &= F(D(x_1), D(x_2), \ldots, D(x_n))
\end{aligned}
\tag{6.6}
$$

Consider a second-order version of the logistic equation, which we define as

$$x(t) = ax(t-1)(1 - x(t-2)) \tag{6.7}$$

We first substitute variables: $x_1 = x(t-1)$ and then $x_2 = x(t)$ using (6.6). Then our equations are

$$x_1 = D(x_2) \tag{6.8}$$

$$x_2 = aD(x_2)(1 - D(x_1)) \tag{6.9}$$

If an nth-order difference equation is not autonomous, it means that there is a term $g(t)$ inside the equation (i.e., a term that is not a function of state variables). The method is to add one extra equation: $x_{n+1} = g(t)$.

We need a way of encoding the state variables $x_1, x_2, \ldots, x_n$ into C code. It is convenient to use two arrays (IN and OUT) as follows: $D(x_1) = OUT[1]$, $D(x_2) = OUT[2]$, $D(x_3) = OUT[3]$, and so on. The IN array stores all of the values to be delayed for 1 time unit. Likewise, an OUT array stores all the state variable values: $x_1 = IN[1]$, $x_2 = IN[2]$, $x_3 = IN[3]$ which are delayed. Now, we can think of the *delay* step for a variable x_i as simply taking the value in cell $IN[i]$, delaying it, and placing this new value into cell $OUT[i]$. This procedure matches the action of the delay block displayed in the block model in Fig. 6.5. The *state* step involves executing equations relating IN to OUT.

Taking the equation $x(t) = ax(t - 1)(1 - x(t - 2))$, we have the following C code for updating the equations of state:

```
a = 1.5;
in[1] = out[2];
in[2] = a*out[2]*(1 - out[1]);
```

To summarize, here is the C code for our three major steps discussed earlier:

1. Set initial conditions.

```
out[1] = 0.8;
out[2] = 0.8;
delta_time = 1;
time = 0;
a = 1.5;
```

2. Update state equations.

```
in[1] = out[2];
in[2] = a*out[2]*(1 - out[1]);
```

3. Delay all state variables.

```
for (i=1;i<=2;i++)
   out[i] = in[i];
time += delta_time;
```

Steps 2 and 3 are repeated in a loop until a termination criterion has been reached (say, until time exceeds 100).

6.2.5 Differential Equation Algorithms

Numerical integration in C. When our model requires continuous time/continuous state descriptions, we will specify that part of the model with differential equations. Differential equations are the natural method of defining physical phenomena that are time varying. Consider the following generic definition for a single differential equation:[2]

$$\frac{dx(t)}{dt} = f'(t, x) = \lim_{\Delta t \to 0} \frac{x(t + \Delta t) - x(t)}{\Delta t} \tag{6.10}$$

This definition is precise and we now need to derive methods that allow us to solve for $x(t)$ for $t \geq 0$.

Forward Euler's method. For very small values of the time interval Δt we can remove the $\lim_{\Delta t \to 0}$ and solve for x as follows:

$$x(t + \Delta t) = x(t) + f'(t, x) \Delta t \tag{6.11}$$

---

[2]Note the following equivalence $\frac{dx(t)}{dt} = \frac{dx}{dt} = \dot{X}$.

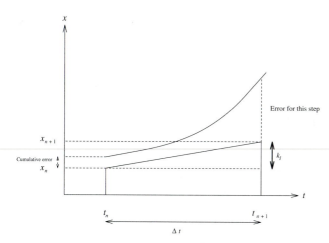

Figure 6.6   RK(1) Method of integration.

Euler's method is a first-order approximation to the Taylor series expansion for the differential equation. Unfortunately, Euler's method is highly unstable for large values of Δt, meaning that the numerical solution diverges from the true analytic solution as $t \rightarrow \infty$.

Runge-Kutta methods.   The Runge-Kutta (RK) methods are quite popular and are characterized as being reasonably stable. In fact, Euler's method is subsumed by Runge-Kutta methods since it is equivalent to the first-order Runge-Kutta method [or RK(1)]. RK1 is defined as

$$k_1 = \Delta t f'(t_n, x_n) \tag{6.12}$$

$$x_{n+1} = x_n + k_1 \tag{6.13}$$

Note that n is defined as the index of iteration, whereas t (the implicit time variable which is not shown) is incremented in fixed time steps of some size Δt which is chosen arbitrarily. A sample value for Δt might be 0.01. The numerical error for RK(1) is of order Δt^2. For $\Delta t = 0.01$ this would mean an error between the true and calculated values of x on the order of 0.0001. In general, a RK(n) has an error on the order of $\Delta t^{(n+1)}$. RK(1) is shown in Fig. 6.6. RK(2) provides a better approximation to x at the expense of computation time. Here is the method for RK(2):

$$k_1 = \Delta t \ f'(t_n, x_n) \tag{6.14}$$

$$k_2 = \Delta t \ f'\left(t_n + \frac{\Delta t}{2}, \ x_n + \frac{k_1}{2}\right) \tag{6.15}$$

$$x_{n+1} = x_n + k_2 \tag{6.16}$$

RK(2) is shown in Fig. 6.7. RK(4) is, perhaps, one of the most standard integration methods used in simulation software. It is simply an extension of our definitions so far:

$$k_1 = \Delta t \ f'(t_n, x_n) \tag{6.17}$$

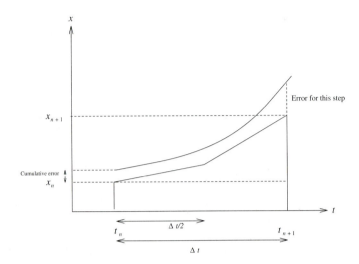

Figure 6.7   RK(2) Method of integration.

$$k_2 = \Delta t \; f'\left(t_n + \frac{\Delta t}{2}, \; x_n + \frac{k_1}{2}\right) \tag{6.18}$$

$$k_3 = \Delta t \; f'\left(t_n + \frac{\Delta t}{2}, \; x_n + \frac{k_2}{2}\right) \tag{6.19}$$

$$k_4 = \Delta t \; f'(t_n + \Delta t, \; x_n + k_3) \tag{6.20}$$

$$x_{n+1} = x_n + \frac{k_1}{6} + \frac{k_2}{3} + \frac{k_3}{3} + \frac{k_4}{6} \tag{6.21}$$

Adaptive step-size control. In certain cases where state variables $x_i(t)$ are fairly linear in some regions, it would be highly beneficial to have a variable step size so that we could make a large step size to "hop" over uneventful areas in state space and then reduce the step size in nonlinear regions. The Runge-Kutta-Fehlberg (RKF) algorithm was constructed for this purpose. The RKF method of order 4 (RKF4) is very similar to RK4:

$$F_1 = \Delta t \; f'(t_n, x_n) \tag{6.22}$$

$$F_2 = \Delta t \; f'\left(t_n + \frac{1}{4}\Delta t, \; x_n + \frac{1}{4}F_1\right) \tag{6.23}$$

$$F_3 = \Delta t \; f'\left(t_n + \frac{3}{8}\Delta t, \; x_n + \frac{3}{32}F_1 + \frac{9}{32}F_2\right) \tag{6.24}$$

$$F_4 = \Delta t \; f'\left(t_n + \frac{12}{13}\Delta t, \; x_n + \frac{1932}{2197}F_1 - \frac{7200}{2197}F_2 + \frac{7296}{2197}F_3\right) \tag{6.25}$$

$$F_5 = \Delta t \; f'\left(t_n + \Delta t, \; x_n + \frac{439}{216}F_1 - 8F_2 + \frac{3680}{513}F_3 - \frac{845}{4104}F_4\right) \quad (6.26)$$

$$x_{n+1} = x_n + \frac{25}{216}F_1 + \frac{1408}{2565}F_3 + \frac{2197}{4104}F_4 - \frac{1}{5}F_5 \quad (6.27)$$

To calculate the next-order RKF of order 5 (RKF5) we must add the additional equations:

$$F_6 = \Delta t f'\left(t_n + \frac{1}{2}\Delta t, \; x_n - \frac{8}{27}F_1 + 2F_2 - \frac{3544}{2565}F_3 + \frac{1859}{4104}F_4 - \frac{11}{40}F_5\right) \quad (6.28)$$

$$x_{n+1} = x_n + \frac{16}{135}F_1 + \frac{6656}{12825}F_3 + \frac{28561}{56430}F_4 - \frac{9}{50}F_5 + \frac{2}{55}F_6 \quad (6.29)$$

The difference between the fourth- and fifth-order RKF methods provides us with an error estimate ε for the integration. Let's say that $x_{(4)}$ is the calculated value of x_{n+1} after a RKF4 integration, and that $x_{(5)}$ is the value after RKF5 (starting over again). We define the error estimate to be $\varepsilon = |x_{(5)} - x_{(4)}|$. This estimate is often calculated in one routine called, say, RKF45 since it calculates the (1) fourth-order solution, (2) fifth-order solution, and (3) error estimate. We also specify two sets of error bounds: one pair for the error estimate $\varepsilon_{min} \le \varepsilon \le \varepsilon_{max}$ and the other for $\Delta t_{min} \le \Delta t \le \Delta t_{max}$. The bounds for the error are specified so that there is a "window" of error tolerance. The bounds for Δt are specified so that we cannot continue dividing or multiplying Δt once the precision of the machine word has been exceeded. The process of adaptive step-size control is now specified in the following algorithm:

Algorithm 6.1. Adaptive Step-Size Control
Procedure Main
> (1) *Given Δt and initial condition for x, call RKF45*
>> *to compute order 4 and order 5 solutions, and ε.*
> (2) *If $\varepsilon_{min} \le \varepsilon \le \varepsilon_{max}$, then*
>> *step size Δt is not altered and the next step is taken to*
>> *compute x (as one would normally do in RK4, for instance).*
> (3) *If $\varepsilon < \varepsilon_{min}$, then $\Delta t \leftarrow 2\Delta t$.*
> (4) *If $\varepsilon > \varepsilon_{max}$, then $\Delta t \leftarrow \Delta t/2$.*
> (5) *If $\Delta t_{min} \le \Delta t \le \Delta t_{max}$, then the step*
>> *is repeated by returning to step 1 with x and the new Δt.*

End Main.

Code for differential equations. Let's consider a simple differential equation, and then go about solving it. Here is a first-order ordinary differential equation (ODE):

$$\dot{x} = x + 3 \qquad \text{where } x(0) = 1 \quad (6.30)$$

Figure 6.8 displays this differential equation in functional block form. Figure 6.8 shows us the feedback inherent within this system (or equation). It is easier to see how we will solve this ODE by studying the block diagram. $\dot{x}$ feeds into the integrator block which produces

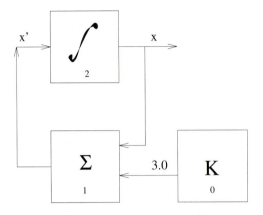

Figure 6.8 Block model for ODE example.

x. Recall that x is really $x(t)$ and $\dot{x}$ is really $\dot{x}(t)$. To solve for this equation we need to start with the initial condition $x(0) = 1$. You can imagine this to be the same as making the block integrator output "line" the value 1 at time zero. Then the value 1 must be propagated through the block diagram or network, so 1 and the constant 3 feed into the addition block to get 4. Now the $\dot{x}$ line has value 4. Now we must integrate to see what the new value of x will be. If we use Euler's method for this example, we know that we can derive the new value for the x line by adding together the old value of the x line to $\Delta t \times \dot{x}$. Let's make Δt some small value like 0.01. Then the new, integrated, value for x is $1 + 0.01 \times 4$ or 1.04. We can continue like this forever or until t or x reaches some specific value. Note that this process of solving the ODE involves three major steps:

1. The *initial condition* step. Specify values for all state variables at time zero. $x(0) = 1$ in this case.

2. The *state* step. Put the current value for x into the state equation to calculate $\dot{x}$.

3. The *integration* step. Integrate $\dot{x}$ to get the new value for x.

Steps 2 and 3 are continued in a loop until the termination criterion has been satisfied.

Most equations are not as simple as the one we just discussed. Assume that we have a single, multiple-order ODE with variable x:

$$\frac{d^n x}{dt^n} = F\left(x, \frac{dx}{dt}, \frac{d^2 x}{dt^2}, \ldots, \frac{d^{n-1} x}{dt^{n-1}}\right) \tag{6.31}$$

This represents an ODE with an arbitrary order n. Initial conditions for (6.31) are as follows (where $t_0 = 0.0$ is common):

$$x(t_0), \frac{dx}{dt}(t_0), \frac{d^2 x}{dt^2}(t_0), \ldots, \frac{d^{n-1} x}{dt^{n-1}}(t_0)$$

We must convert this equation to a set of first-order differential equations by creating new variables as follows:

$$x_1(t) = x, x_2(t) = \frac{dx}{dt}, \ldots, x_n(t) = \frac{d^{n-1} x}{dt^{n-1}}$$

Then the generated set of equations is

$$\dot{x}_1 = x_2$$

$$\dot{x}_2 = x_3$$

$$\vdots \tag{6.32}$$

$$\dot{x}_{n-1} = x_n$$

$$\dot{x}_n = F(x_1, x_2, \ldots, x_n)$$

For our previous example, we let $x_1 = x$; we need not produce more equations since we will have n equations for an nth-order system, and in this case $n = 1$. Our generated equation is $\dot{x}_1 = x_1 + 3$. One can, therefore, consider as a primary step (before solving the equation numerically) the job of converting the ODE set to a first-order ODE set. Now, consider the third-order equation:

$$x''' + x' = 2 \tag{6.33}$$

We first substitute variables using (6.32): $x_1 = x$, $x_2 = x'$, and $x_3 = x''$. Then our equations are

$$\dot{x}_1 = x_2 \tag{6.34}$$

$$\dot{x}_2 = x_3 \tag{6.35}$$

$$\dot{x}_3 = 2 - x_2 \tag{6.36}$$

If an nth-order ODE is not autonomous, it means that there is a term $g(t)$ inside the equation (i.e., a term that is not a function of state variables). The method is to add one extra equation: $\dot{x}_{n+1} = 1.0$ with the initial condition $x_{n+1}(0) = t_0$, where t_0 represents the starting time for the system (0.0, for instance).

We need a way of encoding the state variables $x_1, x_2, \ldots, x_n$ into C code. It is convenient to use two arrays as follows: $\dot{x}_1 = IN[1]$, $\dot{x}_2 = IN[2]$, $\dot{x}_3 = IN[3]$, and so on. The IN array stores all of the derivative values. Likewise, an OUT array stores all the state variable values: $x_1 = OUT[1]$, $x_2 = OUT[2]$, $x_3 = OUT[3]$. Now we can think of the *integration* step for a variable $\dot{x}_i$ as taking the value in cell $IN[i]$, integrating it and placing this new value into cell $OUT[i]$. This procedure matches the action of the integration block displayed in the block diagram in Fig. 6.8. The *state* step involves executing equations relating OUT to IN. Taking the equation $x''' + x' = 2$, we would have the following C code:

```
in[1] = out[2]
in[2] = out[3]
in[3] = 2 - out[2]
```

To summarize, here is the main C code for our three major steps discussed earlier:

1. Set initial conditions. We will assume 1.0 for simplicity.

```
out[1] = 1.0;
out[2] = 1.0;
out[3] = 1.0;
time = 0.0;
delta_time = 0.01;
```

2. Update state equations.

```
in[1] = out[2];
in[2] = out[3];
in[3] = 2 - out[2];
```

3. Integrate all state variables.

```
for (i=1;i<=3;i++)
  out[i] += in[i]*delta_time;
time += delta_time;
```

Steps 2 and 3 are repeated in a loop until a termination criterion has been reached (say, until time equals 5.0). Figure 6.9 displays the output from the RK4 method integration for equation $x''' + x' = 2$.

6.2.6 Predator/Prey Ecological Model

The Lotka-Volterra equations express birth, death, competition, and interaction among species. Consider lions and gazelles in Africa and the dynamics of the species population levels. Lions are carnivores and prey on gazelles, which are herbivores. The equations for such a model are based on the following balance equations:

1. Number of Prey = New Births of Prey Species − Number of Prey Killed by Predators.
2. Number of Predators = New Births of Prey Species + Number of Prey Killed by Predators.

We will go through the entire model design process since this is our first application of models of the ODE type. Our first task is to determine the objects. We see that there are several possible objects: (1) population of lions, (2) individual lions, (3) population of gazelles, and (4) individual gazelles. We could create an object per individual animal but we must look at the level of detail that we are trying to achieve. If, in our modeling pursuit, individual animals have different characteristics which will be important for the dynamics, we should create individual animal objects. If our goal, though, is to model *populations* of animals rather than caring about individual animal dynamics, we will create an object for a population of the same type of animal. Let's choose the latter method. Our class and instance models are shown in Figs. 6.10 and 6.11. For Fig. 6.10, we will want to keep

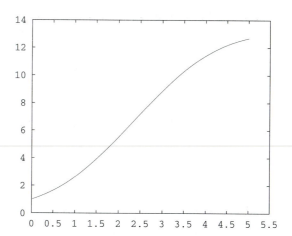

Figure 6.9   Solving for x.

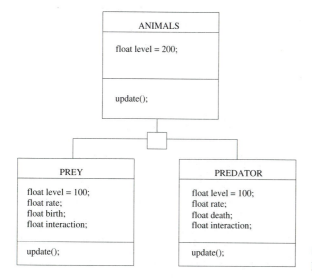

Figure 6.10   Class hierarchy for population dynamics.

track of the current level of the population, the rate of growth, and the interaction between the species. The method update() updates the state equations and integrates the state variable(s). For the prey, update() is presented as pseudo-code:

```
prey.rate = prey.birth - prey.interaction;
prey.birth = k1 * prey.level;
prey.interaction = k2 * prey.level * predator.level;
prey.level = integrate(prey.rate);
```

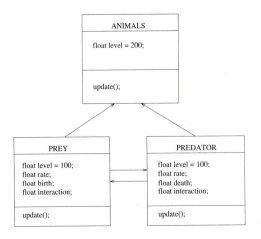

Figure 6.11 Model for population dynamics.

For the predator population, we have similar code:

```
predator.rate = predator.interaction - predator.death;
predator.interaction = k2 * prey.level * predator.level;
predator.death = k3 * predator.level;
predator.level = integrate(predator.rate);
```

For the total animal population, `update()` is coded as `level = prey.level + predator.level;`. The simulation proceeds by sweeping through the population objects while invoking update to update the state variables.

Now, let's observe the dynamics. Using x_1 for prey and x_2 for predator, we develop the equations of balance which correspond to the object design discussed previously:

$$\frac{dx_1}{dt} = k_1x_1 - k_2x_1x_2 \tag{6.37}$$

$$\frac{dx_2}{dt} = k_2x_1x_2 - k_3x_2$$

The k_i are rates specific to the species involved and how they interact. In our model, we are saying that prey increase their population by themselves but predators do not. Predators grow more based on the interaction with prey rather than simply on their own population level. At first, the interaction term x_1x_2 in (6.37) may be confusing and may not seem intuitive, but an example helps. Consider the situation where we have a current population of 1000 gazelles and 1 lion. Now contrast this against a scenario with 1 gazelle and 1000 lions. In either case, the rate of prey death (or predator growth) is close to being equal. It is the combination of both species, with the interaction term representing this combination, that dictates the prey death rate and predator growth. Figures 6.12 through 6.14 show the results of the simulation for this system. Figure 6.12 is the population of gazelles (prey) over time, and Fig. 6.13 is the population of lions (predators). Figure 6.14 shows the interaction between the two species in the phase plane. Note how the system shows a cycle occurring between two species—as the number of gazelles increase, we obtain a

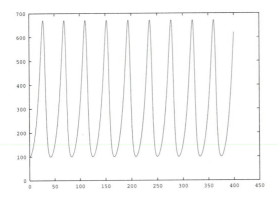

Figure 6.12 Gazelle population x_1 vs. t.

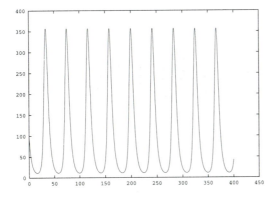

Figure 6.13 Lion population x_2 vs. t.

corresponding increase in lions, but as the number of lions increases, this decreases the gazelle population, which, in turn, causes the lion population to decay.

6.2.7 Ballistics

In Ch. 1, we gave a *ballistics* example of simulation in its many forms. Let's see how we might model the flight of a ball. Even though we could model the flight of a ball with declarative and functional approaches, the most elegant approach is to use Newton's second and third laws of motion. Specifically, all forces acting on a body will be equal and opposite, or put more quantitatively, all forces acting on a body will sum to zero. Consider a ball thrown by a person; the ball has the following forces acting upon it:

1. Force initially provided to the ball as a result of having been thrown (**f**).
2. Force resisting the ball's forward motion, which is opposite in direction to **f**: air resistance (**r**).
3. Force provided by the earth in the form of gravity (**g**).

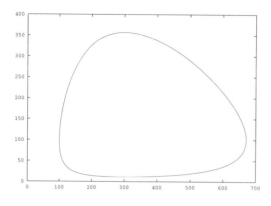

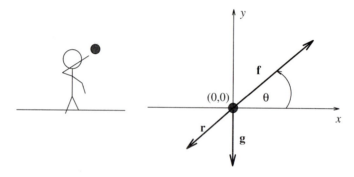

Figure 6.14 Phase plane population cycle x_1 vs. x_2.

Figure 6.15 Person throwing a ball.

Figure 6.15 shows a schematic of the person throwing the ball along with the vectors showing the forces acting on the object at the time that the ball is released.

The constraints on the object specify that:

$$m\mathbf{a} = \sum_{i=1}^{n} \mathbf{f}_i \tag{6.38}$$

More specifically, the acceleration vector ($\mathbf{a}$) for the ball times the mass (m) will equal the sum of all n forces ($\mathbf{f}_i$) acting on the ball. In our example (where $n = 2$), all forces are in directions opposite to the ball's flight, so that

$$m\mathbf{a} = -\mathbf{r} - \mathbf{g} \tag{6.39}$$

We now need to attach a reference frame to this vector equation. Letting the release point of the ball be $(0, 0)$ with a rectangular coordinate system, we obtain, using(6.39),

$$m\ddot{x} + v^2 r \cos\theta + 0 = 0 \tag{6.40}$$

$$m\ddot{y} + v^2 r \sin\theta + g = 0 \tag{6.41}$$

where m is the mass of the ball, v is the velocity, r is the coefficient for air resistance, θ is the angle formed by the horizontal x axis and $\mathbf{f}$, and g is the gravitational constant of 9.8

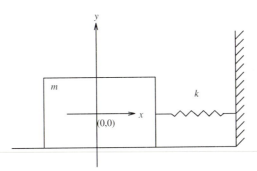

Figure 6.16   Linear harmonic oscillator.

m/s². Noting that $\dot{x} = v \cos \theta$ and $\dot{y} = v \sin \theta$, and dividing by m (where $m = 1$) yields:

$$\ddot{x} = -rv\dot{x} \tag{6.42}$$

$$\ddot{y} = -rv\dot{y} - g \tag{6.43}$$

where $v = (\dot{x}^2 + \dot{y}^2)^{0.5}$. Equations (6.42) and (6.43) can now be solved using the methods of numerical integration.

6.2.8 Harmonic Motion

Linear systems representing physical phenomena are normally categorized as first or second order. First-order systems have exponential solutions where the state variable rises or falls proportional to ke^{at}, where k is most often negative for growth or decay asymptotic to a limit. First-order equations are good at representing things such as growth or decay, whereas second-order systems are characterized by oscillatory behavior; their solutions are sinusoidal. The classic example of oscillatory behavior comes in one of two forms: (1) the linear oscillator and (2) rotational oscillator. The linear oscillator is shown in Fig. 6.16. This oscillator is composed of a mass m resting on a surface and tied to a wall via a spring with stiffness coefficient k. We have created our coordinate system so that the origin $(0, 0)$ is located at the center of mass of m, while the spring is in its rest (i.e., minimal energy) position. The formula for the system is

$$m\ddot{x} = -kx \tag{6.44}$$

where $-kx$ is the magnitude of the force which operates in an opposite direction to the velocity of m. Therefore, if m moves to the right, the spring pushes back with a force that increases linearly with its displacement from the origin. Real springs have nonlinearities, but the linear oscillator serves as a starting point for studying periodic behavior.

The rotational oscillator is better known as a pendulum. Figure 6.17 shows a planar pendulum. m is the mass of the ball on the end of the pendulum and the mass-less rod (length l) connecting the bob to the origin $(0, 0)$ is of length l. The tension on the rod is T. If s is the arc length of the bob, then $s = l\theta$, $\dot{s} = l\dot{\theta}$, and $\ddot{s} = l\ddot{\theta}$. We will show three ways of modeling the pendulum:

1. *Angular Cooordinates.* Using $\theta(t)$ to define the angular displacement of the bob from its rest position where $\theta = 0$.

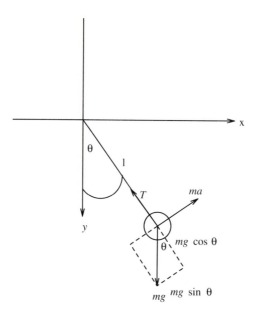

Figure 6.17   Planar pendulum.

2. *Rectangular Coordinates.* Using $x(t)$ and $y(t)$ to define the position of the bob over time.

3. *Energy.* Using the Lagrangian to define the dynamics of the pendulum using potential and kinetic energy and the *principle of least action*.

As before, we can determine the formula for the pendulum using an equation of balance of forces acting on the bob. For the first equation, we will consider a spring to be acting on the pendulum at the origin $(0, 0)$. This provides the term $-(k/ml)\theta$. Also, there is a resistive force due to friction: $-(b/ml)\dot{\theta}$.

$$\underbrace{ml\ddot{\theta}}_{\text{arc accelaration}} = \underbrace{-mg\sin\theta}_{\text{gravity}} \underbrace{-b\dot{\theta}}_{\text{damping}} \underbrace{-k\theta}_{\text{spring}} \qquad (6.45)$$

Dividing by ml yields

$$\ddot{\theta} = -\frac{g}{l}\sin\theta - \frac{b}{ml}\dot{\theta} - \frac{k}{ml}\theta \qquad (6.46)$$

We can integrate for θ to determine the angle over time for the swinging pendulum. θ is converted into rectangular coordinates, if desired, by using the equivalences $x = l\sin\theta$ and $x = l\cos\theta$. For the remaining two approaches to pendulum simulation, we will remove the damping and spring terms for simplicity.

The rectangular approach is not as convenient as the angular method since it results in formulating two equations instead of one and having to insert the tension T provided by the rod into the equations. The following equations balance the forces in the x and y directions:

$$m\ddot{x} = -T\sin\theta \qquad (6.47)$$

$$m\ddot{y} = mg - T\cos\theta \qquad (6.48)$$

$$x = l \sin \theta \tag{6.49}$$

$$y = l \cos \theta \tag{6.50}$$

Substituting for x and y using (6.49) and (6.50) yields

$$\ddot{x} = -\frac{T}{ml}x \tag{6.51}$$

$$\ddot{y} = mg - \frac{T}{ml}y \tag{6.52}$$

The reason for having to use the tension T in the rectangular approach but not in the more familiar angular approach is that T is orthogonal to θ in a polar coordinate system (i.e., for the angular method), while in a rectangular system x and y are functions of T.

Classical mechanics was more expressively formalized in the nineteenth century using principles by Hamilton and Lagrange. The principle of least action (by Lagrange) states that nature operates so as to minimize the time accumulated difference between kinetic and potential energy. From such energy principles it is possible to express mechanical relations using a scalar (energy) and a general set of coordinates. The Lagrangian is defined by $L = T - V$, where T is kinetic energy and V is potential energy. The equation of Lagrange is

$$\frac{d}{dt}\left(\frac{\partial L}{\partial \dot{q}}\right) - \frac{\partial L}{\partial q} = 0 \tag{6.53}$$

where q is termed a generalized coordinate, and so it can be either a linear or an angular displacement. We let $q = \theta$ and define energy values T and V as

$$T = \frac{1}{2}mv^2 = \frac{1}{2}ml^2\dot{\theta}^2 \tag{6.54}$$

$$V = -mgl \cos \theta \tag{6.55}$$

Using $L = T - V$, $L(\theta, \dot{\theta}) = \frac{1}{2}ml^2\dot{\theta}^2 + mgl \cos \theta$ and the substitutions for T and V, we reduce (6.53) to

$$\frac{d}{dt}(ml^2\dot{\theta}) - (-mgl \sin \theta) = 0 \tag{6.56}$$

This reduces further to $\ddot{\theta} = -(g/l) \sin \theta$, which is the same as our original equation [Eq. (6.46)] without the damping and spring terms.

6.2.9 Chaotic Behavior: The Lorenz System

You can see how you can now take any set of nonlinear ODEs and solve them using the integration methods and C code discussed. Consider the often-cited Lorenz system, which can easily be placed within our `contrk.c` framework:

$$\dot{x}_1 = \sigma(x_2 - x_1)$$

$$\dot{x}_2 = (1 + \lambda - x_3)x_1 - x_2 \tag{6.57}$$

$$\dot{x}_3 = x_1 x_2 - bx_3$$

with parameter values of $\sigma = 10$, $\lambda = 24$, and $b = 2$, and initial conditions of $x_i(0) = 1.0$. The resulting chaotic attractor [a projection of (x_2, x_3)] for (6.57) is shown in Fig. 6.18.

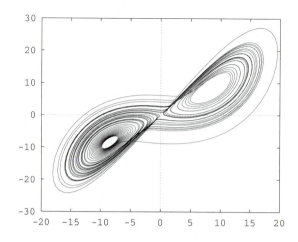

Figure 6.18   The Lorenz system.

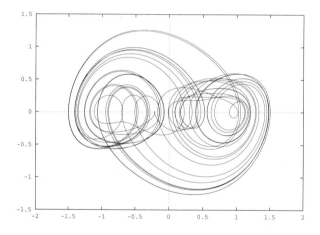

Figure 6.19   Silnikov's homoclinic orbit.

Another example of a chaotic attractor is the homoclinic orbit (an orbit which moves toward an equilibrium point in forward and reverse time) of the Silnikov equations:

$$\dot{x}_1 = x_2$$
$$\dot{x}_2 = x_3 \qquad\qquad (6.58)$$
$$\dot{x}_3 = -ax_3 - x_2 + bx_1(1.0 - cx_1 - dx_1^2)$$

A phase trajectory for Eq. 6.58 was created using parameters found in [124] and is displayed in Fig. 6.19.

6.2.10 Mathematical Expression Handling

In SimPack there is a program called DEQ which parses differential equations expressed in one of two forms:

1. *Form 1.* A single equation with high-order terms (derivatives). Note that the system may be homogeneous or nonhomogeneous, and that the left-hand side of the equation must be the higher-order term. Note the following notation: use $x[1]$ for the first derivative of x with respect to time (i.e., x' or dx/dt), $x[2]$ is the second derivative, and so on. $x[0]$ is simply x.

2. *Form 2.* A set of first-order equations for an n-dimensional system. The user is prompted for each equation in turn. Here $x[1]$ means the first state variable, $x[2]$ means the second state variable, and so on.

An example of a single equation (using form 1) is to enter

```
%deq
x[2] = -16*x[0] + 10* cos (3.7*t)
```

This equation is $x'' = -16x + 10\cos(3.7t)$ using standard notation. We plot $x[0]$ over time as shown in Fig. 6.20.

An example of the Lorenz system phase plot in (6.57) is to enter

```
%deq
x[1]' = 10*(-x[1] + x[2])
x[2]' = 28*x[1] - x[2] - x[1]*x[3]
x[3]' = -2.67*x[3] + x[1]*x[2]
```

(*Note:* The left-hand sides of equations are prompted by DEQ with form 2.)

6.2.11 Delay Differential Equations

Overview. Within the overall domain of differential models, we may incorporate specific time delays that do not involve dispersion of a state variable. That is, a state variable's value will stay fixed for a time period as it does for models using difference equations. An example of such a model is

$$\dot{x}(t) = ax(t-1) + bx(t-2) \tag{6.59}$$

where a and b are parameters. This type of model is sometimes called a "delay differential equation"; however, in engineering, the new item that we have added to the existing differential equation capability is usually termed a *transport delay*. The transport delay can be contrasted against other forms of delay, such as a first-order (exponential) delay that one obtains when a capacitor is charged. The difference is that a transport delay represents a true delay of a signal, whereas a delay as a result of integration results in a response variable (output) *gradually* reaching the forcing value (input) that was input to the system after some period of time. A delay differential equation can be seen as a hybrid

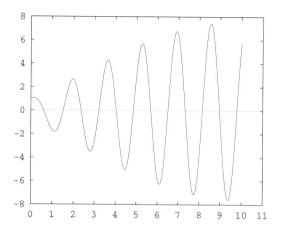

Figure 6.20   Plot of $x'' = -16x + 10\cos(3.7t)$.

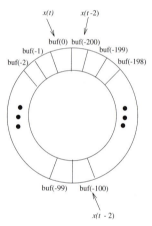

Figure 6.21   Circular queue for delay differential equation.

or combined model composed of both difference and differential terms. For the delays denoted by $x(t - 1)$ and $x(t - 2)$ it is important to note that x, since it is a continuous variable, needs to be stored at many points between $x(t - 2)$ and $x(t)$. It is not sufficient just to keep track of $x(t - 1)$ or $x(t - 2)$ as we did for difference equations since x is integrated using a small time slice. In the following example, we will use 200 points; however, this is sufficient to see how one can use any number of points to store the delayed values of the state variable. Figure 6.21 displays how we store the circular queue values.

Regulation of Hematopoiesis.   Hematopoiesis is the process of blood creation in the body. White and red blood cells are produced in bone marrow. From the marrow they enter the blood circulatory system. As the oxygen level decreases in the body, there is a feedback back to the bone marrow—which produces more cells. While this process can be

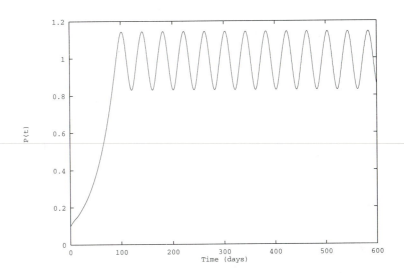

Figure 6.22   Cell concentration vs. time for delay $T = 6$ days.

modeled with the usual differential equation techniques, we need a method for incorporating the delay T between the initialization of cellular production in the bone marrow and the release of mature cells into the bloodstream. Mackey and Glass [138] provide a delay model for hematopoiesis of the following form:

$$\frac{dP(t)}{dt} = \frac{\lambda \theta^m P(t-T)}{\theta^m + P^m(t-T)} - gP(t) \tag{6.60}$$

This form, along with accompanying analysis, is also considered by Murray [155]. Depending on the maturation delay T, we can generate different solutions. Figure 6.22 displays the time trajectory for the total concentration of blood cells between times 0 and 600 when the delay $T = 6$ days. As the delay moves upward to $T = 20$ days, we find a nonperiodic, chaotic trajectory in Fig. 6.23. A phase graph depicting $P(t)$ against $P(t-20)$ is shown in Fig. 6.24.

6.3 GRAPH-BASED APPROACH

Unlike declarative or functional graphs, constraint graphs have no directionality and therefore no directed arcs. The most general type of constraint graph useful in simulation is one based on *energy*. This is because all networks containing physical components are governed by energy conservation laws; what flows into a component must flow out, and if something flows into a junction via several input lines, the same amount of *stuff* must come out on the output lines.

We begin with electrical networks since they are quite common. Electrical networks involve the flow of electrical power and they contain two key variables that allow us to model all dynamic behavior: voltage drops and current levels in different parts of the network.

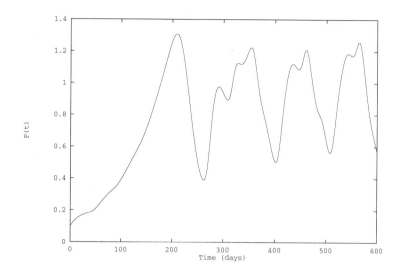

Figure 6.23   Cell concentration vs. time for delay $T = 20$ days.

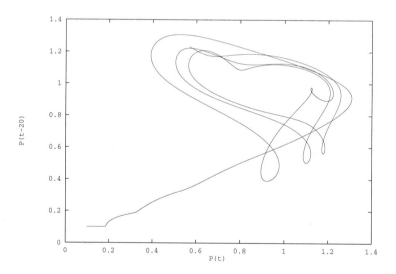

Figure 6.24   Phase graph for delay $T = 20$ days.

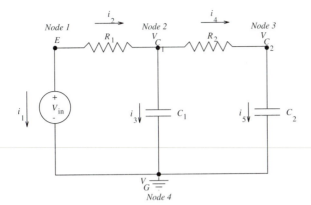

Figure 6.25   Sample electrical circuit.

Then we cover a very general type of constraint graph technique called *bond graphs* which subsume electrical networks and all other networks containing elements which are subject to fluctuations in power and energy.

6.3.1 Electrical Networks

Electrical networks have components with flow and electric potential. We are now covering only analog networks since digital networks are designed to be directional and are consequently better modeled using functional approaches. Flows can occur across a component in *either direction,* so we must base our dynamics on conservation of flow and potential. Fig. 6.25 displays an electrical circuit with a voltage source, two resistors, and two capacitors. The resistors are labeled R_1 and R_2 and the capacitors, C_1 and C_2. This type of graph—with nodes and links—is a constraint graph because there is no simple flow relationship that would cause the R_1, for instance, to behave as a simple function within an input X and output Y. Instead, three types of constraints serve to characterize the lumped dynamics[3] of all electrical and electronic circuits:

1. *Power Relation.* One must keep track of two variables, current and voltage, for a circuit. The product of these two variables forms the power associated with each part of a circuit. A simple algebraic relationship exists for resistors since $V = RI$ defines the voltage drop across a resistor to be equal to the resistance times the current flowing through the resistor. This is a *linear* model since the relationship involves a linear combination of variables, although certain resistors can be defined by more complex, nonlinear relationships such as FETs (field-emitting transistor) that contain a region of approximate linear (I, V) behavior.

2. *Kirchhoff's Current Law (KCL).* The value of all currents entering a node equals the value of all currents exiting that node. Since currents are denoted positive or negative

---

[3]The term *lumped* means that we are building a model whose parameters and state variables are assigned to components rather than to subcomponent (device) pieces. An example of a nonlumped model (often called a *distributed parameter* model) is one where we break up a component such as a transistor into many smaller pieces, where each pieces has its own properties, parameters, and state.

depending on whether they flow to or from a node, the law is more precisely defined so that all currents for any node add to zero. KCL is a *node law* since it focuses on conservation of flow where links meet (at nodes).

3. *Kirchhoff's Voltage Law (KVL).* For all loops in a circuit, the voltage drops around each loop add to zero. KVL can be thought of as a *loop law.*

There are three types of components in Fig. 6.25: voltage source, resistor, and capacitor. The voltage source outputs a signal with voltage V_{in} (or E). The power relationship for the resistor is $v = iR$ and the relationship for the capacitor is $v = (1/C) \int i \, dt$. The last relationship can also inversely be represented as $i = C\dot{v}$.

We will formulate KCL and KVL equations for the circuit in Fig. 8.2. First, let's define some terms. Lowercase letters will be used to define *link* values so that the current *through* component j will be i_j and the voltage *across* component j will be v_j. An uppercase letter V_j will refer to the *node* voltage. E, V_{C1}, and V_{C2} are node voltages, which just means that they represent the voltage value at a node relative to ground V_G. E and V_{in} represent the source voltage.

Using KCL, we derive the equations for current flow in and out of the nodes V_G, V_{C1}, and V_{C2}:

$$i_1 + i_3 + i_5 = 0 \tag{6.61}$$

$$i_2 - i_3 - i_4 = 0 \tag{6.62}$$

$$i_4 - i_5 = 0 \tag{6.63}$$

KVL provides similar equations for two of the loops:

$$-v_1 + v_2 + v_3 = 0 \tag{6.64}$$

$$-v_1 + v_2 + v_4 + v_5 = 0 \tag{6.65}$$

From the state relationship for capacitance and Eq. (6.63), we derive

$$V_{C2} - V_G = \frac{1}{C2} \int i_5 \, dt = \frac{1}{C2} \int i_4 \, dt \tag{6.66}$$

Let $V_G = 0$ and note that $i_4 = (V_{C1} - V_{C2})/R_2$. With this result, Eq. (6.66) yields

$$V_{C2} = \frac{1}{C2} \int \frac{V_{C1} - V_{C2}}{R_2} \tag{6.67}$$

which simplifies to

$$\dot{V}_{C2} = \frac{1}{C2} \frac{V_{C1} - V_{C2}}{R_2} \tag{6.68}$$

A similar derivation for $\dot{V}_{C1}$ can be made by starting with $V_{C1} - V_G$ (see the Exercises). We obtain state equations for either the currents or the voltages. The state equations for the two nodes V_{C1} and V_{C2} are

$$\dot{V}_{C1} = \frac{1}{C1} \left(\frac{E - V_{C1}}{R_1} + \frac{V_{C2} - V_{C1}}{R_2} \right) \tag{6.69}$$

$$\dot{V}_{C2} = \frac{1}{C2} \left(\frac{V_{C1} - V_{C2}}{R_2} \right) \tag{6.70}$$

6.3.2 Bond Graphs

There is a more general network formulation than the one presented previously for electrical networks. Bond graphs were developed to allow a system to be defined in terms of a group of elements with bonds connecting each element to the rest of the network. Systems can contain components from many domains, such as mechanical, electrical, pneumatic, thermal, and hydraulic. Even though the domains are different, each domain has an analogous component in every other domain. All domains are concerned with two primary variables: *effort* and *flow*. The product of effort and flow is *power* and the time integral of power is *energy*. So, for instance, a resistor in the electrical domain performs the same basic function as a dashpot in the mechanical domain. In fact, the relationship between effort and flow is the same: $e = fR$, where R is the damping coefficient or the resistance value, depending on the domain. This *law* is Ohm's law in the electrical domain ($V = IR$), and the basic law for resistance [that is, a resistive force (effort) is proportional to velocity (flow)]. By building a modeling method founded on the basic variables *effort* and *flow*, we derive a powerful technique that has some advantages over using domain-specific modeling approaches:

- In instruction, the student learns analogies among circuits in different domains.
- All dynamics in a domain is seen as stemming from the same domain-independent terms of *effort* and *flow*, their product *power*, and the time integral of power: *energy*.
- When modeling a hybrid circuit containing parts from different domains, the bond graph approach creates a holistic modeling methodology; systems with multiple domain-specific components can easily be designed.

Table 6.1 lists three example domains (mechanical/translational, mechanical/rotational, and electrical) along with the generic variable applicable to each:

Our first concern is to specify a clear method for designing bond graphs given a conceptual model (usually a schematic) of a system. Some bond graph texts such as [191, 219] provide good discussions on how to do this. We will first adapt an algorithm by Breedveld [30], who focuses on the translation problem from schematic to bond graph by associating a template to handle all point efforts (relative velocities). The algorithm uses terms for nonmechanical domains while placing terms for mechanics in parentheses. For

TABLE 6.1   VARIABLES FOR EXAMPLE DOMAINS

		Mechanical		
Variable	Generic	Translational	Rotational	Electrical
Effort	$e(t)$	F, force	τ, torque	e, voltage
Flow	$f(t)$	V, velocity	ω, ang. vel.	i, current
Momentum	$p = \int e\,dt$	P, momentum	H, ang. mom.	λ, flux
Displacement	$q = \int f\,dt$	X, distance	θ, angle	q, charge
Power	$P(t) = e(t)f(t)$	$F(t)V(t)$	$\tau(t)\omega(t)$	$e(t)i(t)$
Energy	$E(p) = \int f\,dp$	$\int V\,dP$, kinetic	$\int \omega\,dH$	$\int_\lambda i\,d\lambda$, magnetic
	$E(q) = \int e\,dq$	$\int F\,dX$, potential	$\int \tau\,d\theta$	$\int_q e\,dq$, electrical

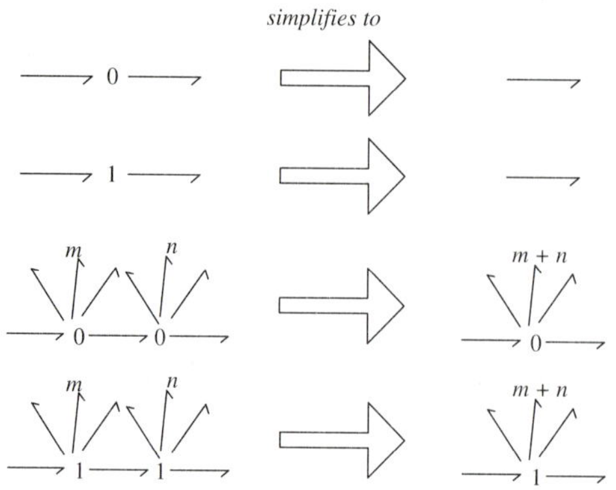

simplifies to

Figure 6.26 Simplification rules for bond graph components.

instance, when we specify efforts by 0-junctions in nonmechanical domains, we will specify relative velocities by 1-junctions for mechanics.

Algorithm 6.2. Junction-Oriented Bond Graph Method

Procedure Main
1. *Draw the schematic of the system*
2. *Identify all physical domains and elements and choose a reference for each domain*
3. *Specify the efforts (velocities) by 0-junctions (1-junctions) except for the the reference effort (reference velocity), and indicate the corresponding variables*
4. *Connect the 0-junctions (1-junctions) together with 1-junctions (0-junctions) The 1-junction (0-junction) will have 3 0-junctions (1-junctions) attached to it Two will be the effort (velocity) nodes and the third will be the the effort difference (relative velocity) Junctions should be connected so that the flow is from the highest potential to ground*
5. *Connect all ports of all elements to the corresponding efforts (or velocities) or effort differences (relative velocities) and indicate the constitutive parameters with a colon*
6. *Simplify the bond graph using the rules in Fig. 6.26 and removing bonds connected to zero-valued references*

End Main

Each bond in a bond graph is labeled with the two power variables: *effort* and *flow*, which are normally represented on either side of a bond. Therefore, a bond from element A to element B looks like: $A \rightharpoonup B$, with e on one side of the half-arrow and f on the other. The product $e(t)f(t)$ defines the power of the bond connection at time t. A "1" junction means *common flow* and a "0" junction means *common effort*.[4] These two junctions route

---

[4]A good way to remember this difference is to think of the "1" looking like the "f" in flow and the "0" to be the closed semicircle in the "e" for effort.

power through the network. Consider junctions with three bonds attached to them. Each bond will have an effort (e_1, e_2, e_3) on one side of the bond and a flow (f_1, f_2, f_3) on the other. For the 1-junction the flow equations are $f_1 = f_2 = f_3$ (i.e., common flow). The effort equation says that all inputs to the 1-junction equals all outputs. Therefore, a 1-junction with one input labeled e_1/f_1 and two outputs labeled e_2/f_2 and e_3/f_3 would have the following two equations:

$$f_1 = f_2 = f_3 \tag{6.71}$$

$$e_1 = e_2 + e_3 \tag{6.72}$$

The 0-junction is the dual of the 1-junction, so for the same inputs and outputs for a 0-junction, we have

$$e_1 = e_2 = e_3 \tag{6.73}$$

$$f_1 = f_2 + f_3 \tag{6.74}$$

Now that we have specified how junctions work, we need elements to add to the junction network. While junctions cause power to be routed, elements specify the precise relationship between two power variables on a bond. For a resistance element (R), the relationship between e and f is $e = Rf$. For a capacitive element (C), the constraint is $e = e_0 + (1/C) \int f \, dt$. For an inductor (I), the equation is the inverse of that for capacitance: $f = f_0 + (1/I) \int e \, dt$, where e_0 and f_0 are initial conditions for effort and flow.

Let's take the bond graph algorithm for a two-mass oscillator with springs joining the masses. We show each step starting with the initial schematic in Fig. 6.27. Figure 6.28 displays the simplification step taken in the two-mass oscillator represented in Fig. 6.27 from step 5 to step 6. We see that in the simplified bond graph, we have the following relationships coming from the 1-junction: $e_1 = e_3$ and $f_1 = f_3$. We need to show that this holds given the left-hand graph in Fig. 6.28. First we write down the equations for the lower 1-junction:

$$e_1 = e_2 \tag{6.75}$$

$$f_1 = f_2 \tag{6.76}$$

Now, for the 0-junction,

$$e_2 = e_3 = e_4 \tag{6.77}$$

$$f_2 = f_3 + f_4 \tag{6.78}$$

Since v_0 is the reference velocity which equals 0, we let $f_4 = 0$. We see that $e_1 = e_3$ from (6.75) and (6.77), and $f_1 = f_3 + 0$ or $f_1 = f_3$ from (6.76) and (6.78).

Let's take our simplified graph of step 6 in Fig. 6.27 and assign power variables as shown in Fig. 6.29. We can finally formulate the equations of state that can be used to simulate the double oscillator. First note that $S_{e1} = e_6 = m_1 g$ and $S_{e2} = e_1 = m_2 g$. The constraint equations for the three junctions from the bottom of Fig. 6.29 to the top are:

$$e_1 = e_2 + e_3 \qquad f_1 = f_2 = f_3 \tag{6.79}$$

$$e_3 = e_4 = e_5 \qquad f_3 = f_4 + f_5 \tag{6.80}$$

$$e_4 + e_6 = e_7 + e_8 \qquad f_4 = f_6 = f_7 = f_8 \tag{6.81}$$

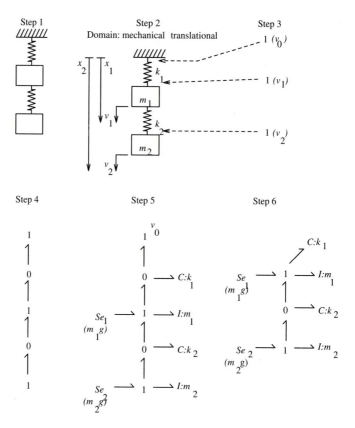

Figure 6.27   Two mass oscillator.

Figure 6.28   Analyzing the simplification in Fig. 6.27.

Now we take each element and show the relationship between each power variable on the respective element bond (again starting with the bottom-most element and working upward):

$$f_2 = \int e_2 \, dt \qquad\qquad (6.82)$$

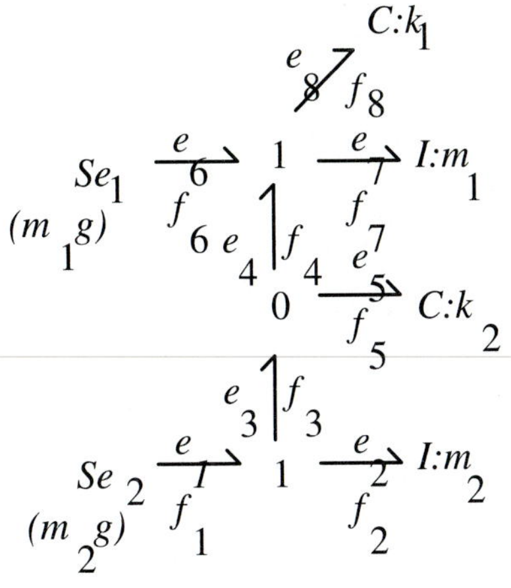

Figure 6.29 Simplified graph with power variables assigned.

$$e_5 = \int f_5 \, dt \tag{6.83}$$

$$f_7 = \int e_7 \, dt \tag{6.84}$$

$$e_8 = \int f_8 \, dt \tag{6.85}$$

Using these substitutions with Eqs. (6.79) through (6.81), we arrive at the formulation

$$m_2 g = m_2 \dot{f}_2 + k_2 \int f_5 \, dt \tag{6.86}$$

$$k_2 \int f_5 \, dt + m_1 g = m_1 \dot{f}_7 + k_1 \int f_8 \, dt \tag{6.87}$$

Now we pick a coordinate system in order to formulate the state equations. From Fig. 6.27, the positive values of x_1 and x_2 are measured from the top (reference point) down. We let the $(0, 0)$ position be the rest position of each mass; this eliminates the necessity to adjust our spring equations based on Hooke's law. We use a rectangular coordinate system and specify the following equalities for the flows: $f_2 = \dot{x}_2$, $f_7 = \dot{x}_1$, $f_5 = \dot{x}_2$, $f_8 = \dot{x}_1$. Using these substitutions, we derive

$$m_1 \ddot{x}_1 = m_1 g + k_2 x_2 - k_1 x_1 \tag{6.88}$$

$$m_2 \ddot{x}_2 = m_2 g - k_2 x_2 \tag{6.89}$$

The previous algorithm allowed us to create a bond graph for a mechanical system by specifying velocities. We now develop a different algorithm by focusing on the objects where objects are defined as components (elements) or junctions. The distinction is, again, made in this algorithm between nonmechanical domains (terms not in parentheses) and mechanical domains (terms in parentheses):

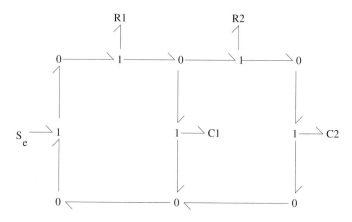

Figure 6.30   Object-oriented bond graph for Fig. 6.25.

Algorithm 6.3. Object-Oriented Bond Graph Method
Procedure Main
    1. Draw the schematic of the system
    2. Identify all physical domains and elements and
        choose a reference for each domain
    3. Build an object model where an object is either an element
        or a junction
    4. For each object, specify the correct junction:
        (a) For C or R objects, use a 1-junction (0-junction)
        (b) For I and jub=nction objects, use a 0-junction (1-junction)
    5. Connect all junctions together Junctions should be
        connected so that the flow is from the highest to lowest potential
    6. Connect all ports of all elements to the corresponding object junctions
    7. Simplify the bond graph using the rules in Fig. 6.26
        and removing bonds connected to zero-valued references
End Main

Let's use this algorithm on two examples. First, consider the electrical circuit in Fig. 6.25. By looking at this circuit, it is clear where the objects are located. If we use the object-oriented bond graph approach, we end up with a bond graph shown in Fig. 6.30. Note the similarity in connectivity between Fig. 6.30 and the original circuit in Fig. 6.25. There is a one-to-one mapping between the graphs.

Rube Goldberg produced a series of cartoons that portray all types of objects, animals, and humans within a cause-effect chain [142]. Figure 6.31 displays one called "Professor Butts' Automatic Garage Door Opener," which he drew in 1928. Goldberg's drawings, aside from being entertaining, are useful when discussing bond graph structure. The drawing in Fig. 6.31 does not have much two-way interaction; most objects receive a *push* from another object, and there is a sort of domino-like chain effect. Due to this directionality, it would be reasonable to use a functional block model; however, bond graphs are useful for their expression of the functionality of elements and the relations among domains. Speaking of

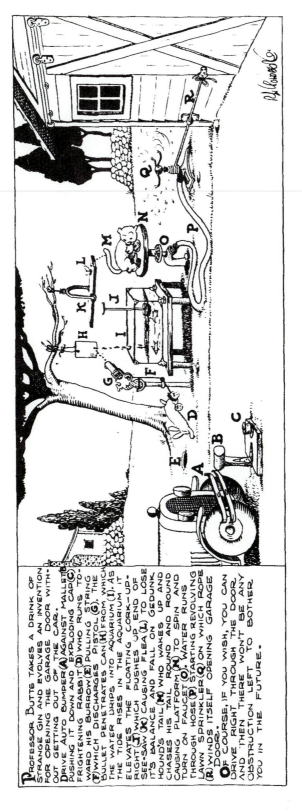

Figure 6.31 Professor Butts' automatic garage door opener. (From Marzio, *Rube Goldberg, His Life and Work* © 1973, pg. 194. Reprinted with permission of Harper and Row, New York, NY.)

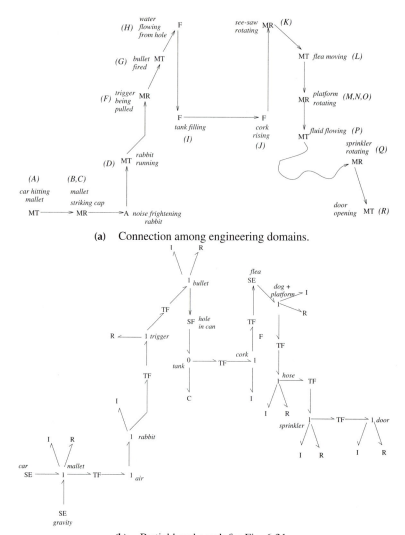

(a)   Connection among engineering domains.

(b)   Partial bond graph for Fig. 6.31

Figure 6.32   Bond graph for Professor Butts' invention.

domains, there are four represented in Fig. 6.31: mechanical translational (MT), mechanical rotational (MR), fluid (F), and acoustic (A). Figure 6.32(a) displays a graph of each domain used in Fig. 6.31 and Fig. 6.32(b) is a partial, simplified bond graph. It is simplified for several reasons, such as (1) complex actions such as the rabbit running, bullet colliding with the can of water, and (3) moving dog can be aggregated to obtain the basic effects of these actions. When studying Fig. 6.32, one notices that Rube's cartoons involving Butts' inventions involve a small number of domains, often mechanical, with frequent transformers which serve to shift power from one form to another (such as translational to rotational). This type of causal chain is present in any complex mechanical device such as a grandfather

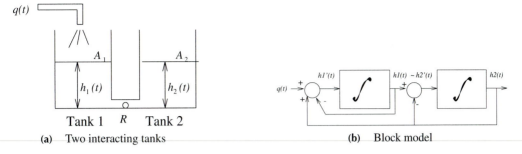

(a)   Two interacting tanks   **(b)**   Block model

Figure 6.33   Interacting tanks.

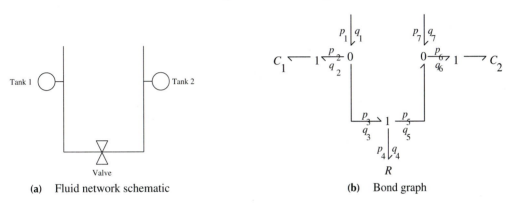

(a)   Fluid network schematic   **(b)**   Bond graph

Figure 6.34   Interacting tanks modeled with a bond graph.

clock with multiple gears (acting as transformers) and weights which convert linear motion to the angular motion of the gears.

Recall that in Ch. 5 we presented two interacting tanks, shown again in Fig. 6.33(a) along with its block representation in Fig. 6.33(b). We see that the model in Fig. 6.33(b) does not bear a structural relationship to the picture in Fig. 6.33(a). This is because functional models, due to their directionality, can show only one power variable (effort or flow) at a time. A bond graph, however, will show the power bonds and maintain a structual mapping to the original schematic. First, we create a more formal schematic for the two-tank scenario and then build a bond graph by recognizing three distinct generic objects: (1) capacitor 1, (2) resistor, and (3) capacitor 2. The capacitors represent storage (or the tank) locations for the flow of liquid and the resistor represents a valve which controls the amount of flow from one tank to the other.

Figure 6.34 shows a more accurate schematic of the two-tank system [than Fig. 6.33(a)] in that junctions are placed where the flow separates: part of the flow goes into the tank and the remaining flow goes to the attached tank via the resistive element. For the fluid regime, we are concerned with the power variables pressure (effort) and volume flow (flow). Pressure is a function of the volume of liquid (area ∗ height) and the volume flow represents the velocity throughout the system. Figure 6.34(b) represents the bond graph with all power variables attached to their respective bonds. Pressure variables are identified with p and flow variables use q. Using the object-oriented approach, we attach the objects (C_1, C_2, R)

to 1 junctions. If our bond graph is correct, we should derive the same equations that we listed in Ch. 5. Let's begin with

$$q_2 = q_1 - q_3$$

which defines the splitting of the liquid flow in tank 1. This leads to

$$A_1 \frac{dh_1(t)}{dt} = q(t) - q_3 \tag{6.90}$$

We proceed to define q_3 as follows:

$$p_3 = p_4 + p_5 \tag{6.91}$$

$$h_1(t) = Rq_4 + h_2(t) \tag{6.92}$$

$$q_4 = q_3 = \frac{1}{R}(h_1(t) - h_2(t)) \tag{6.93}$$

Substituting this value for q_3 in Eq. (6.90) yields

$$A \frac{dh_1(t)}{dt} = q(t) - \frac{1}{R}(h_1(t) - h_2(t)) \tag{6.94}$$

Equation (6.94) correctly represents the dynamics of the liquid in tank 1. For tank 2, we use a similar approach(with $q_7 = 0$ since there is no external flow into tank 2):

$$q_6 = q_7 + q_5 \tag{6.95}$$

$$A_2 \frac{dh_2(t)}{dt} = 0 + q_5 = q_3 = \frac{1}{R}(h_1(t) - h_2(t)) \tag{6.96}$$

6.4 EXERCISE

1. Write a short program that simulates the difference equation $y(t) = y(t-4) + y(t-6)$ without using arrays or other data structures (i.e., just using simple variables).

2. Specify the previous difference equation in canonical first-order form.

3. Specify the equation $2x' - 3x'' + x + 5t = 0$ in canonical first-order form and draw a functional model for this equation.

4. Formulate an equation similar to Eq. (6.3) that encodes the dynamics for a rabbit population with a gestation period of 3 months, rather than 2, and a 6-month life span rather than an infinite one.

5. Difference and differential equations can be used to model the growth of populations. Under what circumstances would you use difference equations?

6. What is the key difference between algebraic equations and difference (or differential) equations? Can both be used to model dynamic systems?

7. Why would someone choose to model a system with equations instead of using a declarative or functional model?

8. A set of equations can be drawn as a dependency graph where an arc from $A \rightarrow B$ means that variable A on the right-hand side of an equation affects variable B on the left-hand side. Draw a dependency graph for the Lorenz system. Does this type of graph resemble any model types that we have already covered?

9. Discrete delays are produced using the block operator D and continuous delays are caused by

integrator blocks ($\int$). Consider a system defined with n D blocks coupled together. How would such a system differ behaviorally from one with n $\int$ blocks?

10. Equations (6.69) and (6.70) define the two state equations for the electrical circuit in Fig. 6.25. Show a complete derivation of Eq. (6.69). Also, show the equations of state in terms of current.

11. Simplify the bond graph in Fig. 6.30 and demonstrate that the bond graph derived from using either of the two "schematic $\rightarrow$ bond graph" algorithms will result in the same graph.

12. Show the bond graph for Fig. 5.22(a).

6.5 PROJECTS

1. Write a program that will draw cobweb diagrams given the logistic equation with different initial conditions and parameter values. A more sophisticated program will allow the user to input any difference equation. Allow an option to let the user step through the drawing process or to have the drawing occur at a specific rate.

2. Build a simulation program for simulating the population changes for rabbits. Allow the user to key in any number for (1) gestation period, and (2) life span. Test out your program using several values for these two variables. Do you notice any qualitatively interesting characteristics for the growth?

3. Choose a method of integration not currently supported in SimPack, and add it to the current selection.

4. Build an expert system that determines the best type of ODE solver to use based on the requirements of the user. This will require a survey of existing techniques and the stability attributes of each method. For instance, some approaches, such as Gear's, are useful for stiff systems. Adaptive time step methods are useful when transients are separated by long stretches of inactivity for a state variable. In the expert system, include help messages and queries that lead the user through the decision process.

5. Take the deq simulator, which accepts differential equations in a noncanonical form, and rebuild it using a parser generator such as *yacc*. Develop a complete grammar that includes a capability to process difference, differential, and delay-differential equations.

6. A spring network (or mesh) is a network of springs where springs are joined with masses at the junctions. The compliance of a spring can model deformations and elasticity in materials, and so when building a network of springs, you can model simple geometric objects with such material properties. Create a program that takes as input a network description and allows the user to animate a spring network under the influence of a force applied to some part of the network.

7. There are three forces acting on a golf ball in flight [100]: (1) lift (L) because of its shape, (2) drag (D), and (3) gravity. Assume an angle to the horizontal of θ representing the current velocity vector. Drag is directly opposed to the velocity vector and lift is perpendicular to the velocity vector. The equations of motion in (x, y) are

$$m\frac{d^2x}{dt^2} = -D\cos\theta - L\sin\theta \qquad (6.97)$$

$$m\frac{d^2y}{dt^2} = L\cos\theta - D\sin\theta - mg \qquad (6.98)$$

Let $D = 3.589 \times 10^{-4}V^2$ and $L = 5.834 \times 10^{-3}V$, where V is the velocity. Build a simulation of this process using the SimPack differential equation routines and illustrate the output for a variety of different conditions and parameter settings.

8. Choose another of Professor Butts' cartoons [142] and create a bond graph for it. Be sure to list the objects, their domains, and their dynamic function (i.e., capacitive, inductive, resistive) within the scene.

6.6 FURTHER READING

The literature on difference equations and differential equations is vast. A text on ODE solving for serial architectures is given in [164]. Korn and Wait [126] and Cellier [34] provide a variety of continuous modeling domains for which ODEs are often the best modeling technique. Several texts on mathematical modeling [19, 97, 114] provide good examples of using ODEs to solve practical problems. A book on mathematical biology [52] provides an excellent introductory treatment on ODEs and PDEs along with a good description of the qualitative theory of differential equations as it is applied to biological problems. For a more in-depth treatment in biological systems, the reader is directed to Murray's text [155]. The two-volume collection by Gould and Tobochnik [90, 91] serves as an excellent introduction to simulation model design and execution (with programs in BASIC, Pascal, and FORTRAN) for physics problems. It provides the best introduction to the use of simulation for basic problems in classical physics (volume 1) and statistical physics (volume 2). Other good texts that include numerical approaches to physics problems are ones by Merrill [149] and Koonin [125]. Heermann [102] also has an overview of simulation methods applied to physics. The text by Horowitz and Hill [110] is a good general introduction to electrical and electronic components. Mastascusa [143] provides a good introduction to algorithms for network theory of analog electrical circuits, and software is included in his text. Other more detailed treatments are available [39, 224]. Bond graphs have been around for almost three decades and have good treatments in [119, 191] and [219]. Basic common analogies among different engineering domains are presented by Shearer et al. [206]. Bond graphs are not the only approach to modeling energy fluctuations and power flow; Odum [159, 160] has made a significant study of *energy graphs* for modeling energy changes in ecological systems.

6.7 SOFTWARE

For numerical software in general and methods of integration, the text *Numerical Recipes* [177] contains a library of software routines and examples in C, Pascal, and FORTRAN. For electronic circuits, a good starting place is the book by Mastascusa [143], who includes FORTRAN examples. His approach is most useful since he provides a staggered approach where the overall network software is gradually augmented as the text describes additional features. The best overall program for dynamic analog circuit analysis is SPICE. This is available on the Internet.

In addition to the SimPack software for difference and differential equations, Koçak's PHASER software [124] is a very good tool for exploring trajectories on PCs. Information about an archive site for nonlinear dynamics software is found bye reading `lyapunov.ucsd.edu/pub/README`. One such program is XPP, which was written by Bard Ermentrout. XPP uses X windows. SimPack includes a subdirectory on constraint modeling which is further described in Ch. 10.

## Chapter 7

Spatial Modeling

The role of physics to some extent is to divide the existing groupings—call them particles—into entities isomorphic or almost isomorphic to each other.[1]

7.1 OVERVIEW

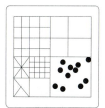

Models that reflect a decomposition of space, usually with straight boundaries, are spatial models. Spatial models are useful when we fragment the world into many small pieces with the idea that if we can specify the behavior of each piece, we can understand the behavior of the whole system.

Here are some heuristics to use when spatial modeling is chosen:

- When the system is seen as many small, regular pieces. The pieces may fit together as in a jigsaw puzzle, or they may be particles which either map-to or approximate a physical phenomenon.

[1]S. Ulam, " Conversation with Gian-Carlo Rota" in [40], p. 308.

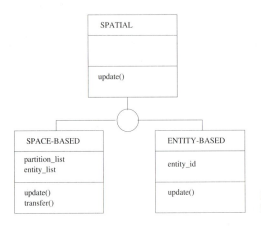

Figure 7.1   Spatial model class hierarchy.

- When you want to study a system in great detail. Normally, most spatial models are composed of a large number of pieces which divide the overall system geometry.

From an object-oriented perspective, we can implement spatial objects in one of two ways: (1) the space is the object, or (2) the entity (in the space) is the object. It is also possible to have some combination of these two approaches. Figure 7.1 displays the class hierarchy for spatial models based on one of these two schemes. Let's first consider the case where we define an object to be the space in which particles or entities move. For such an object, our attributes would include a representation of the data structure for the space and any auxiliary structures. The methods would include update(), which updates the position of all entities located within the space object. For the entity-based OO approach, our attributes will include position and orientation information and our methods will focus on what can happen to the entity over time. We can envision two types of entity methods: (1) a method to update the entity state, and (2) a method to permit the entity to grow or decay over time. A growth method is unnecessary in the space-based approach because entity structure can grow or decay without a change in the structure of space.

7.2 SPACE-BASED APPROACH

7.2.1 Overview

Space-based models contain rules that say "Given a lattice with the pattern X, perform operation f on all cells in X to produce pattern Y." The models involve convolution of a template over an entire lattice. Where rules are applicable, they are invoked on the matching part of the lattice. Due to the *sweeping* method of invoking rules, the best model execution approach is to use massive parallelism[2] since there are usually only local interactions present.

---

[2]To the level constrained by the number of processors.

7.2.2 Percolation

One of the simplest declarative spatial models relates to the phenomenon of *percolation*. Percolation is defined by the kind of clustering that occurs in a material when heated. If we heat a flat pan containing a large number of disconnected pieces of material, the pieces will expand and coalesce to form *clusters*. We can define a cell space where there is a two-dimensional lattice containing sites each of which contains the material or does not contain material. Moreover, we specify a critical threshold p_c between zero and one. An algorithm for percolation is:

Algorithm 7.1. Percolation Simulator
For each row
    For each column
        Pick a random number (X) between 0 and 1
        If $X < p_c$ then let F(i,j) = 1 otherwise 0
    End for
End for

The simulation of percolation is a matter of sampling a uniform distribution using a Monte Carlo approach. The main goal of percolation is to determine *connectivity* among clusters that arise depending on the value of p_c. At what value of p_c does a phase transition occur where many independent pieces of material suddenly become connected? Forest fires can be simulated with percolation methods. Critical to any fire is the connectivity associated with tree clusters; if there is a line of trees where each tree is in close proximity to its neighbors, fire will flash from one tree to the next, causing a chain of burning trees. By design, the effective way of halting such burning is to have a *fire wall*, or an area that the fire cannot span. Percolation simulations are not true simulations in the sense that time is not a factor; the "simulation" outputs a steady state for the system, not a sequence of states comprising a temporal state trajectory. However, percolation plays the following roles in simulation:

- Percolation can be used to create the initial spatial configuration for occupied sites. Then a simulation algorithm can be applied. For the forest fire situation, a very simple three-state automaton at each site serves to illustrate such a simulation. Each tree can have three states during a fire: *safe*, *burning*, and *burnt*. Therefore, percolation is used to try out various scenarios and determine at what critical value of p_c, forests would burn.

- If we make the parameter p_c time dependent, we can gradually increase time and watch for when clusters start appearing within forests. In the simplest case, we could assume a linear function: $p_c(t) = t$ and increment t by an arbitrary amount.

There are two types of percolation simulation: site percolation and bond percolation depending on which features (i.e., sites or the bonds that connect the sites together) one wishes to focus.

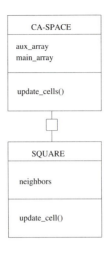

Figure 7.2 Object constructed to represent CA space.

7.2.3 Cellular Automata

Overview. A cellular automaton is conveniently described in terms of a set of rules where each rule $A \rightarrow B$ has a precondition A and a postcondition B. These rules define how the system state evolves by checking a substate of part of the system, and then firing the rules where the precondition patterns match the substate. Many cellular automata (CA) are equivalent to partial differential equations (PDE) discussed in Sec. 7.2.5; however, CAs represent a larger set of dynamical system since the rules need not be algebraic in form.

CAs are dynamical systems that operate upon a grid that can be of any geometry, including rectangular, triangular, or hexagonal. For our purposes, we will assume a finite square tessellation of a plane with $n \times n$ squares (like a chess or checker board). The simplest object representation for a CA involves methods for updating all entities in the space object where the primary object attribute is an array defining the cell space and where entities are located in the space.

Let's discuss the implementation of a CA using object-oriented methods. Since the CA modeling approach is *space-based*, our object will be the space over which individual cells multiply and decay. The productions will be applicable to all parts of the space at any time, so a single object is created of the form shown in Fig. 7.2. There are two attributes shown: (1) `main_array` for holding the cellular automaton lattice or grid, and (2) `aux_array` for holding cells that have been updated but which cannot yet replace old cells in the main array. The method `update_cells()` is defined using entity-based models of the production system type discussed in Sec. 7.3.2.

Game of Life. The Game of Life is a classic CA developed by John Conway which serves as a good introduction to the basic CA method. First consider a square board with $n \times n$ squares. Each square is either empty or contains a cell. Here are three rules for the Game of Life:

1. *Birth Rule.* A cell is born in an empty square if the square is touching exactly three cells.

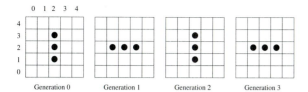

Figure 7.3   A blinker CA.

2. *Death Rule.* A cell dies if it is touching less than two cells (isolation) or if it is touching more than three cells (overcrowding).

3. *Life Rule.* A cell lives to the next generation if it touches either two or three cells.

Rule 3 is, strictly speaking, unnecessary since it is the counterpart to rule 2; however, we include it to emphasize that the history of a cell encompasses one of three phases: birth, life, and death.

Let F be the array used. If we consider any location on F as (i, j), we define one of two types of commonly employed neighborhoods of (i, j), where i is a row and j is a column:

- *Moore Neighborhood.* $\{(i-1, j-1), (i-1, j), (i-1, j+1), (i, j-1), (i, j), (i, j+1), (i+1, j-1), (i+1, j), (i+1, j+1)\}$.
- *Von Neumann Neighborhood.* $\{(i-1, j), (i, j-1), (i, j), (i, j+1), (i+1, j)\}$.

A new value for $F(i, j)$ depends on surrounding neighbors and, possibly, on the old value of $F(i, j)$. This permits the simulation of systems with memory of one past state. A new value of $F(i, j)$, denoted $F'(i, j)$, depends on its von Neumann neighbors according to relation f:

$$F'(i, j) = f(F(i-1, j), F(i, j-1), F(i, j), F(i, j+1), F(i+1, j))$$

f is a relation that can be algebraic (where the CA may be approximating a PDE), or it can be rule-based as in the Game of Life.

We begin the Game of Life by putting down any number of cells onto the initially empty board. This is called *generation* zero since it represents a starting population of cells (or initial condition). Starting with generation zero, we apply each of the three rules to determine generation one. After doing that, we create generation two. This repeated procedure will yield a surprising amount of visual complexity where cells grow and die, forming an intricate pattern over time. Consider the *blinker* in Fig. 7.3. We specify a two-dimensional array F which ranges from 0 to 4 in both dimensions. When a square (i, j) has no cell, we denote this with $F(i, j) = 0$. A square containing a cell is $F(i, j) = 1$. Therefore, the initial condition for this system is defined as $F(1, 2) = F(2, 2) = F(3, 2) = 1$, and $F(i, j) = 0$ for all remaining i and j. We have now specified generation zero. To obtain generation one, we use each of the three rules where appropriate. To do this, we scan every square on the board and apply each of the three rules. Let us ignore the boundary cells for the moment (where i or j equal 0 or 4). Consider square $(3, 1)$. Can a cell be born here? A cell is born on an empty square if it touches exactly three cells. We see that square $(3, 1)$ touches only two cells [at $(3, 2)$ and $(2, 2)$], so no new growth occurs there. The only two squares where growth occurs are at $(2, 1)$ and $(2, 3)$. We cannot immediately fill in

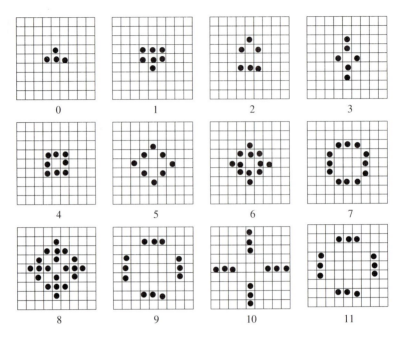

Figure 7.4   A CA that evolves into four blinkers.

these squares until we have processed all other squares on the board. To see why this is so, consider cell $(1, 2)$. This cell dies because it touches only 1 cell at $(2, 2)$, but it would have survived had we prematurely added $(2, 1)$. One way to think of the square processing is that *all squares are processed in parallel.* The blinker always loses its two outermost cells and rotates 90 degrees.

Figure 7.4 displays twelve generations (labeled 0 through 11) of a second CA with four cells in generation zero. The pattern grows steadily until parts of the population are too separate from one another. At that time (generation nine), four blinkers emerge. An algorithm for simulating cellular automata is shown below.

Algorithm 7.2. Cellular Automata Simulator
For each generation
        Let F be the array for the current generation
        For each row
                For each column
                        Let the current square be (i,j)
                        Apply all rules to F(i,j)
                        Write new and surviving F(i,j) to new array G(i,j)
                End for
        End for
        Copy array G to F
End For

This algorithm is simple, but it is not well optimized. We can improve it by noting:

1. There is no reason to copy array F. We just need to alternate between F and G, where F would become the primary array at even-numbered generations, and G would become the primary at odd generations.

2. If space is an issue, we need not use two full-sized arrays if we assume that the updating of cells is performed in an orderly fashion from bottom-up/left-right sequence for instance. Consider that we update array location $F(i, j)$ using a Moore neighborhood; what other locations will need to know the old value of $F(i, j)$ before we are permitted to update it? If we are moving left to right, then location $F(i - 1, j - 1)$ [where value $(0, 0)$ is in the lower left corner of the cell space] is the last cell that is required to update the new value of $F(i, j)$. All remaining (i, j) locations prior to $(i - 1, j - 1)$ will move beyond the range. This means that we need to have a temporary array of size $n + 1$ in addition to array F.

3. A *lookup table* can be constructed for all potential neighborhood patterns given a certain collection of rules. Consider the use of a von Neumann neighborhood. If we lay down a template onto the cell space where the template center is at (i, j), we note that there are 2^5 possible values of that template assuming a cell space with ones [a cell is present in square (i, j)] and zeros (a cell is not present in the square). A value of 00100, for instance, denotes the situation where $F(i, j) = 1$ and its surrounding neighbors are equal to 0. For the set of life rules, using a von Neumann neighborhood, the cell at (i, j) would die of isolation since it would be surrounded by four zeros. This means that given an input to the table of 00100, we would obtain an output of 0. The lookup table replaces the functional calculation or rule application with a hash table, thereby improving the efficiency of generating updates.

Margolus neighborhood.   A problem with the rules as we have defined them is that they are not robust enough to model physical principles such as conservation of mass or energy or the concept of uniform motion of cells across the cell space. To see why, consider an typical rule where $F(i, j)$ depends on its neighborhood. How can we ensure with each application of f that a certain subarea will maintain the same total number of cells? What we are forced to say with our existing method is "The new value of (i, j) is a function of its neighbors." What we want to be able to say is "The new value of a *substate* of cell space is a function of its old value." Thus, we want to operate on several squares at once so that we can use the *spatial context* to specify new cells. Spatial context plays the same role as it plays for context-sensitive L-Systems described in Sec. 7.3.2. The cellular automaton specified in Ch. 1 within Fig. 1.6(e) is such a CA; the rules specify operations on *blocks* instead of a single square.

The *Margolus neighborhood* is defined in terms of *blocks* as follows [221]:

1. The cell space is partitioned into a collection of finite, disjoint, and uniformly arranged pieces called *blocks*.

2. A *block rule* is given that looks at the contents of a block and updates the whole block rather than a single cell as in an ordinary CA. The same rule is applied to every block.

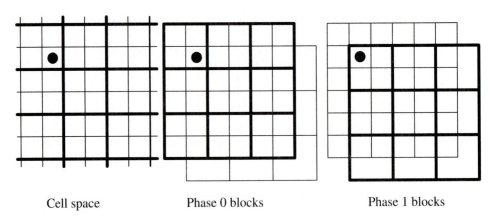

Cell space   Phase 0 blocks   Phase 1 blocks

Figure 7.5 Margolus neighborhood with two phases.

Blocks in the same phase do not overlap, and no information is exchanged between blocks.

3. The partition is changed from one step (phase) to the next, so as to have some overlap between the blocks used at one step and those used at the next one.

Consider 2×2 blocks created on a cell space. Two overlapping partitions are created in Fig. 7.5. The left-hand subfigure in Fig. 7.5 shows the original cell space. The remaining two figures show the phases or partitions that, when overlaid on one another, define the cell space. Successive generations alternate between the two grids (or phases). Note how the single cell is located in the lower right-hand corner of a block in phase 0, but it is in the upper left-hand corner of a block in phase 1. Given this type of neighborhood, a cell can approximate the motion of a particle undergoing uniform motion. A particle placed anywhere in the cell space will move at a constant rate diagonally across the space. For instance, the particle shown in Fig. 7.5 will travel downward and to right at each time step.

Diffusion.   CAs can model uniform motion using a Margolus neighorhood and alternate clockwise/counterclockwise block rotations. If we make these rotations random instead of regular, it results in a diffusion effect (or 2D random walk). The spread of heat or particles by diffusion is shown in Figs. 7.6 through 7.8 for the following times: 1, 8, 16, 32, 64, 128.

Lattice gas dynamics.   The type of particle motion afforded by the Margolus neighborhood is used to define the motion of gas particles using a CA approximation to the Navier-Stokes equation for fluid flow. The use of CAs permits a *bottom-up* approach to modeling spatial phenomena by creating simple rules that capture the basic conservation laws of mass, energy, and momentum. Consider the case where particles move diagonally and then meet each other. Figure 7.9 shows a set of block rules for diagonal motion. When a particle meets another one within the same block, it swaps diagonally with the other particle. The block transitions shown in Fig. 7.9 look very much like the production systems

(a)   Time 1   **(b)**   Time 8

Figure 7.6   Diffusion using a 2D CA random walk (times 1 and 8).

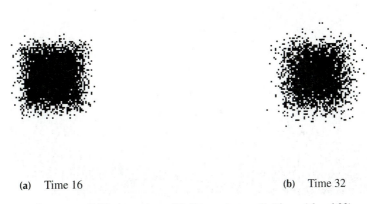

(a)   Time 16   **(b)**   Time 32

Figure 7.7   Diffusion using a 2D CA random walk (times 16 and 32).

that we saw in Ch. 4. By using rotational invariance, we can create fewer productions that represent the original set. Specifically, if we look at the second column of block rules, we see four productions, each with blocks containing a single particle. These four rules, however, all point to the same behavior—given a particle in a block, move the particle at the next time step to the opposite corner of the block. All four productions are identical if we rotate each block by increments of 90 degrees. In this sense, we say that they are *rotationally invariant*, and actually represent one rule—not four. The condensed set of six rules is shown in Fig. 7.10. The seventh production rule in Fig. 7.10 provides for what happens in a collision. There are two interpretations in this case: (1) two particles that are in diagonally touching squares will move right past each other, or (2) two particles will bounce off each other moving in the reverse directions. The visual effect is the same regardless of the interpretation. An alternate collision rule is shown in the second subfigure within Fig. 7.10, where the particles, after colliding, move off in 90-degree directions from

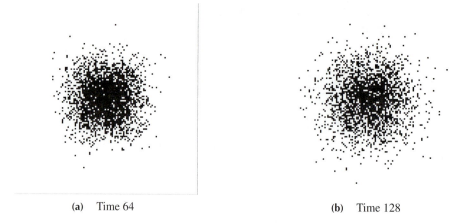

(a)   Time 64   **(b)**   Time 128

Figure 7.8   Diffusion using a 2D CA random walk (times 64 and 128).

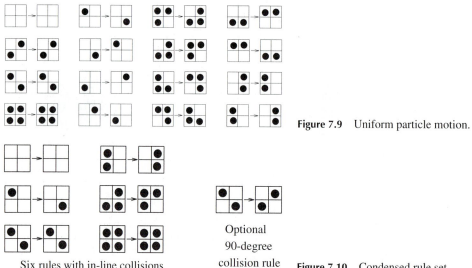

Figure 7.9   Uniform particle motion.

Six rules with in-line collisions

Optional
90-degree
collision rule

Figure 7.10   Condensed rule set.

their original paths. At first, one wonders why we might want to make the particles move off at an odd angle. Due to the regularity in the lattice, we will obtain unrealistic dynamics of colliding particles if they are not permitted to *scatter*. By having the colliding particles move off at different directions, we provide the sort of scattering found in real particles where the collision is not precisely *head on*.

Are these methods physically accurate? CAs can be as accurate as the laborious calculations necessary using other schemes. A key point is that the CAs provide an exploratory vehicle for simulating spatial processes, and rules are easily manipulated to answer "What if" scenarios. The rules covered satisfy the basic conservation laws as well, in that mass and energy can be approximated by the number of particles in the simulation. As long as the number of particles does not change, energy is conserved. Momentum is also conserved

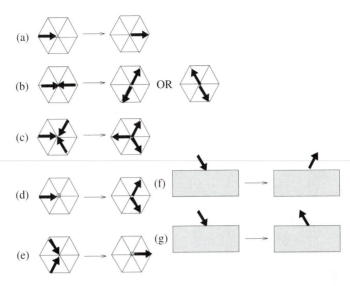

Figure 7.11   Rules for discrete fluid flow on a hexagonal lattice.

in that particles (which have the same "weight") react in physically realistic ways due to the simplification of uniform rate applied to all cell space particles.

The main difference between using a CA technique to simulate physical phenomena and using a fluid flow approach such as Navier-Stokes is that the CA method is *bottom up* or microscopic, whereas the Navier-Stokes approach is *top down* or macroscopic. The Navier-Stokes equation defines how fluid flows through volumes of a liquid, whereas CA methods begin with very simple interactions. As you require more realism in your model, you go beyond the basic CA to include particles with multiple speeds and other attributes. This provides more realism but at the expense of simulation time. The qualitative, macroscopic features of actual fluid flow can be better characterized by a CA constructed on a hexagonal lattice. This is because the hexagonal lattice provides for a good characterization of vorticity at a microscopic level. Instead of a rectangular space with cells, consider a hexagonal space with arrows where each arrow represents uniform motion in the stated direction. Figure 7.11 (derived from [47]) shows rules of motion for the arrows. Note the following meanings for Fig. 7.11:

(a) Rule of transport at constant speed.

(b) Scattering rule for two head-on particles. One of two possible scattering rules is chosen at random.

(c) Scattering rule for three particles.

(d) Rule for a particle striking a stationary particle.

(e) Rule for two particles. The outcome is that one particle stays fixed (denoted by the hollow circle) and the other proceeds at an angle equal to the average of the incoming particles.

(f) Rule for a particle against a fixed boundary (rebound).

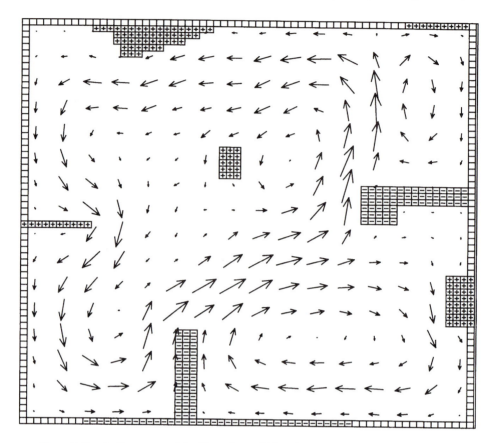

Figure 7.12 Lattice gas model of heat flow within a box containing sources and sinks. (Reprinted with permission of Harlan W. Stockman, Sandia National Laboratories.)

(g) Same as (f) except that the particle obeys the "angle of incidence equals angle of reflection" rule.

Figure 7.12 displays a lattice gas automaton in action—convection in a box. Boxes with plus (+) signs represent heat sources and boxes with minus (−) signs represent heat sinks. Empty boxes represent heat insulators.

7.2.4 Ising Models

In the study of statistical physics, the Ising model provides a technique using vast numbers of magnetic *spins*. An array (or lattice) of spins models a magnetic material. The Ising model contains a set of N sites where each site S_i is labeled as an up arrow or down arrow. Up arrows are $+1$ and down arrows are -1. The state of the system at any one time is shown as a lattice of arrows. When all arrows in a system point in the same direction, the system is said to be in a stable, minimal energy state. Figure 7.13 displays three material configurations or states. The first two states are minimal energy and the third reflects an in-between state that was achieved through the external introduction of heat. The energy E

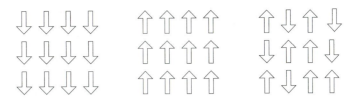

Figure 7.13   Three sample Ising models.

of the system comprising a material object is defined as

$$E = -\frac{1}{2} \sum_i \sum_j J_{ij} S_i S_j - B \sum_{i=1}^{N} S_i$$

where B is a uniform magnetic field and J_{ij} represents the connectivity of the material. In our case, we will define J_{ij} to be 1 where spin S_i is connected (i.e., physically adjacent) to spin S_j and zero otherwise. This means that with the term $\sum_i \sum_j J_{ij} S_i S_j$ we are really interested in the spin value products of all spins connected to one another. In a one-dimensional site space, a spin will have two neighbors, and in a two-dimensional site space, a spin will have four neighbors. In accordance with the Boltzmann distribution, we define a probability $P(S_i)$ that spin S_i remains in its current state (whether -1 or $+1$):

$$P(S_i) = \frac{1}{1 + \exp[-(\sum_j J_{ij} S_i S_j / k_B T)]}$$

where T is temperature in Kelvin and k_B is Boltzmann's constant. Let's assume, for the moment, that $T = 0$ and that we are concerned with the one-dimensional Ising model $\uparrow\downarrow\uparrow$. In this model there are three spins: $S_1 = +1$, $S_2 = -1$, and $S_3 = +1$. Now, let's compute the probability that S_2 will change its spin. We need to multiply the spin value of S_2, -1, with each of its two neighbors and then add them together: $(-1)(+1) + (-1)(+1) = -2$. The probability is defined:

$$P(S_2) = \frac{1}{1 + \exp[2/K_B T]}$$

With $T = 0$, the probability converges to $1/(1 + \exp[+\infty]) = 0.0$. This means that the probability is infinitely small that spin S_2 will remain in the same direction. Instead, the spin wants to change its direction and sign to minimize the configuration energy. This direction change will yield $\uparrow\uparrow\uparrow$. An analogy with springs will intuitively validate this calculation. Consider that when two spins are connected their arrowheads are connected with a small spring. With such a configuration, the middle spin S_2 in model $\uparrow\downarrow\uparrow$ will have two stretched springs extending to the arrowheads of S_1 and S_3. An extended spring represents a positive energy, and the system can lower its energy by flipping, thereby relaxing the two springs. Conversely, S_2 within a system such as $\uparrow\downarrow\downarrow$ can flip arbitrarily since, by flipping direction, the total energy is not increased. This guess is made firm by calculating the probability of

$$P(S_2) = \frac{1}{1 + \exp[-0/K_B T]} = \frac{1}{1 + 1} = 0.5$$

As $K_B T$ approaches zero, we can create production rules for a one-dimensional, three-spin material as shown in Fig. 7.14.

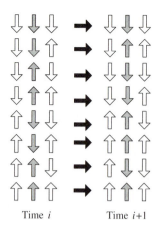

Figure 7.14   Eight possible production rules for spin flip dynamics.

Time i Time $i+1$

7.2.5 Partial Differential Equations

Partial differential equations (PDEs) are an extension to ordinary differential equations where time is not the only independent variable in the equation set. In ODEs, we saw that there were perhaps several variables—all dependent upon time. For PDEs, we will continue to have a time dependency but also spatial dependencies as well. Most PDEs specify a relation between rate with respect to time and rate(s) with respect to position. To solve PDEs in simulation, we use the finite difference method over time and the space of independent variables. For instance, note the following discrete approximations for a function Φ that is dependent on two variables t and x:

$$\frac{\partial \Phi}{\partial x} = \frac{\Phi_{i+1}^n - \Phi_i^n}{\Delta x} \tag{7.1}$$

$$\frac{\partial^2 \Phi}{\partial x^2} = \frac{\Phi_{i+1}^n - 2\Phi_i^n + \Phi_{i-1}^n}{(\Delta x)^2} \tag{7.2}$$

$$\frac{\partial \Phi}{\partial t} = \frac{\Phi_i^{n+1} - \Phi_i^n}{\Delta t} \tag{7.3}$$

To solve equations use these equivalences, substitute all terms with their differences and then solve the equation using explicit or implicit means. In simulation, we are concerned with time dependency so we will want to consider equations that contain t. Usually, our systems will include one of two terms: $\partial \Phi / \partial t$ or $\partial^2 \Phi / \partial t^2$. Consider the heat equation in two dimensions (x, y):

$$\frac{\partial^2 \Phi}{\partial x^2} + \frac{\partial^2 \Phi}{\partial y^2} = \frac{\partial \Phi}{\partial t} \tag{7.4}$$

Equation (7.4) represents the time-dependent behavior of variable Φ (where Φ is temperature) for a flat plate. Heat is applied to the plate using boundary conditions which assign initial temperature values to different parts of the plate. For instance, one side of the plate might initially be hot and the rest of plate at ambient temperature. By translating (7.4) into

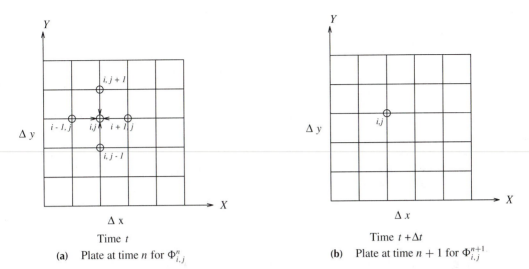

(a)   Plate at time n for $\Phi_{i,j}^n$

(b)   Plate at time $n + 1$ for $\Phi_{i,j}^{n+1}$

Figure 7.15   Finite Difference discretization of a flat plate

differences, we arrive at

$$\frac{\Phi_{i-1,j}^n - 2\Phi_{i,j}^n + \Phi_{i+1,j}^n}{(\Delta x)^2} + \frac{\Phi_{i,j-1}^n - 2\Phi_{i,j}^n + \Phi_{i,j+1}^n}{(\Delta y)^2} = \frac{\Phi_{i,j}^{n+1} - \Phi_{i,j}^n}{(\Delta t)} \qquad (7.5)$$

Rearranging terms in (7.5) yields the solution for time $n + 1$ ($\Phi_{i,j}^{n+1}$) in terms of the past value $\Phi_{i,j}^n$ and the four nearest von Neumann neighbors of $\Phi_{i,j}^n$. The term $\Phi_{i,j}^n$ is multiplied by -4 like terms are added together. Figure 7.15 displays the plate and the indexes for Φ at time n and $n + 1$. In this example, we see a similarity between cellular automata (CA) and PDEs. Namely, we have an effective CA rule that says "Let the new value of cell (i, j) be equal to the old value times -4 plus the values in the von Neumann neighborhood of the cell." This suggests a relationship between CAs and PDEs. PDEs can be considered an algebraic subclass of CA in that PDEs are involve equations with algebraic operations. CAs can be defined in a similar way, but they are more general in that CAs support rule-based definitions. Rules permit discontinuities and allow for nonlocal interactions which are not amenable to PDEs since the PDE state space is continuous.

Vector calculus.   The language of mechanics is vector calculus. Refer to Sec. 7.6 for references with good in-depth coverage of vector methods. We defined vector quantities for ODE models and our systems became balances for vectors and derivatives of vectors. For PDEs, however, our reliance on vector methods is more substantial since we will introduce new vector concepts that are essential for cleanly formulating balance equations. For our needs, a function of several variables will include the variable for time t along with spatial variables. Since we must visualize in a maximum of three dimensions, we will limit our discussion to a function of the form

$$\Phi = f(x, y) \qquad (7.6)$$

Φ is a function of x and y, so Φ could be the temperature or pressure at any point x on a flat plate, for instance. Φ is called a *scalar field* because for any point in the domain (x, y) there is a single scalar value defined by (7.6). The *gradient* of a scalar field Φ is denoted either grad Φ, or more commonly, $\nabla\Phi$, and it is defined as follows:

$$\nabla\Phi = \left(\frac{\partial\Phi}{\partial x}, \frac{\partial\Phi}{\partial y}\right) \tag{7.7}$$

A gradient is an operator which takes as input a scalar field and produces a vector field. Each vector at any point (x, y) reflects the gradient at that point. This means that there is a vector whose tail is located at a value (x_0, y_0) whose length represents the length of the hypotenuse for the orthogonal lengths $|\partial\Phi/\partial x|$ and $|\partial\Phi/\partial y|$. The length of a vector $\mathbf{v}$ at an arbitrary location (x_0, y_0) is

$$|\mathbf{v}| = \left(\left|\frac{\partial\Phi}{\partial x}\right|^2 + \left|\frac{\partial\Phi}{\partial y}\right|^2\right)^{0.5} \tag{7.8}$$

The gradient is a vector field with dimension one less than the scalar field dimension. This means that our 3D scalar equation in Eq. (7.6) results in a 2D vector field after we take $\nabla\Phi$. Figure 7.16 shows the gradient for a scalar field defined as $z = f(x, y)$.

Diffusion.   We already saw an example of diffusion using CAs in Sec. 7.2.3. Let's cover it from the *top-down* perspective of PDEs. Consider a one-dimensional balance equation for particles entering and leaving an arbitrarily small volume as shown in Fig. 7.17. In Fig. 7.17, we are viewing particles entering and leaving *at the same time*. The variables are defined:

1. ΔV is the volume and A the cross-sectional area. For simplicity, let $A = 1$ cm^2. This makes things easier since now $\Delta V = \Delta x$, where x is the sole independent spatial variable.

2. $\mathbf{J}(x, t)$ is the *flux*[3] of particles at location x and time t. Flux is measured as the number of particles per unit area per unit time. Therefore, we might say that the flux into the small volume in Fig. 7.17 is 5000 particles per square centimeter (cm^2) per second. Flux is therefore the same as rate (i.e., 5000 particles per second) with the addition of a unit of area within which the rate applies.

3. $c(x, t)$ is the concentration of particles at location x and time t. Concentration is measured as number of particles per unit volume (cm^3).

4. $s(x, t)$ is the number of particles created or destroyed per unit volume at location x and time t.

Given these functions, we build mass balance formula using: rate of change of the number of particles = rate of input to ΔV − rate of output from ΔV ± rate of creation/destruction in ΔV. The formula is defined as

$$\frac{\partial c(x, t)}{\partial t} = -\frac{\partial\mathbf{J}(x, t)}{\partial x} \pm s(x, t) \tag{7.9}$$

---

[3]Flux represents more than *flow rate* since flux encapsulates flow across an *area*.

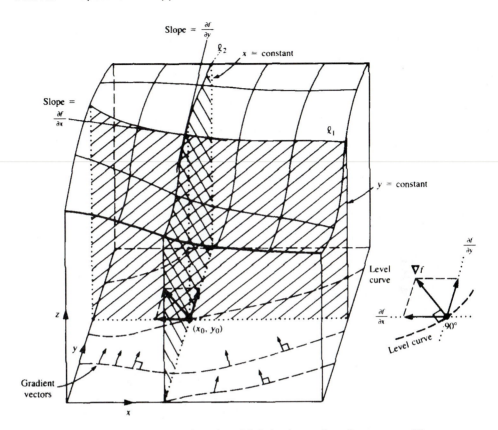

Figure 7.16   An interpretation of partial derivatives and gradient vectors. The surface intersects planes for which y is constant along curves l_1 and l_2. The slope of a tangent to l_1 is $\frac{\partial f}{\partial x}$, and the slope of a tangent to l_2 is $\frac{\partial f}{\partial y}$. (From Edelstein-Keshet, *Mathematical Models in Biology* ©1988, pg. 389. Reprinted with permission of Random House, New York, NY.)

We let $s(x,t) = 0$ for the following discussion. Equation (7.9) can also be obtained by starting with the number of particles N in ΔV. The number of particles between positions $x = a$ and $x = b$ is

$$N = \int_a^b c(x,t)\,dx \qquad (7.10)$$

Then the conservation equation states that the rate of change in number of particles equals the difference between the flux values at $x = a$ and $x = b$:

$$\frac{\partial N}{\partial t} = \mathbf{J}(a,t) - \mathbf{J}(b,t) \qquad (7.11)$$

Combining Eqs. (7.10) and (7.11) yields the integral conservation law:

$$\frac{\partial}{\partial t}\int_a^b c(x,t)\,dx = \mathbf{J}(a,t) - \mathbf{J}(b,t) \qquad (7.12)$$

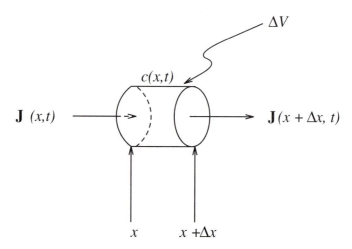

Figure 7.17 Particles entering and leaving volume ΔV at time t.

After further simplification of Eq. (7.12), by equating $a = x$ and $b = x + \Delta x$ as Δx approaches zero, we obtain

$$\frac{\partial c(x,t)}{\partial t}\Delta x = \mathbf{J}(x,t) - \mathbf{J}(x + \Delta x, t)$$

which, in turn, yields the same mass balance formula in Eq. (7.9).

   Now that we have a basic mass balance formula, we can define one common method of transport. A *method of transport* describes the way that fluid or particles move around in the system. Let's study a common form of transport called *diffusion*. With diffusion, particles of a liquid or gas move according to local gradients in particle concentration. Diffusion is omni-directional in that particles are not carried via a transporting medium such as blood through arteries. Instead, diffusion represents a kind of *random walk* spreading of the particles outwards from their initial state. Humans use diffusion to transport needed oxygen molecules to the bloodstream via the alveoli in the lungs. Diffusion is useful for transporting particles short distances in every direction.

   Flux $\mathbf{J}$, within transport by diffusion, is defined as

$$\mathbf{J} = -k\nabla c \tag{7.13}$$

Equation (7.13) defines that particle flow will tend to move down the concentration gradient, or from a higher concentration down to a lower concentration. From the one-dimensional balance equation, we may now construct the one-dimensional diffusion equation using the above substitution:

$$\frac{\partial c}{\partial t} = k\frac{\partial^2 c}{\partial x^2} \tag{7.14}$$

Convection.   While diffusion is appropriate for omnidirectional, short-distance flows, convection is appropriate for more narrowly focused, long-distance transport where the goal is to get particles from point A to point B quickly. Convective transport (convection) is the way that hot air moves around in a convection oven to heat food. It also defines the movement of fluid that is contained in pipes or blood vessels; blood moves in human bodies by conduction using arteries, veins, and capillaries. Consider traffic flow along a road in one direction. Modeling traffic flow in this way is no different than modeling particles in a blood flow stream; cars are carried along at the same rate as the vector field.

Flux $\mathbf{J}$, within transport by convection, is defined as

$$\mathbf{J} = c\mathbf{v}$$

where $\mathbf{v}$ represents the velocity vector field. In convection the flow of particles through a cross-sectional area ($\mathbf{J}$) is directly proportional to the velocity field. That is, particles flow along at the same rate as the velocity of the transporting fluid which carries the particles in its stream. From Eq. (7.9) we may now construct the one-dimensional convection equation using the above substitution:

$$\frac{\partial c}{\partial t} = -\frac{\partial c\mathbf{v}}{\partial x} = -\nabla \cdot (c\mathbf{v})$$

Fluid mechanics. Two equations define the behavior of *particles* of fluid: the conservation of mass and the conservation of momentum:

- *Mass.* Rate of change of mass in volume ΔV = rate at which mass enters ΔV − rate at which mass leaves ΔV.
- *Momentum.* Rate of change of linear momentum in volume ΔV = rate at which momentum flows into ΔV − rate at which momentum flows out of ΔV.

By encoding the terms in these formulas of constraint, we can derive the Navier-Stokes equations for fluid flow where *fluid* can be gas, liquid, or solid particulate matter.

The mass balance equation [Eq. (7.9)] is also known as the *conservation of mass* or *continuity equation* in fluid mechanics. Instead of thinking in terms of particle flow with concentration in one dimension $c(x, t)$, we refer to density $\rho(x, t)$ where density is a measure of mass per volume rather than particle count per volume. For our discussion we will assume an equivalence between these two measures. Also, we will treat $\mathbf{J}$ as a velocity vector field $\mathbf{v}$. Equation (7.9) is then more cleanly represented in an n-dimensional system as

$$\frac{\partial \rho}{\partial t} + \text{div}(\rho\mathbf{v}) = 0 \tag{7.15}$$

where "div" refers to divergence of the field and is equivalent to $\partial\mathbf{v}/\partial x$ in one dimension. div $\mathbf{v}$ is also equivalent to $\nabla \cdot \mathbf{v}$. For incompressible flows, ρ = constant, so (7.15) simplifies to div $\mathbf{v} = 0$.

The left-hand side of the *conservation of momentum* represents the rate of change of momentum. There are components that affect the left-hand side: (1) the velocity of the fluid particle with respect to time, and (2) the velocity of the particle due to the change in velocity potential as a function of position. The latter term is called *advection*[4]. On the right-hand side, we have the forces acting on the fluid particle: (1) the driving force representing the pressure gradient (pressing on the particle on all sides), and (2) a diffusion

---

[4]Horizontal transport of fluid.

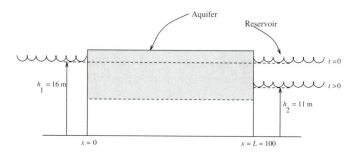

Figure 7.18 Reservoir with one dimensional flow. (Adapted From Wang and Anderson, Fig. 4.1 *Introduction to Groundwater Modeling* ©1982, pg. 71. Reprinted with permission of W. H. Freeman and Company, San Francisco.)

term representing diffusion based on velocity. The complete Navier-Stokes formulation is

$$\underbrace{\rho \frac{\partial \mathbf{v}}{\partial t}}_{\text{Rate of Change}} + \underbrace{(\mathbf{v} \cdot \nabla)\mathbf{v}}_{\text{Advection}} = \underbrace{-\nabla P}_{\text{Driving Force}} + \underbrace{\nu \nabla^2 \mathbf{v}}_{\text{Diffusion}} \qquad (7.16)$$

Groundwater modeling.   Many systems that involve fluid flow use simplified versions of (7.16) by assuming, for instance, that there is no advection or diffusion. Consider the response of an aquifer to a sudden change in reservoir level (from Wang and Anderson [227]). Figure 7.18 represents a one-dimensional problem where the reservoir (to the right of the aquifer) drops its level from 16 m to 11 m. What is the response, in terms of head height, in the aquifer? The equation is determined based on mass balance: the amount of water present in volume ΔV within the aquifer is proportional to the amount of water leaving. This is represented as

$$\frac{\partial^2 h}{\partial x^2} = \frac{S}{T} \frac{\partial h}{\partial t} \qquad (7.17)$$

with boundary conditions $L = 100$ m, $h(0, t) = h_1$ and $h(L, t) = h_2$ for $t > 0$, where $h_1 = 16$ m, $h_2 = 11$ m. The initial condition is $h(x, 0) = h_1$ for $0 \le x \le L$.

In an n-dimensional groundwater problem, we use a more general formulation. We have added a new term $R(x, y, t)$ which represents the amount of water added or removed to/from the reservoir. This is similar to the $s(x, t)$ term used in Eq. (7.9).

$$\nabla^2 h = \frac{S}{T} \frac{\partial h}{\partial t} - \frac{R(x, y, t)}{T} \qquad (7.18)$$

7.3 ENTITY-BASED APPROACH

7.3.1 Overview

Entity-based models focus on describing the dynamics of entities moving around in space rather than how the (local) space changes as a result of entity dynamics. A simple example of an entity-based model type is random walk of a particle or the simulation of particle systems with nonlocal interactions. One does not sweep over the entire lattice to engage such particles. Instead, each particle behaves according to its own set of constraints, which might be different from neighboring particle functions. Spatial models of the constraint type are most often based on equational models where the equations are either ordinary or partial differential equations. In both types of models, there are normally a great number of spatial entities: particles or small segments created by tesselating or fragmenting a region of space.

7.3.2 L-Systems

Overview. Lindenmeyer Systems (L-Systems) represent a declarative type of spatial model used for modeling plant life and development [135]. L-Systems are entity-based spatial models in that we begin with an object whose state is defined by an initial condition ω, and continue simulation by growing an aggregation tree composed of objects. The recursion defined in the productions provides for a tree of objects which grows and is constructed as the method pN() is applied (where pN refers to production number N). After a tree has grown, other state updating methods can be applied for modifying the object states; however, in the majority of cases, growth and decay methods will continue to be applied in parallel with methods that, for instance, change the state of a plant or tree as a result of wind, water, or other environmental effects. Figure 7.19 displays a sample plant with branches and flowers and its corresponding object aggregation hierarchy at this time in the plant's development. The hierarchy is abbreviated to show the objects along with the relevant productions used for growth. The L-System production model for simulating the plant in Fig. 7.19 is defined below:

$$p_1 : \qquad \omega : a$$
$$p_2 : a \rightarrow I[A][A]A$$
$$p_3 : \qquad A \rightarrow I B$$
$$p_4 : \qquad B \rightarrow [C][C]$$
$$p_5 : \qquad C \rightarrow I K$$

Production p_1 represents the starting point of the L-system. p_2 provides the basic support structure for the plant with an internode, two angle branches, and a straight branch. Each branch is constructed with p_3. p_4 provides a two-branch structure at the end of each of the three branches just created, and each of these new branches contains a flower (via production p_5).

Productions p_1 through p_5 represent a declarative model. The objects in such a model will generally form a tree due to the recurrences in the productions. The tree is composed of several objects as seen in Fig. 7.19. The base (or root) object is the main trunk, where the tree starts its growth. As it grows it has the potential to generate several types of possible objects: leaves (L), internodes (I), and flowers (K). The growth in Fig. 7.19 is composed of internodes and flowers. The methods `grow()` and `update()` are defined

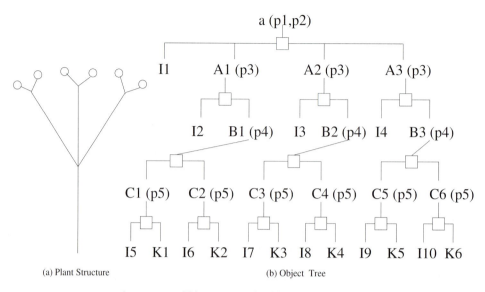

(a) Plant Structure   (b) Object Tree

Figure 7.19   Object aggregation hierarchy for a plant.

with production rules. The main difference between production rule methods as applied with a global working set, and our OO object-based approach is that only those productions associated with changing the state of a particular object are included within an object. This can be seen in Fig. 7.19 by noting the production numbers relevant to a particular part on the plant. Other productions that do not change an object's state are located elsewhere in those objects where the productions are applicable.

L-Systems are based on production systems described in Sec. 4.2.3 and represent simulation over time where time equals the generation number. A generation is defined as the rewriting of expression (i.e., string) using as many productions as can be fired, but not firing any one production more than once. L-Systems have been used for modeling many types of biological growth, including flowering plants, branching, tree growth, phyllotaxis, and cellular development. We follow the approach in [180, 181] by focusing on the geometric interpretation of L-Systems.

Basic L-System.   Consider the following L-System:

$$\omega : \qquad F - F - F \qquad\qquad (7.19)$$

$$p_1 : F \rightarrow F + F - F - F + F \qquad\qquad (7.20)$$

where ω represents the initial condition of the system defined by production p_1. Production p_1 says that wherever we see symbol F, we can replace it by $F + F - F - F + F$. ω is called the *initiator* and p_1 is the *generator*. One begins, at time 0, with the initiator and then recursively generates patterns by replacing each F at the previous time with the new

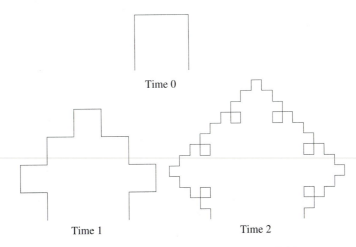

Figure 7.20   Three generations for L-System 7.19.

pattern. So, at time 1, we have a new pattern formed by taking each F^5 and replacing it with $F + F - F - F + F$. '+' means that we rotate counterclockwise by an angle δ, and '−' means that we rotate clockwise by δ. Figure 7.20 shows the first three generations 0, 1, 2 for this system where $\delta = 90°$. One can imagine a turtle [1] drawing a straight line while moving forward. This is equivalent to F. Then the turtle rotates $90°$ and moves forward again F. After a third time, the first generation is complete. Even though one turtle was used, this gives the impression that time moves according to the turtle's movement—which is incorrect. For our purposes, the system evolves with generation zero representing the *spatial system state* at time zero. At times one and two, the system appears as in Fig. 7.20. As in constraint systems based on difference equations, the system is defined only at discretely spaced intervals. One can imagine a continuous system where there is a continuous mapping between generations. Such a transformation is shown in Fig. 7.21 with some potential intermediate states.

   Features.   We have covered a basic L-System, and now we specify features of more fully developed L-System models that are necessary for modeling plant development.

   1. *Higher Dimensions.* For modeling in 3D, we need to specify rotational movement in two additional planes. If the original 2D plane is defined as $X \times Y$, then the symbols − and + represent rotation around the Z axis, which is perpendicular to $X \times Y$. Other symbols are used to denote rotations around the X and Y axes.

   2. *Branching.* The development of branches corresponds to the recursive act of creating self-similar substructures. What we need is a way of *saving state* while the branch geometry is defined. This is done using brackets '[' (save state) and ']' (return to saved state). A stack data structure can be used as a mechanism for state saving. The

---

[5] F represents a move forward of one step while drawing.

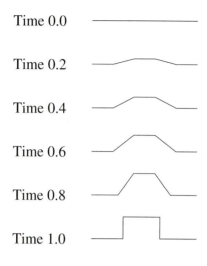

Time 0.0

Time 0.2

Time 0.4

Time 0.6

Time 0.8

Time 1.0   **Figure 7.21** A smooth transformation
                                            between system state at times 0 and 1.

form $F[-F]FFF$, for example, specifies that (a) a line segment is drawn, (b) another
line segment is drawn clockwise at an angle δ, and (c) three additional line segments
are drawn starting with the end of the first line segment, and not from the end of the
second line segment (i.e., the branch).

3. *Stochastic Processes.* A stochastic L-System uses probability values attached to
   production rewriting arrows. This is similar in form to what is done for Markov models
   and other stochastic processes. For instance, the following represents a situation
   where L may be rewritten in one of two possible ways (p_1 and p_2), each with a 50%
   chance of being applied:

$$\omega : \qquad F$$
$$p_1 : F \underset{0.50}{\rightarrow} F[+F]F \qquad\qquad (7.21)$$
$$p_2 : F \underset{0.50}{\rightarrow} F[-F]F$$

4. *Context Sensitivity.* A production can be fired only when specified within a context.
   $p_i : a < b > c \rightarrow d$ means that given a word or string with b contained within it,
   b is rewritten as d only when b occurs within the specified context where a and c
   surround b as in the word $ccabca$.

5. *Parametric Systems.* Parameters provide the same sort of flexibility available within
   the Prolog language in the form of first-order calculus. So far, all L-Systems can
   be seen as formulas specified in a purely propositional calculus. The expression
   $F(i) \rightarrow F(i + 1)$, where $F(0)$ is an initial condition, causes $F(0)$ to be rewritten
   as $F(1)$, which, at the following generation, is rewritten as $F(2)$, and so on until a
   termination criterion is reached. The termination criterion must be provided as an
   additional production, as in $F(i) : i = 10 \rightarrow G$.

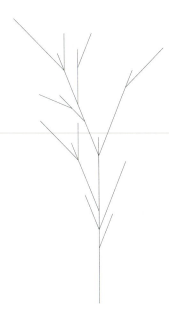

Figure 7.22   Plant growth at time 15.

Plant Models.   Hogeweg and Hesper [105] provide several examples of plant branching structure. We show a particular plant based on the following L-System as it forms, selecting generations 15, 20, and 26 for display in Figs. 7.22 through 7.24:

$$
\begin{aligned}
\omega : \quad & F1F1F1 \\
p_1 : \quad & 0 < 0 > 0 \to 0 \\
p_2 : \ 0 < 0 > & 1 \to 1[-F1F1] \\
p_3 : \quad & 0 < 1 > 0 \to 1 \\
p_4 : \quad & 0 < 1 > 1 \to 1 \\
p_5 : \quad & 1 < 0 > 0 \to 0 \\
p_6 : \quad & 1 < 0 > 1 \to 1F1 \\
p_7 : \quad & 1 < 1 > 0 \to 1 \\
p_8 : \quad & 1 < 1 > 1 \to 0 \\
p_9 : \quad & * < + > * \to - \\
p_{10} : \quad & * < - > * \to +
\end{aligned}
\tag{7.22}
$$

The branch angle parameter δ is set to $22.5°$. For purposes of rewriting, $+$, $-$, and F are ignored. The symbol $*$ is a *wildcard* symbol which can match either 0 or 1.

More interesting, and realistic, time-based sequences of growth are found in *developmental models* which include the features of *phase transition* and *signal propagation*. Real plants exhibit phased development where qualitative features for leaf growth, branching, and exhibit flowering occur in phases as the plant matures. Consider the following

Figure 7.23   Plant growth at time 20.

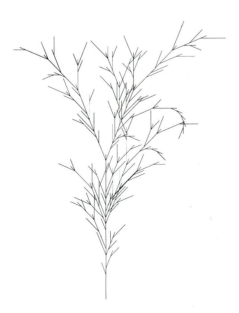

Figure 7.24   Plant growth at time 26.

L-System:

$$
\begin{aligned}
\omega : \quad & D(1)a(1) \\
p_1 : \quad & a(i) \quad : i < m \rightarrow a(i+1) \\
p_2 : \quad & a(i) \quad : i = m \rightarrow I[L]a(1) \\
p_3 : \quad & D(i) \quad : i < d \rightarrow D(i+1) \\
p_4 : \quad & D(i) \quad : i = d \rightarrow S(1) \\
p_5 : \quad & S(i) \quad : i < u \rightarrow S(i+1) \\
p_6 : \quad & S(i) \quad : i = u \rightarrow \epsilon \\
p_7 : \quad & S(i) < I \quad : i = u \rightarrow I S(1) \\
p_8 : \quad & S(i) < a(j) : * \rightarrow I[L]A \\
p_9 : \quad & A \qquad : * \rightarrow K
\end{aligned}
\tag{7.23}
$$

As terms are rewritten, the following sequence occurs from times 0 through 9 with $d = 4$, $m = 2$, and $u = 1$.

$$
\begin{aligned}
time\ 0 :\ & D(1)a(1) \\
time\ 1 :\ & D(2)a(2) \\
time\ 2 :\ & D(3)I[L]a(1) \\
time\ 3 :\ & D(4)I[L]a(2) \\
time\ 4 :\ & S(1)I[L]I[L]a(1) \\
time\ 5 :\ & I S(1)[L]I[L]a(2) \\
time\ 6 :\ & I[L]I S(1)[L]I[L]a(1) \\
time\ 7 :\ & I[L]I[L]I S(1)[L]a(2) \\
time\ 8 :\ & I[L]I[L]I[L]A \\
time\ 9 :\ & I[L]I[L]I[L]K
\end{aligned}
\tag{7.24}
$$

I refers to internode growth, which represents the growth, of the main shoot. L is the growth of a leaf, A is the apex of the plant, and K is a flower. It takes m time units to grow a single internode segment, or one can consider $1/m$ of a segment to grow at each time unit. The interval of time taken to grow the main shoot is termed the *plastochron*. Once the internode segment I has grown, it grows a leaf $[L]$, making $I[L]$. Leaves can all grow on the same side of the plant, or they can grow alternately from left to right. The method chosen is arbitrary since this facet is not specified in the model. So far we have grown a single primary shoot with a single leaf. Parameter d represents the delay. After the plant begins to grow, a signal is transmitted from the plant's base to its apex. In the previous time sequence, the signal starts at time 4 and gradually is shifted to the right of the developing string [$S(1)$ moves from left to right]. The signal propagates along the main axis at a rate of u time steps per internode. Eventually, since $u < m$, the signal pushes through the apex and causes a flower to grow.

Figure 7.25 displays a fully grown tree at time 8 using an L-System with ternary branching [181]:

- Tree segments are straight and their girth is not considered.

Figure 7.25   Tree growth at time 8.

- A mother segment produces two daughter segments through one branching process (i.e., ternary).
- The lengths of two daughter segments are shortened by ratios r_1 and r_2 with respect to the mother segment.
- The mother segment and its two daughter segments are contained in the same branch plane. The daughter segments form constant branching angles, a_1 and a_2, with respect to the mother branch.
- The branch plane is fixed with respect to the direction of gravity as to be closest to a horizontal plane. An exception is made for branches attached to the main trunk. In this case, a constant divergence angle α between consecutively issued lateral segments is maintained.

To demonstrate the realism and potential for declarative plant modeling, Fig. 7.26 displays a more complex model of a canopy of oil palms using stochastic modeling and incorporating 3D geometry for tree parts with ray tracing for final image rendering.

7.3.3 Gas Dynamics with Time Slicing

As a gas heats and expands, gas molecules move faster; gas molecules move slower when a gas cools. We can model a gas composed of an arbitrary number of molecules by using a square box of dimensions $0 \to 1$ on each side as shown in Fig. 7.27. The box will contain many molecules, each of which has some initial position and velocity in both the x and y directions. We will ignore interparticle collisions, gravity, and other short-range forces for

Figure 7.26   Oil palm tree canopy. (From de Reffye et al., *SIGGRAPH '88* ©1988, pgs. 151-158. Reprinted with permission of the Association for Computing Machinery, New York, NY.)

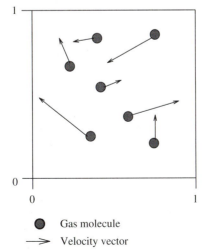

    ●   Gas molecule

    →   Velocity vector   **Figure 7.27**   Motion of gas molecules.

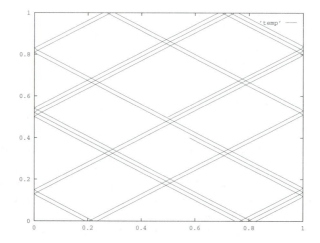

Figure 7.28   Gas molecule trace for one time unit of motion.

this model and specify that each molecule will travel in a straight line until it strikes one of the four walls. When a molecule hits a wall, it bounces off in the same manner as photons, striking a perfectly reflecting surface (i.e., angle of reflect equals angle of incidence). Using the time slice method, we can simulate the system by keeping track of position and velocity for each ball—in fact, position and velocity will serve as our state variables for this model. State variables will be incremented over time by using the following equation:

$$x_{new} = x_{old} + v_x \times \Delta t$$

This is simply the differential equation $dx/dt = v_x$ solved using Euler's method of integration. Normally, we will sample a uniform distribution to produce random initial positions and velocities for the molecules; however, we will trace the action of a single molecule by pre-setting its x, y position to $(0.5, 0.5)$ and its initial velocity vector to $(15.0, 10.0)$. Figure 7.28 displays the motion of the molecule for 1 time unit with a delta time of 0.001 time unit.

7.3.4 Gas Dynamics with Event Scheduling

We discussed in Sec. 7.3.3 a molecular dynamics example using the time-slicing approach to simulation. Is there a way that we can take this same model and formulate the simulation problem using an event scheduling approach? How would we define the events? We will define an event as the time when a wall is hit by a molecule. Let's consider the motion of a single molecule. The molecule travels in a straight line, and therefore its time until collision with a wall can be calculated without having to explicitly simulate all in-between positions as was done in the time-slicing method. Once the intersection time is calculated, the molecule moves directly to the wall, where it intersects the wall with its straight-line path. The molecule—if plotted over time—would appear to "jump" from one wall to another. Now, if we add more molecules into the picture, we will take the molecule that will intersect a wall in the shortest amount of time. We will not need to perform explicit sorting to determine

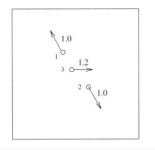

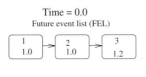

Figure 7.29   Step 1: Time to impact for 3 molecules.

this molecule since the sorting occurs as a natural result of initially scheduling all molecules to strike their respective walls at certain times. When next_event is called, we remove an event from the FEL. This event will be the molecule that is next in line to hit a wall. The simulation time is incremented, as usual, in the standard way and all molecule positions are updated to reflect this passage of time.

The advantage to the event scheduling method for simple molecular dynamics is that—if we are only interested in when molecules hit walls for counting purposes—the simulation time is decreased dramatically since we are not doing computation whenever all molecules are simply moving in space, each one not yet having hit a wall.

There are two key aspects of the algorithm to perform event scheduling simulation on gas molecules:

1. When one checks to see if the molecule has hit a wall, it is important to specify an error tolerance. For instance, the right wall is specified at $x = 1.0$; however, it is unwise to use the code if (x == 1.0) ··· since when the positional variable x is updated, it is very possible for x never to precisely equal the value 1.00000. Instead, x may take on the value 1.00001. In a program such as this, *numerical errors* play a vital role, and it is paramount that we adjust for them when checking for positional and temporal boundaries.

2. If, at any one time (given tolerance), no more than one molecule hits a boundary, then our algorithm is straightforward (see below): an event will occur which signifies a gas molecule striking a wall. All positions are updated accordingly, and the next event is considered. However, a consideration arises when two or more molecules strike a wall simultaneously. At first, it is not obvious that this should cause a problem with our general algorithm; however, consider the sequence of events pictures in Figs. 7.29 through 7.31.

Algorithm 7.3. Molecular Dynamics Event-Scheduling Simulator
Procedure Main
       Initialize SimPack
       Set initial positions and velocities for all molecules
       Initially schedule each molecule to strike its respective wall
       While more events to process do
              Get next event
              For event MOVE:
                    Determine time to impact for this ball

> *Check if this event time is the same as the previous*
> *event time*
> *If the two times are different:*
> > *Update all molecule positions based on this*
> > *molecule time to impact*
> > *Determine new time to impact for this*
> > *molecule*
> *End if*
> *Schedule this molecule to impact the next wall*

> *End While*
End Main

Note the progression from step 1 (Fig. 7.29) through step 3 (Fig. 7.31). In step 1, three gas molecules are scheduled to strike their respective walls. Molecules 1 and 2 are scheduled to hit opposite walls at the same time (within a tolerance for equal times). Molecule 1's position is updated along with the other two molecules (Fig. 7.30). The next event to occur is the updating of molecule 2's position. Since molecule 2 is also at a boundary, its subsequently scheduled time to impact will be *for the left wall* and not *for the bottom wall*. This means that regardless of the time taken to hits its next wall, molecule 2 will be scheduled accordingly. Molecule 3, meanwhile, is only a distance of 0.2 away from the right wall. If molecule 2 initiates an updating of position now, molecule 3 will move through the right wall instead of impacting it. The problem is that while molecule 3 is closer to a wall, molecule 2 gets priority since it was scheduled to occur before molecule 3. Note how this situation could have been avoided. Suppose that the initial times to wall impact in Fig. 7.29 had all been different, say (1) molecule 1 to strike in time 1.0, (2) molecule 2 to strike in time 1.1, and (3) molecule 3 to strike at time 1.2. Then there would be no problem in position updating; after molecule 1 updates all positions, molecule 2 has not yet hit the

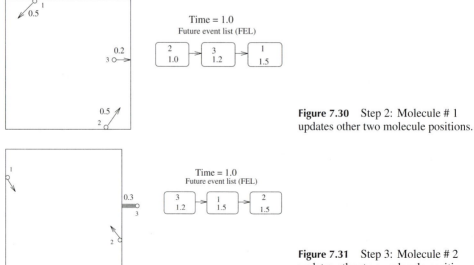

Figure 7.30   Step 2: Molecule # 1 updates other two molecule positions.

Figure 7.31   Step 3: Molecule # 2 updates other two molecule positions.

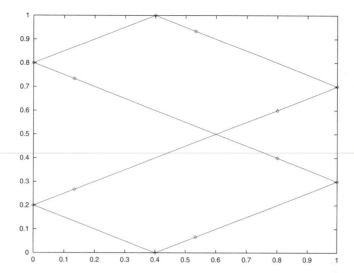

Figure 7.32   Molecule 0 trajectory with discrete event points.

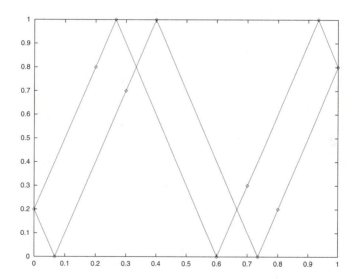

Figure 7.33   Molecule 1 trajectory with discrete event points.

wall (it has a time of 0.1 to go). Therefore, molecule 2 updates all molecule positions by a time to impact of 0.1. But what if we do have two of the same event times in the FEL? Then we should update positions only *for the first event at that time*. Subsequent events scheduled at the same time should not cause a position update. This issue of *duplicate* events caused related problems in Ch. 5 when we discussed digital logic simulation and the scheduling of the same logic block at the same time.

Figures 7.32 through 7.35 display the events and molecular trajectories of four molecules with initial conditions:

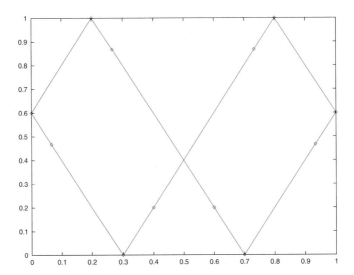

Figure 7.34   Molecule 2 trajectory with discrete event points.

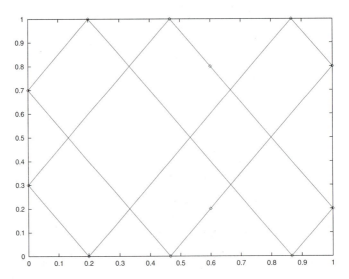

Figure 7.35   Molecule 3 trajectory with discrete event points.

1. Position: $0.4, 0.4$, velocity: $-2, -1$.
2. Position: $0.4, 0.6$, velocity: $-1, 3$.
3. Position: $0.6, 0.6$, velocity: $1, 2$.
4. Position: $0.6, 0.4$, velocity: $2, -3$.

The trajectories for each molecule represent simple patterns; events occur at the marked points specified in the trajectories.

TABLE 7.1   PLANETARY ORBITAL DATA

Number	Planet	Mass	Distance to Sun
1	Mercury	0.055	0.418
2	Venus	0.805	0.726
3	Earth	1.000	0.987
4	Mars	0.106	1.508
5	Jupiter	314.031	5.275
6	Saturn	94.026	10.038
7	Uranus	14.325	19.431
8	Neptune	16.944	30.259
9	Pluto	0.002	29.789

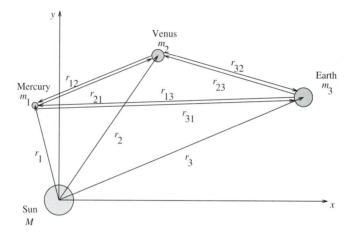

Figure 7.36   Solar system with three planets.

7.3.5 Planetary Orbital Mechanics

Newtonian mechanics are sufficient to build a model of the solar system. The basic three laws provide all that is necessary in order to construct a simulation. Let us consider the Sun along with the innermost three planets: Mercury, Venus, and Earth. Table 7.1 displays some basic planetary data for planets orbiting the Sun. Figure 7.36 displays the configuration of the Sun and the first three planets where mass values are defined relative to Earth's mass and shown in parentheses: Sun (M), Mercury (m_1), Venus (m_2), and Earth (m_3). Forces for each planet with the Sun are shown pointing toward the planet $\{r_1, r_2, r_3\}$ since the Sun will be stationary for our reference frame, where the Sun occupies the $(0, 0)$ spot on the axes. We proceed as we did for the ballistics example in Ch. 6 by specifying the net forces on m_1, m_2, and m_3:

$$\mathbf{f}_1 = -\mathbf{r}_1 - \mathbf{r}_{12} - \mathbf{r}_{13} \qquad (7.25)$$

$$\mathbf{f}_2 = -\mathbf{r}_2 - \mathbf{r}_{21} - \mathbf{r}_{23} \qquad (7.26)$$

$$\mathbf{f}_3 = -\mathbf{r}_3 - \mathbf{r}_{31} - \mathbf{r}_{32} \qquad (7.27)$$

$\mathbf{f_1}$, $\mathbf{f_2}$, and $\mathbf{f_3}$ are the three vectors representing the net force vectors for each of the three planets. The only forces acting upon a planet are the gravitational forces exerted upon it by the other planets and the Sun. The law of gravitation states that the force f between two masses m_1 and m_2 is defined as $f = m_1 m_2 G \mathbf{r}/r^3$, where $\mathbf{r}$ is the vector pointing from one mass to the other and G is the gravitational constant ($G = 6.67 \times 10^{-11}$ m^3/ kg $\cdot$ s^2). Given this equivalence and $\mathbf{f} = m\mathbf{a}$, our equations become

$$m_1 \mathbf{a_1} = -\frac{m_1 M G \mathbf{r_1}}{r_1^3} - \frac{m_1 m_2 G \mathbf{r_{12}}}{r_{12}^3} - \frac{m_1 m_3 G \mathbf{r_{13}}}{r_{13}^3} \qquad (7.28)$$

$$m_2 \mathbf{a_2} = -\frac{m_2 M G \mathbf{r_2}}{r_2^3} - \frac{m_2 m_1 G \mathbf{r_{21}}}{r_{21}^3} - \frac{m_2 m_3 G \mathbf{r_{23}}}{r_{23}^3} \qquad (7.29)$$

$$m_3 \mathbf{a_3} = -\frac{m_3 M G \mathbf{r_3}}{r_3^3} - \frac{m_3 m_1 G \mathbf{r_{31}}}{r_{31}^3} - \frac{m_3 m_2 G \mathbf{r_{32}}}{r_{32}^3} \qquad (7.30)$$

If a unit vector $\hat{\mathbf{r}}$ is used instead of $\mathbf{r}$, the force takes on the inverse square realization.

7.3.6 Particle System Algorithms

The N-body problem has a long history in science. First, the problem is intractable from a closed-form perspective (i.e., direct computation) for systems with greater than two bodies. Therefore, simulation is used to determine the effects of applying forces and velocities to bodies.[6] We will discuss N-body simulation under the rubric of particle systems. Second, bodies and particles are used virtually everywhere. For instance, a particle can be a soil particle, gas particle, stellar object, quantum particle, or fluid particle. In the gas dynamics simulation, our particles were not under the force constraints of other particles; we ignored them since, at high temperatures, gas molecules move quickly and are more unlikely to be affected by other particles (for long-range forces) than they would be if the particles were stars orbiting in galaxies. Figure 7.37 shows a particle simulation in progress depicting the formation of a spiral galaxy.

In Sec. 7.3.5 we provided the equations for orbital mechanics with a small number of bodies given force conservation. We did not explicitly go over an algorithm for the orbital mechanics problem, but we will provide one here: for each body, add up all forces exerted from other bodies. This is quite straightforward, but it is expensive if we wanted to analyze anything other than a small number of bodies. If there are N bodies, we will be performing $N(N-1) = N^2 - N$ calculations to determine the effect of force on each body. Therefore, the direct approach is $O(N^2)$. The following algorithm specifies the "particle-particle" approach. In this algorithm, we improve the efficiency slightly by noticing the symmetry of forces between any two objects, resulting in half the calculations of the naive approach.

Algorithm 7.4. Particle-Particle Method
Procedure Main
  Compute Forces
  Perform Integration Step

---

[6]We will interchange the terms *body* and *particle*.

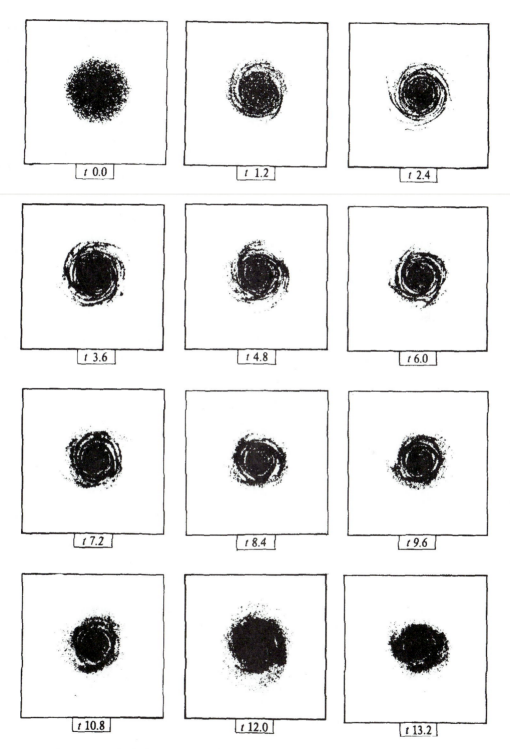

Figure 7.37 Evolution of a "cold disk" galaxy stabilized with a heavy diffuse halo showing the development of filamentary ring and spiral structures. (From Hockney and Eastwood, *Particle Simulation* ©1988, pg. 433. Reprinted with permission of Adam Hilger, Philadelphia, PA.)

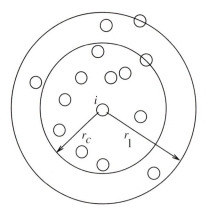

Figure 7.38 Building a Verlet neighbor list based on distance.

> *Update Time Counter*
>
> *End Main*
> *Procedure ComputeForces*
> > *Clear force accumulators*
> > *Sum up forces as follows:*
> > > *For $i = 1$ to $N - 1$ do*
> > > *For $j = i + 1$ to N do*
> > > > $F_i = F_i + F_{ij}$
> > > > $F_j = F_j - F_{ij}$
>
> *End ComputeForces*
>
> *Procedure IntegrationStep*
> > *Integrate using one of the methods in Ch. 6*
> *End IntegrationStep*

The particle-particle method can be enhanced in several ways since it has a bottleneck with regard to the shear number of interparticle interactions which must be calculated. Since the particles for a simulation usually represent an aggregation of a lower-level physical component (such as a molecule particle aggregate representing a set of atomic particles), we can use the concept of aggregation to further reduce the number of particles. This is proposed in the particle-mesh method to be discussed. If the force is electrical or gravitational proportionality, we can limit our *search* for particles which can affect particle i by using a *cutoff*. That is, we use only those particles within a circle of radius r whose center is located at particle i. All outside particles are deemed to have less of an impact on the force calculation, due to their distance and the inverse square relationship between force and distance which exists for gravitational and Coulombic forces. Let's discuss two methods that can help reduce the time complexity: *neighbor lists* and *adaptive time scaling*. The first type of neighbor list is the Verlet neighbor list, which involves storing a linked list for each particle in the system. Figure 7.38 shows two radial vectors (r_c and r_l). After the neighbor list is constructed for particle i, we need only search its linked list neighbors. The linked list is just a way of bounding the search for other particles which have an influence

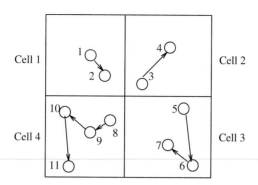

Figure 7.39 Building a neighbor list based on partition.

or interaction with particle i. Any particle outside the r_l radius will not be on the particle i's list. Now, there is the issue of when to update the lists because an update on every time slice would be expensive. When the sum of the magnitudes of the two largest particle displacements within a cell is greater than $r_l - r_c$, the list is updated. Intervals of between 10 and 20 steps are reported in [4].

A related approach uses a rectangular partitioning where each partition is known as a *cell*. Each cell has eight neighbors in its neighborhood, and each cell contains a linked list of its current contents. This approach is shown in Fig. 7.39. The arrows in Fig. 7.39 do not represent forces—they are there to illustrate the internal data structure of the linked list. The *time scaling* approach applies the concept of distance partitioning, as shown in Fig. 7.38, to time. Each particle position will be updated by integrating along the velocity vector. This integration uses a time slice, and we can use smaller time slices for calculating forces between a given particle i in Fig. 7.38 and particles in the immediate neighborhood of i (inside the r_l circle). Larger time slices are used for any outside particles considered during the force calculation.

We will now discuss the "particle-mesh" approach and the "tree-code approach." There are two important aspects to large groups of particles. The first aspect is that we can think of all the particles, along with their forces, as we have just done in the "particle-particle" method. The other approach is to represent, not the particles themselves, but rather the potential field that results from particles being placed inside it. A potential field occurs in many areas, such as gravitation and electromagnetics. For gravitation problems in two dimensions it is easiest to imagine a three-dimensional surface, where the first two dimensions are the x and y positions of the particle in space and the z dimension is the value of the potential. The potential is a *scalar field*. An example potential is where z is the height of a hill for some x, y area of land. In this case, all particles fall vertically toward the largest particle (Earth). Consider creating a potential in space from Earth, Moon, and Sun. This can be done intuitively by letting the potential be a flat sheet of rubber. Onto the sheet, you then place representative masses for the Sun, Moon, and Earth. The Sun would cause a deep depression since it has the greatest mass, whereas the moon would hardly cause any depression at all. If one places a new mass on the sheet, it will run downhill toward a nearby mass. Mathematically, *downhill* means away from the gradient. So that if our potential is ϕ, we move toward $-\nabla\phi$. In fact, the forces between masses are directed along the gradients. The "particle-mesh" approach utilizes these gradients to determine the forces for a set of masses. Given the forces, we can then use $\mathbf{F} = m(d\mathbf{v}/dt)$ to integrate new positions for each particle. The potential field changes since the masses have moved, and the process repeats itself. The algorithm using the nearest grid point (NGP) to assign charges (in the case of electrodynamics) or masses (in the case of gravitational fields) is defined as:

Algorithm 7.5. Particle-Mesh Method (NGP Scheme)
Procedure Main
        Assign charge (mass) to mesh by assigning it to
              the nearest grid point (called the NGP scheme)
        Solve the field equation: $\nabla^2\phi = -\rho/\epsilon_0$
              where ϕ is the potential scalar field
        Calculate force field using $F = -\nabla\phi$
        Interpolate to find forces on particle
        Integrate for new position
End Main

Tree-code approaches *group* particles and redefine these groups to be new pseudoparticles that will contain a mass and a center. The trees most often used are the quad-tree for 2D situations and oct-trees for 3D. At the tree leaves, we have single particles. At each internal tree node, we have a pseudoparticle that derives its mass from its children's combined mass and its position from a center of mass (COM) calculated from the children's positions. The idea behind tree-codes is that when calculating all forces on particle p, sometimes it is possible to add in a force between p and a pseudoparticle's COM rather than to all children of the pseudoparticle. For forces that depend on $1/D^2$, where D is distance between two particles, one can use higher nodes in the tree when particle p is far away from the other particles. This results in much shorter searches and, on average, means that for any p, the search for the right pseudoparticle to use will be $O(N \log N)$ rather than $O(N^2)$ as in the particle-particle approach. An algorithm by Barnes and Hut [15] defines a procedure using this approach:

Algorithm 7.6. Tree-Code Method
Procedure Main
      BuildTree
      CalculateForces
End Main
Procedure BuildTree
      Top Down Pass:
           Begin with one root cell, and load particles into
              the root cell
           If two particles fall into the same cell then
              divide the cell into 4 square cells (for 2D), or
              divide the cell into 8 cubicle cells (for 3D)
           A cell will hold mass and center of mass information

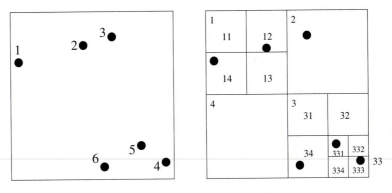

Figure 7.40   Labeled bodies in 2D space with partitioning.

and information for its children
Bottom Up Pass:
Calculate the center of mass for cells based on children
This creates a tree where cells are pseudoparticles
End BuildTree

Procedure CalculateForces
The force on any particle p is calculated as follows
using a top-down tree search
Define: *l = the length of a cubicle cell,*
D = the distance from the cell's COM to p,
θ = *fixed accuracy parameter,*
If l/D < θ then include this interaction between cell
and particle p in the total force being accumulated for p
Else
Resolve current cell into its children and recursively
examine each child using same criteria above
End CalculateForces

Figure 7.40 illustrates the tree-code method in 2D space. Each filled black dot on the left-hand square represents a body in space. There are six labeled bodies. In the first phase, the tree is built by performing two passes. The first pass creates a tree by recursively partitioning the space. The root node represents the original unpartitioned square. New nodes are added to the tree at the root. Therefore, consider body 1 as being the first one entered as the root node. Then another body, say 3, is added. This causes a tree to develop with one root node and four subnodes each representing a quarter of the original space. Since body 3 found another body (1) in the root node position, this caused the recursive subdivision. The first subnode would now contain body 1; the second subnode would contain the body 3; and the third and fourth subnodes would be empty. As new bodies are added top-down to the tree this procedure continues: if a body encounters another body in

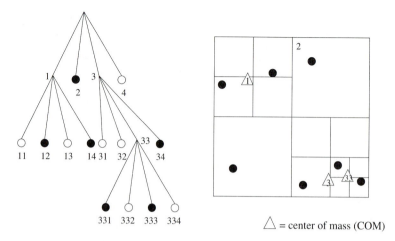

$\triangle$ = center of mass (COM)

Figure 7.41   Quad-tree and COM locations.

a treenode, it causes a new subdivision. The left-hand figure in Fig. 7.41 displays the tree created after *all* six bodies have been entered in the tree. At this point, we must start from the leaf nodes and proceed upward to create COM values for all internal tree nodes. The right-hand figure in Fig. 7.41 shows the positions of the COMs using triangular markers. Note that node 33 represents a COM from bodies 4 and 5, but that node 3 represents a COM calculated from body 6 and the COM created from bodies 4 and 5. Finally, let's consider the force calculation which uses the constructed quad-tree. Say that we are updating the position of body 3. Normally, this would require a search of all remaining five bodies to obtain the total force on body 3; however, given the tree-code algorithm we only need visit to a particular depth based on the distance of body 3 from the other bodies. In general, the bodies nearest body 3 will necessitate a deeper descent into any subtrees which house those bodies. Even though COM 1 (Fig. 7.41) stores the COM for bodies 1 and 2, we would need to descend this subtree to calculate 3 $\leftrightarrow$ 1 and 3 $\leftrightarrow$ 2 forces because body 2 is very close to COM 1 (i.e., l/D is not less than θ). However, the only remaining force calculation would be between body 3 and COM 3 since this distance is large enough to avoid a recursive search.

7.3.7 Rigid Body Mechanics

The dynamics of particles are based on Newton's laws of motion. If we choose to study a *system of particles*, we can choose to do so by using the particle methods previously discussed. However, there is a special case where all the particles remain a relative distance to one another even though the particle system—as a whole—moves. This kind of particle system is called a *rigid body*. A rigid body is one that does not deform if we create loads using forces (linear) and torques (angular). Figure 7.42 displays a rigid body whose center of mass is C and whose local reference frame is $x'y'z'$. The frame with origin O is called the inertial reference frame since it can be translated with constant velocity but not accelerated.[7]

---

[7]The inertial frame xyz cannot rotate since this implies a force and, therefore, an acceleration.

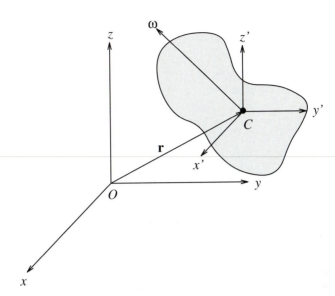

Figure 7.42   Free rigid body in local reference frame $x'y'z'$.

The rigid body in Fig. 7.42 has an angular velocity of ω. The ω vector is the spin axis, passing through C, for the body. The dynamics for a free rigid body are given as follows:

$$m\ddot{\mathbf{r}} = \sum_{i=1}^{n} \mathbf{F}_i \qquad (7.31)$$

$$I\ddot{\theta} = \sum_{i=1}^{n} \mathbf{r}_i \times \mathbf{F}_i \qquad (7.32)$$

Equation (7.32) represents the conservation of linear momentum for the rigid body. To apply this equation, the body can be considered to be a particle with all of its mass centered at C. Equation (7.32) says that the change in linear momentum for a rigid body is equal to the applied forces directed toward the center of mass. Equation (7.33) represents the conservation of angular momentum. I plays a similar role to m in Equation (7.32) except that since the body mass is distributed, a matrix (inertia tensor) must be specified. I is the inertia tensor. The term $\mathbf{r}_i \times \mathbf{F}_i$ is the torque τ for force i. The radii $\mathbf{r}_i$ are vectors whose base is at C and whose tip is located somewhere within or on the body. Angular velocity $\omega = \dot{\theta}$ and angular acceleration $\alpha = \ddot{\theta}$.

Simulation of a rigid body proceeds in a similar way to the simulation of a system of particles: (1) the forces are specified, (2) the differential equations are set up given these forces, and (3) values for vectors $\mathbf{r}$ and θ are calculated through numerical integration. Equations (7.32) and (7.33) are sufficient for simulating rigid bodies that are translating and spinning through a medium such as air, but what happens when a body strikes a boundary object or another body? In these cases, momentum must be conserved as usual. Figure 7.43 illustrates a collision of two bodies, each with a local reference frame. There is a point of

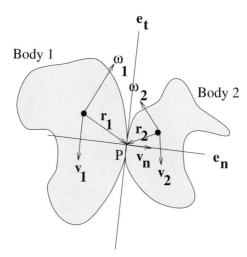

Figure 7.43 Two colliding rigid bodies.

first contact P and two key features: e_t the tangent plane and e_n the normal line which is orthogonal to e_t. The following scenario captures the situation in Fig. 7.43. Body 1 and body 2 are initially moving within a system. Each body is spinning and translating. Forces and torques may be applied at any time. At time t_c an event occurs at point P, where one body comes into contact with another. This event is called a collision. At that precise time, body 1 will have certain characteristics: it will be spinning at an angular velocity of ω_1 and it will have a linear velocity of $\mathbf{v}_1$. The vector $\mathbf{r}_1$ points from the center of mass to the point of collision P. Body 2 has similar characteristics at time t_c. There is another vector coincident with e_n, which is the velocity vector normal $\mathbf{v}_n$. This vector is defined in terms of the vectors in Fig. 7.43:

$$\mathbf{v}_n = \mathbf{n} \cdot ((\mathbf{v}_1 + \omega_1 \times \mathbf{r}_1) - (\mathbf{v}_2 + \omega_2 \times \mathbf{r}_2)) \qquad (7.33)$$

The sign of $\mathbf{v}_n$ reflects the current state of the two bodies:

1. $\mathbf{v}_n < 0$: separation
2. $\mathbf{v}_n = 0$: contact
3. $\mathbf{v}_n > 0$: collision

The second condition is difficult to achieve computationally since there is always a round-off error, so we could state that a collision is occurring while $|\mathbf{v}_n| < \epsilon$, where ϵ is a small quantity. An elastic collision can be modeled using Hooke's law for springs along $\mathbf{v}_n$.

Figure 7.44 demonstrates 15 rendered animation frames from a simulation of rigid bodies after a ball drops and then rolls down a ramp toward a stack of cans (i.e., cylinders). A stereo pair of one frame of the falling cans simulation is shown in Fig. 7.45 along with an additional pair for a *house of cards* (Fig. 7.46). One can view each stereo pair without the use of external viewing aids, by diverging the eyes.[8]

---

[8]Divergence can be achieved in a variety of ways. Try focusing on an object 3 or 4 feet from your eyes. Then place these figures between your eyes and the object. You will see three frames, with the middle frame being the combined left-right stereo frame.

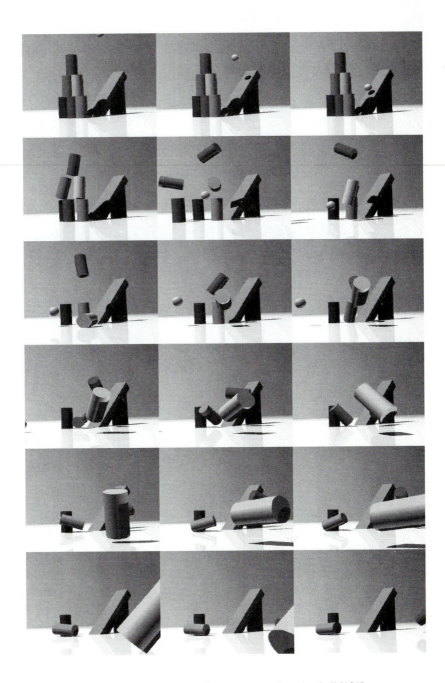

Figure 7.44   Falling cans in response to dropping ball [121].

Figure 7.45   Falling cans stereo pair [121].

Figure 7.46   House of cards stereo pair [121].

7.4 EXERCISES

For all spatial problems, assume periodic boundary conditions.

1. How does one specify recursion in L-Systems and CAs? What does this recursion physically represent in the plant phenomena modeled by L-Systems?

2. Figure 7.47 represents the same initial condition for three different CAs, each with different rules. For each CA, the first three generations are displayed (letting the initial condition be generation zero). Give rules defining each CA.

3. For the CAs in Exercise 2, formally define them using the canonical systems formalism defined in Sec. 2.2.3.

4. Reconsider the two-jug problem in Sec. 4.2.3. Express the dynamics of the two-jug system as a CA by adding the concept of *input* to CAs.

5. For the first L-System defined in Sec. 7.3.2, output the first five generations, starting with ω.

6. There are many ways to search for the critical parameter value during percolation studies. One way is to start with $p_c = 0$ and gradually increase p_c by an arbitrary amount. Can you use methods from the study of search within computer science to shorten the search time?

7. Board games such as Chess and Go involve taking turns and moving pieces on a squares or lattice points. Compare these games against CAs or PDEs by clearly defining the concept of *turn*. Do Chess or Go serve as potential simulations of anything?

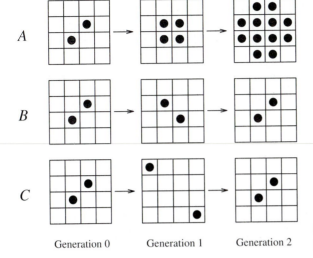

A

B

C

| Generation 0 | Generation 1 | Generation 2 |

Figure 7.47   Three CAs for three generations.

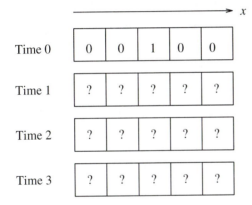

x

Time 0 | 0 | 0 | 1 | 0 | 0 |

Time 1 | ? | ? | ? | ? | ? |

Time 2 | ? | ? | ? | ? | ? |

Time 3 | ? | ? | ? | ? | ? |

Figure 7.48   Rod with heat applied in the center.

8. Given the rules of Life, what are the maximum number of cells that will all die, simultaneously, on an $N \times N$ space?

9. Spatial models can be entity-based or space-based. Consider the motion of a particle in a lattice. The particle's movement can be described by its change in state over time, or it can be defined declaratively by convolving a template over the whole space. In the convolution approach, the particle's motion is described by its spatial context (i.e., the spaces around the particle). Briefly overview these two different approaches. Are there advantages to using one method versus the other?

10. The data structure used for CA simulation is an array. Would it ever be more appropriate to use a linked list? When?

11. The heat equation (7.14) represents the flow of heat by varying space and time. Consider a discretized rod with an initial condition in Fig. 7.48. If we let T stand for temperature, our equation is

$$\frac{\partial T}{\partial t} = \frac{\partial^2 T}{\partial x^2} \qquad (7.34)$$

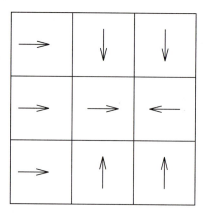

Figure 7.49 Discretized fluid vector field.

where T is defined as 0 except at the center of the rod where we have set $T = 1$. Using a discretized form of the heat equation, specify how the heat changes in the rod for the first three time steps. Let $\Delta X = 1$ and $\Delta T = 0.25$. If we let the change in T over time be equal to zero, we obtain the Laplace equation $\partial^2 T/\partial x^2 = 0$.

12. For the plant growth example in Sec. 7.3.2, draw two state trajectories representing the position of the tip of the shoot and the position of the signal over time. Hand simulate the system for the following values: $d = 6, m = 3$ and $u = 1$.

13. Consider the blinker CA in Fig. 7.3. Specify an FSA model which matches this behavior. Add one cell to the initial configuration so that it breaks the periodicity in the first FSA. That is, the revised FSA should not involve a cycle. In how many places can you add one cell so that this steady-state condition holds?

14. What is the difference between a CA and a PDE when the underlying semantics are the same? As an example, one could define the Laplace equation semantics using a PDE, $\nabla^2 \phi = 0$, or using CA rules. Is there a difference in approach or modeling philosophy?

15. The 8-puzzle is a game where there is a 3×3 square with eight square tiles and one empty square space. The tiles can be moved around horizontally and vertically by moving the empty space around. Specify how this problem relates to fluid mechanics as far as its conservation laws are concerned. Can the puzzle be an analog simulation for fluid flow? How can the puzzle be altered to conflict with conservation laws?

16. Equation (7.15) represents the equation of continuity for a fluid. We create a vector field where vectors are all of the same length, with one of four possible directions (Fig. 7.49). Consider that this system has a uniform density ρ set equal to one particle per discrete cell in Fig. 7.49. Answer the following questions:
    (a) Intuitively, what will the particles do given this velocity field?
    (b) Prove that your intuition is correct by manually calculating the first three time steps.

17. Suppose that you are required to build a CA, with rules, so that a cell moves in a circular path around the space. Can you do this with a declarative modeling approach such as a CA? What are the problems that you encounter?

18. A *reversible CA* is one where, given a set of CA rules, one may go backward as well as forward in time. Create and define a new CA that is reversible. Is the Game of Life reversible? Why or why not?

7.5 PROJECTS

1. A key aspect of percolation is to determine the value p_c so that there is a linking span of occupied sites (or bonds) that bridges one edge of the lattice to the opposite edge. Perform a Monte Carlo simulation by determining many values of p_c by using different random number generator seeds each time you begin a sweep to increase p_c gradually. Build a probability density function based on your observations of p_c at which the span occurs.

2. Build a visual forest fire simulation based on the percolation method to set up the initial condition and the three-state automaton mentioned in Sec. 7.2.2. Use colors if you have them available: green for the initial state of a tree, red for a burning tree, and black for a burnt tree.

3. Build a CA simulator that can operate over an arbitrary surface, instead of only a rectangular grid. Example surfaces include (1) circular, (2) hexagonal, and (3) spherical. Take a selection of standard CA rules and see how they execute over these surfaces. What are the qualitative differences among the results? Can you classify the CAs into groups?

4. Construct a 3D visual simulation for one of the following types of models: (1) percolation, (2) Ising, (3) CA. Review the literature on scientific visualization to determine what techniques you can use to improve the 3D effect. For instance, it would be useful to use texture mapping for variety or transparency so that users can see *through* the 3D lattice. An extension to this project is to permit nD models where $n > 3$ while allowing the user to explore this type of model using projection from nD to 3D space.

5. Build a graphical simulation of a piston in a cylinder. In this project, you will have to model (1) the rise and fall of the piston due to gravity, (2) collisions of gas particles with the piston forcing it to rise, and (3) the motion of particles. This extends the simulation discussed in Secs. 7.3.3 and 7.3.4.

6. Make an extension of the existing DEQ program in SimPack so that it can handle PDEs. The output from the extended program will provide 2D or 3D state variable information that can be fed into a scientific data visualizer.

7.6 FURTHER READING

Most of the work on percolation [50, 95, 212] is theoretical, with emphasis on proving properties of percolation models. Percolation is a random process and so is often discussed along with other stochastic process, such as random walks. L-Systems have continuous-time models called `timed DOL Systems` [181]. In a timed DOL (tDOL) system, time is advanced using an arbitrarily small time increment. Work on cellular automata originally began with the work of Stanislaw Ulam [40] and John von Neumann [226]. Recent activity in CA is based on workshops [57] and papers written in the journal *Complex Systems* edited by Stephen Wolfram, who also published a collection of key CA-related papers [230]. There is a good overall book on cellular automata by Toffoli and Margolus [221], and two other texts [228, 202] introduce the reader to the concept of *complex systems*, which is defined as the study of systems with a large number of small interacting components using simple behaviors. Frisch et al. [84] introduce lattice gases based on CA rules.

Most spatial simulation approaches are discussed within the context of algorithms for *Supercomputing* [120]. This is not too surprising when you consider that, of all modeling approaches covered, spatial models—involving thousands or millions of elements—will require the greatest amount of computing power. The dynamic gas simulation was inspired by Dewdney [45]. Algorithms for particle system problems are discussed by Hockney and Eastwood [104] and the tree-code approach was developed by Barnes and Hut [15]. The use of quad-trees and oct-trees is covered by Samet [197, 196]. Allen and Tildesley [3] overview methods for simulating liquids composed of particles. Moreover, the fast multiple method (FMM) [94, 93] is reported to be $O(N)$ in complexity. There are numerous texts on partial differential equations. Hopf [107] provides a concise treatment of the basic spatially oriented models for physics. One of the better explanations of spatial phenomena using PDEs is in Edelstein Keshet's text [52], which is oriented toward biological systems and provides a readable account of PDEs. Farlow's book [56] is excellent in its use of examples and figures. Al-Khafaji and Tooley [2] also present a clear account of PDEs for engineering problems. For a lighter, but edifying treatment of spatial models in biology and about the use of diffusion versus convection in organisms, consult Vogel's text [225]. Haberman [97] also has an excellent coverage of PDEs within the domain of traffic flow. Many other PDE texts exist; however, most focus on theoretical analysis of PDEs rather than how to model with PDEs by intuitively deriving the conservation laws from scratch and then proceeding to applications. Schey [198] provides an excellent introductory treatment of vector calculus for PDE applications. Denn [43] provides a thorough treatment of fluid mechanics in that the reader is taken from the conservation equations gradually toward the Navier-Stokes formulation. Also, the compendium containing Ulam's contributions [40] presents the relationship between the top-down Navier-Stokes approach to fluid flow dynamics and the *discrete fluids* approach using cellular automata.

7.7 SOFTWARE

Lsys is a program for simulating L-Systems, and was written by Jonathan Leech of the University of North Carolina. We used *lsys* for generating the following Postscript figures in this chapter: 7.22, 7.23, 7.24, and 7.25. *lsys* can be ftp'd using `ftp.cs.unc.edu:/pub/lsys.tar.Z`. For cellular automata, CELLULAR, a program written by J. Dana Eckart can be obtained by sending e-mail to `dana@rucs.faculty.cs.runet.edu`. CELLULAR version 3.0 was used to generate the 2D diffusion examples in Figs. 7.6, 7.7, and 7.8. Another CA program, CELLSIM, is available for Unix workstations and was written by Chris Langton (Los Alamos National Labs) and David Hiebeler (Thinking Machines, Inc.). CELLSIM supports color and black-and-white 2D cell spaces, and can be ftp'd from directory `think.com:/pub/cellular-automata/cellsim`. A CA simulator based on [221], called CAM, can be ftp'd from `turing.com:/pub/CAM.tar.Z`. Rudy Rucker's CALAB [194] is also an excellent CA simulator for the IBM/PC. For a more theoretical study of artificial life (whose models are based on spatial modeling principles discussed in this chapter), TIERRA can be obtained at `tierra.slhs.udel.edu`. The finite element method is an alternative to using finite differences. FElt, a finite element program with X windows output, was developed by Jason Gobat (`jgobat@ucsd.edu`) and Darren Atkinson (`atkinson@ucsd.edu`). Barnes has tree-code software for bodies in a gravitational potential field at `hubble.ifa.hawaii.edu:/pub/barnes/treecode`. The frames found in Figs. 7.44, 7.45 and 7.46 were provided courtesy of Thomas Bräunl of the University of Stuttgart, Germany.   AERO [121] may be obtained from `ftp.informatik.uni-stuttgart.de/pub/AERO`. AERO provides a useful tool, using the X windows system, for exploring rigid body dynamics.

## Chapter 8

Multimodeling

The test of abstractions then is not whether they are "high-level" or "low-level" abstrac-tions, but whether they are referable to lower levels.[1]

    Six blind men from Indostan. And each did want to know. "What is this thing called an Elephant?" And so the men of Indostan argued loud and long. Though each was partly in the right, they were all in the wrong.[2]

8.1 OVERVIEW

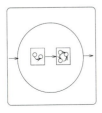

Models that are composed of other models, in a network or graph, are called *multimodels*. Most real-world models are multimodels since the models that we have discussed so far are good at portraying only a subset of the overall device or system behavior. Moreover, a normal model, by itself, can aid the analyst in answering a fairly small class of questions about model prediction and diagnosis. Multimodels, on the other hand, employ a number of abstraction perspectives and can,

[1]S. I. Hayakawa, *Language in Thought and Action* [101], p. 93.

[2]J. G. Saxe, *"An Elephant?"* [12].

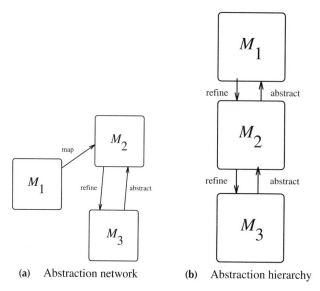

(a) Abstraction network   **(b)** Abstraction hierarchy

Figure 8.1 Simple multimodel structures.

therefore, accommodate a larger variety of questions. Most large, complex systems will have phases where different models define the activity at different stages of the simulation.

Figure 8.1 displays two potential *abstraction networks* created for reasoning and simulating multimodels. Figure 8.1(a) shows a multimodel composed of three models: M_1, M_2, and M_3. M_2 and M_3 represent distinct levels of abstraction, whereas M_1 and M_2 are at the same level of abstraction, although they represent different perspectives of the system. Figure 8.1(b) displays a strict hierarchy (i.e., total ordering) of models—we will use this approach to defining abstraction levels since there is a clear relationship among levels. When moving down a hiearchy, one *refines* a model; when moving up, one *abstracts*. The relation *map* in Figure 8.1(a) defines that M_1 and M_2 are different perspectives of the same model; however, neither one is viewed as an abstraction of the other.[3]

Multimodeling is a modeling process in which we model a system at multiple levels of abstraction. From the previous definitions of aggregation and abstraction we have seen that with aggregation the simulation of such a system need only look at the lowest level since there is where we find the actual functional semantics associated with the model. With abstraction we are concerned with many levels—each of which has the potential to be *independently simulated*. Consider the three-state tea system introduced in Ch. 2. Suppose we want to map the concept of "Heating water" to a lower-level state space? This lower-level space can be used in conjunction with lower-level models that further refine the vague concepts used in the definition of the automaton. For abstraction to take place, we must first create mappings from system components in each model.

---

[3]There is a partial ordering represented by the *map* relation in Fig. 8.1(a).

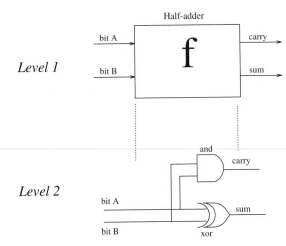

Figure 8.2   Aggregation of a half-adder circuit.

8.2 AGGREGATION AND DECOMPOSITION

In a programming language or a digital circuit design, we use *aggregation* frequently to simplify our model using levels of representation. Aggregation is defined by a high level model component decomposed into several lower level components. An integrated circuit can be decomposed into gates which can be further decomposed into discrete components such as transistors, resistors, and capacitors. When we aggregate, we cluster components together to form another, higher-level component. We create levels of aggregation where each level depends on the lower levels for a complete definition of the functional semantics. Consider the circuit in Fig. 8.2. The circuit represents a half-adder circuit composed of two subcomponents: an *and* gate and an *xor* gate. The half-adder is used in series with full-adder circuits to create *n*-bit adder circuits.

The inputs and outputs for levels one and two are identical; the only difference is that level one does not contain any functional information—it is a black box with input and output. Normally, one cannot simulate this system at anything but the lowest level; the other levels serve as "place holders" for purposes of structured representation. These place holders are no small artifact since they can represesent linguistic terms or identifiable, geometrically disparate subcomponents. We see, then, that aggregation enables us to create hierarchical models where:

- There is no information passed up from lower levels to higher levels (as in abstraction).
- The bottom level of decomposition contains all of the functional detail.
- Other levels will contain at least one black box (i.e., input/output) component whose functional identity is concealed at that level and revealed at lower levels. However, the input/output behavior at higher levels is not specified independently of the behavior at the lowest level of aggregation.
- Aggregation and decomposition permit us to create models with fewer components since the functional identities (coupling among components) of those components are defined at a lower level.

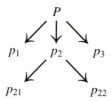

Figure 8.3   A program *calling hierarchy* with subroutines.

Other than being associated with a functional clustering, aggregation is often associated with new names for clusters since clusters form new categories. In Fig. 8.2, a combination of three level two components (two gains and one summer) form a "filter."

In programming languages, hierarchical decomposition is often drawn as a tree (Fig. 8.3). Program P calls subroutines p_1, p_2, and p_3, and p_2 calls p_{21} and p_{22}. What is not shown in Fig. 8.3 is the intercomponent coupling at each aggregation level; for instance, the output of p_1 might be fed into the input of p_2, but we have no way of knowing this from the diagram. However, the *calling hierarchy* is valuable for displaying the overall hierarchical structure of a program without crowding the picture with semantic information on how inputs relate to outputs (i.e., the algorithm).

8.3 ABSTRACTION AND REFINEMENT

We have seen that a system is described by a change of state over time, and that the change of state is associated with an event. *Abstraction* is more general than *aggregation* in that information is passed from one multimodel level to another. A system can be in a state for an arbitrary period of time; however, an event is specified as a point in time. Can an event ever be associated with a period or interval of time? Yes, states and events are both relative to the *level of abstraction* at which the system is viewed. This means that events that are points in time at one level can be defined by several events at the next-lowest abstraction level. For instance, in the Ch. 2 tea example, we could call event 1 "Start inserting" and event 2 "Start steeping." Event 1 is being treated as if it were timeless. Even though this might bother us, remember that models are simply abstractions of reality and that we use a model to answer a set of questions about the world; no single model can serve all purposes, and models may seem unrealistic in some situations. At a lower level of abstraction, we can let the second state, "Inserting bag in water," be divided by four substates: (1) walking to get the tea box, (2) taking the bag from the box, (3) walking from the box back to the water, and (4) placing the bag in the water. Each of these substates will have subevents, such as "Start walking to get to the tea box" or "End placing the bag in the water."

In terms of system behavior, we can view the process of abstraction as a kind of compression of states or events; likewise, refinement is a kind of tension. Consider Fig. 8.4(a) and (b). The methods shown in these two figures provide a convenient way of viewing abstraction and refinement from a state trajectory viewpoint. Compression is actually a homomorphism where we map (or compress) a large set of elements into a smaller set of elements . The use of compression or tension is a geometric tool for viewing homomorphisms where the items to be compressed or tensed are contiguous. With this in mind, we can create

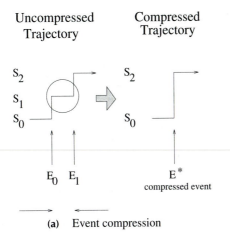

(a)   Event compression

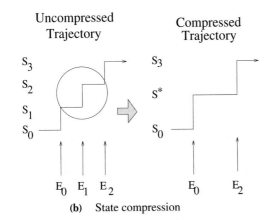

(b)   State compression

Figure 8.4   Abstraction by compression.

two maps f and g as follows: $f\colon \{E_0, E_1\} \to E^*$ for Fig. 8.4(a) and $g\colon \{S_1, S_2\} \to S^*$ for Fig. 8.4(b). While one component (state or event) is compressed, the orthogonal component is a subset of the original, uncompressed set. For instance, state compression involves a reduction in events; prior to compression the event set $E = \{E_0, E_1, E_2\}$ and after compression $E' = \{E_0, E_2\}$. $E' \subset E$. The resulting compressed state or event $\{S^*, E^*\}$ is not a member of the original state S or event set E, respectively. This is due to the nature of abstraction and to the fact that when we form abstraction hierarchies, the child nodes will be different from the parent (abstract) node. This means that abstraction implies a loss of information and refinement involves an increase in information.

8.4 ABSTRACTION AS HOMOMORPHIC SIMPLIFICATION

Abstraction, and its dual method, refinement, are performed by mapping a model structure, say M_1 to another M_2. First, let us briefly review mapping concepts. Systems in Ch. 2

were defined as tuples with components. Even though we have found utility in studying different modeling techniques, the task of refinement and abstraction is made easier when we consider each model in the same structural way. For instance, a general time-invariant system is defined as the tuple: $\langle I, O, S, \delta, \lambda \rangle$, where I is the set of inputs, O is the set of outputs, S is the set of system states, δ is the state transition function, and λ is the output function. Systems, therefore, are generically defined as sets and functions. This is important when discussing abstraction, since for a model M_1 to be refined into M_2, we must form functional mappings between the respective set and function components for each model.

Besides the key types *set* and *function*, we introduce three other concepts. The concept of "symmetry breaking" in complex systems is defined by the qualitative changes to system behavior after a bifurcation occurs in phase space. When a system has a relatively simple pattern of behavior, there are a small number of attractors and basins of attraction constituting phase space. The small number of basins characterizes the system as being symmetric (or homogeneous) with respect to operations that affect the system. However, as certain parameters or external inputs are modified, systems become more complex by exhibiting a variety of distinct qualitative behaviors. Eventually, a system undergoing symmetry breaking will continue to evolve into chaos, where sensitivity to initial conditions and parameter changes predominate. Although bifurcations are known to be at the heart of many nonlinear systems, the partitioning of phase space, in general, is central to the way we can model *all* systems. We begin with a short discussion of partitioning and system morphisms.

Let A and B be sets. A is said to be a *subset* of B ($A \subset B$) if each element of A is an element of B. Therefore, if $A = \{0, 2\}$ and $B = \{0, 1, 2, 3\}$, then $A \subset B$ since $0 \in A \Rightarrow 0 \in B$ and $2 \in A \Rightarrow 2 \in B$. The definition of *partition* uses the subset concept. A partition π of a nonempty set A is a collection of nonempty subsets of A such that (1) for all $X \in \pi$ and $Y \in \pi$, either $X = Y$ or $X \cap Y = \phi$, and (2) $A = \bigcup_{S \in \pi} S$. Partitions of sets are best viewed in pictorial form as in Fig. 8.5(a), where we have set A, which is a two-dimensional subset of $\mathcal{R}^2$, where $\mathcal{R}$ is the set of real numbers. A is partitioned into four subsets. Depending on the connectivity in phase space, we can create a finite state automaton (FSA) to control[4] motion through the phase space using a forcing function. An automaton reflecting control through phase space is shown in Fig. 8.5(b).

Let's consider the tea example in Fig. 2.6. One might regard the two states "inserting bag in water" and "steeping tea," to be clustered into one state at a higher abstraction level called "making tea." We make the following definitions:

- Let states S_1, S_2, S_3 represent the states "heating water," "inserting bag in water," and "steeping tea," respectively.
- Let the input set be $\{1\}$, specifying the control of the system.
- We will call the system in Fig. 2.6 level 2 since we will abstract a new system to be called level 1.
- The state space for level 2 is $A = \{S_1, S_2, S_3\}$, and for level 1 it will be $B = \{T_1, T_2\}$, where state T_1 is "preparations" and state T_2 is "making tea."

---

[4]The term *control* is used somewhat loosely since the FSA serves to "organize" lower-level behavior.

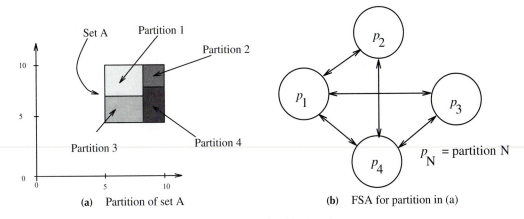

(a)   Partition of set A   (b)   FSA for partition in (a)

Figure 8.5   Partitioning phase space.

- Level 2 has event space $Y = \{E_1, E_2\}$ and level 1 has event space $Z = \{E^*\}$.
- The transition functions for levels 1 and 2 are δ_1 and δ_2. These define state-to-state transitions. For instance, in level 2, $\delta_2(S_2, 1) = S_3$.

The two levels (i.e., models), along with their respective state trajectories, are depicted in Fig. 8.6. An "abstraction" of a system is a homomorphism that maps the identifying sets of that system to sets of an abstract system. The "sets" of a system are its set of possible inputs, set of possible states, set of possible events, and set of possible outputs. A "homomorphism" is a mapping that preserves the behavior of the lower-level system under the set mappings; this ensures that our new abstract model will be an exact abstraction—by using a behavior-preserving morphism. First we create two homomorphic functions h_1 and h_2 which will map our state and event spaces. h_1 is defined as $h_1: A \rightarrow B$:

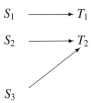

and the event spaces are mapped similarly with $h_2: Y \rightarrow Z$:

A commutative diagram displaying the preservation of behavior between the two

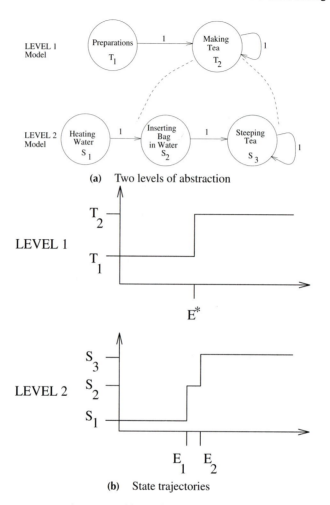

(a)   Two levels of abstraction

(b)   State trajectories

Figure 8.6   Abstracting the tea system.

systems is shown below.

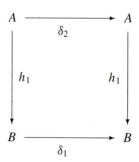

The primary features of homomorphism require:

1. Two systems (i.e., models), called models 1 and 2, that will be used to define two different multimodel levels.
2. Each system must be converted to the canonical system definition in Ch. 2.
3. A set of homomorphic functions. Each function maps each successive component of model 1 to the respective component of model 2.
4. The test of homomorphism is behavior preservation. That is, the mapping must take place so that the behavior of the source model is preserved as shown in the previous commutativity diagram.

Let's create a third level of abstraction for the tea system. For instance, if we create a lower-level state space defined as the the cross product of two variables T (temperature of water/tea) and C (concentration of tea in the water), we can map subsets of this two-dimensional state space to the state space containing the three states: water heating, inserting bag in water, and tea steeping. We will define our current three-state system as model M_1, and we'll create three lower-level models (all at the same abstraction level) to further refine the high-level states:

1. (M_2) HEATING WATER: $\dot{T} = k_1(100 - T)$, $C = 0$.
2. (M_3) INSERTING BAG IN WATER: $T = 100$, $C = 0$, $\Delta t = 0.1$.
3. (M_4) STEEPING TEA: $\dot{T} = k_2(\alpha - T)$, $\dot{C} = k_3(k_4 - C)$.

M_2 and M_4 contain differential equations that represent first-order exponential behaviors for heating, cooling, and diffusion of tea into water. M_3 represents a state that holds for 0.1 time units ($\Delta t = 0.1$). The k_i are heating and cooling constants, with the exception of k_4, which represents the maximum concentration of tea in water. Figure 8.7 displays both abstraction levels for the tea multimodel. The multimodel system in Fig. 8.7 has a total of four models: $M_1, \ldots, M_4$. If the system is simulated using M_1, the simulation would proceed by shifting state according to an input string given by the user.[5] Alternatively, a simulation can be performed at the lowest level (level 2) while using level 1 to control or coordinate the activities at level 2. In this case, the equations of $M_2, \ldots, M_4$ would be solved as time (t) is incremented. Here we see the difference between the two types of inputs to M_1. One input is associated with the agent's activities in moving the tea cup and placing the bag in the boiling water; this input is external to system M_1, so it is termed an "external event." The other input comes from level 2 and is associated not with activities outside the system, but with state changes at level 2. These two input types provide the necessary state transition at level 1: We transit from state "Heating water" when the agent moves the cup and tea bag and places the tea bag in the boiling water. Then, after a period of 0.1 time unit, the state "Steeping tea" is entered.

The homomorphic relationship of level 2 to level 1 is most easily displayed using a "phase graph," as shown in Fig. 8.8. This graph shows the mapping between state spaces

---

[5]An internal input would also be required to drive the system from state "Inserting bag in water" to "Steeping tea."

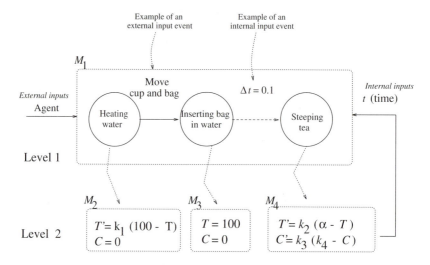

Figure 8.7   Tea multimodel with four models.

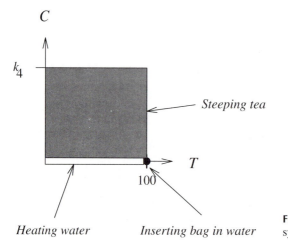

Heating water   *Inserting bag in water*

Figure 8.8   Three phases of the tea system.

associated with each abstraction level. Given the two abstractions presented (Figs. 8.6 and 8.7) we can form a more comprehensive multimodel composed of three levels, as shown in Fig. 8.9.

8.5 REFINING AGGREGATE VECTORS

Many simulations involve the motion of some number (n) of bodies in a d-dimensional space. Partitioning space—for these kinds of systems—is, perhaps, the most straightforward type of state space partitioning. We will take an aggregate vector (which can be either a single number or a vector of numbers), such as the center of mass for n bodies

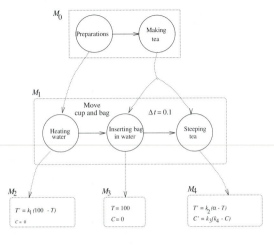

Figure 8.9   Multimodel with three levels and five "nodes."

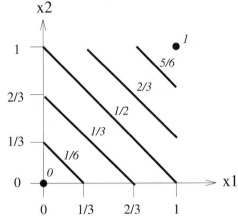

Figure 8.10   Selected diagonal line partitions of state space.

and then partition a lower-level state space accordingly. For instance, consider a rod (i.e., one dimension) that permits two bodies (i.e., spheres) to slide along it. The rod is 1 meter in length. Furthermore, we will assume that two bodies can occupy the same location on the rod. Call the position of body 1: x_1, and the position of body 2: x_2. We can then imagine the space defined by $[0, 1] \times [0, 1]$ to be the space in which we will find points (x_1, x_2) representing the combined position of each of the two spheres. If we aggregate the positions of the two spheres by calculating the center of mass (COM), the center mass value can be seen as partitioning the state space with partitions that are diagonal lines. Figure 8.10 displays the diagonal lines defining constant center of mass for masses that can be positioned at: 0, 1/3, 2/3, and 1 along the rod. The COM locations are specified to the right of the respective diagonal line. In the normal situation where the spheres can be placed anywhere (in a continuum) along the rod, the diagonal lines would partition the state space completely, creating a filled square.

We have accurately represented state space as being a cross product of individual mass locations. However, it is often convenient to let the state space be equal to the state space for a single object, where multiple objects simply overlay the existing state space rather than to increase the state-space dimension. In this case one can easily partition state space using a clustering or nearest-neighbor method. For instance, if n particles are moving in a two-dimensional space, we can partition the space much as we did in Fig. 8.5(a). This kind of partitioning was performed as a matter of course in Sec. 7.3.6.

8.6 DISCONTINUITY AND INTEGRATION

Multimodels involve the switching of models as we progress from one phase to another. This is true, for instance, when we have an FSA-controlled multimodel, where each FSA state represents a lower-level model. Consider a two-state FSA where we switch from state s_1 to s_2 when $s_1 > a$, where a is a real number. This condition represents an internal (state) event. Consider that the level being controlled by the FSA is modeled by ODEs. This means that we will numerically integrate the solutions to s_1 and s_2 as we execute the FSA model. The model below s_1 is $M1$, and the model below s_2 is $M2$. In the ODE controlled by s_1 in $M1$, let the lower-level ODE state variable be $x(t)$. When integrating the equation containing this variable, we will have the situation in Fig. 8.11(a). The integration step is shown in Fig. 8.11 to be large, so that the effect of the internal event can be highlighted. Figure 8.11(b) represents the possible effects of moving to model $M2$ when the condition $x(t) > a$ arises. Four potential trajectories (labeled 1 through 4) are shown in Fig. 8.11(b). While the integration error can be made equivalent to a power of the integration step Δt, a discontinuity caused by an event a time t^* can wreak havoc with our accuracy. Therefore, it is important to locate the time t^* so that (1) integration takes place to time t^* and then, as the model is switched from $M1$ to $M2$, and (2) integration continues with $M2$. The location of internal events can be equivalenced to the problem of *root finding* (or locating zeros). Locating t^* where $x > a$ is performed by locating the root of $x(t) - a$.

A brief discussion of terminology is warranted before overviewing root-finding methods. A root is *bracketed* when a bracket (or interval with a left and right value) can be positioned around the root value. Most methods employ bracketing to ensure that the search for the root does not stray. A root is bracketed in an interval (a, b) if $x(a)$ and $x(b)$ have opposite signs. With continuous functions and a basic result in calculus (the intermediate value theorem), we can assert that at least one root must be present in the interval.

Four methods for root finding are shown in Fig. 8.12. The two points that are known, 1 and 2, are obtained during the integration of model $M1$. Other points are identified in Fig. 8.12 as they are calculated in accordance with the appropriate method. Each method has its strengths and weaknesses. We describe them as follows:

1. *Bisection Method.* This is the continuous equivalent of binary search for discrete structures. First we take the two points, $x(t1)$, and $x(t2)$. Then we determine the point $x((t2 - t1)/2)$. While maintaining a bracket around the root, we continue in this fashion. This method is guaranteed to locate the root, but it can be slow.

2. *Secant Method.* By using secant chords, we converge on the root; however, bracketing is not applied. In the case, where the chord passes through two points on the same

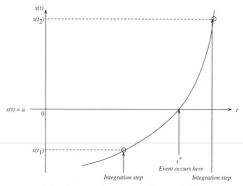

(a)   Integration according to M1

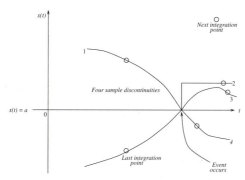

(b)   Four possible discontinuities at time t^*

Figure 8.11   State discontinuity given an internal event.

side of $x(t) = a$, we extrapolate the chord until it reaches the line $x(t) = a$ and then use this position. The last two positions are always used to draw the secant.

3. *Regula Falsa Method.* This is similar to the secant method except that bracketing of the root is always done.

4. *Newton-Raphson Method.* This method is distinctly different from the other three methods in that first derivative information about the solution curve is used to draw tangent lines. Since our model is an ODE and not simply an algebraic equation, this kind of information is easily obtained. At the point where the tangent line intersects $x(t) = a$, we draw another tangent and hopefully, approach the root. If the solution is fairly linear in the region of the search, this method works well.

The only method guaranteed to work is the bisection method. The other three methods depend on the shape of the curve in the region of the root.

8.7 REFINING DECLARATIVE AND FUNCTIONAL MODELS

We first introduce a notation for two types of events (or *transitions*) that can occur within a multimodel:

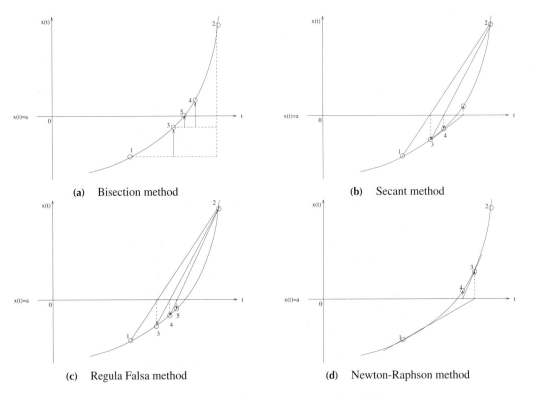

(a)   Bisection method   (b)   Secant method

(c)   Regula Falsa method   (d)   Newton-Raphson method

Figure 8.12   Four methods of root finding.

1. *External event* for model M_i at abstraction level i is an input that is generated outside M_i and is represented as a signal that is input to M_i.

2. *Internal event* for model M_i at abstraction level i is an input that is generated from a refined model M_{i+1} (at abstraction level $i + 1$) of a component of M_i. This signal is input to M_i, as are the signals containing external events.

Figure 8.13 displays a phase graph with three phases A, B, and C. The word "phase" simply refers to a state that has lower-level semantics attached to it.

A different model is associated with each phase; for instance, in phase B, the underlying model M_1 will generate the state trajectories. External events are denoted in Fig. 8.13 with a solid arc. If in phase A, for instance, an input of i_1 is received, the phase moves immediately to B and the model is M_1 starting in a state determined by the state of M_5 when the input i_1 occurred; otherwise, the phase remains A. Internal (also called state) events occur when the state of a model satisfies specific transition conditions. They are denoted by dashed lines from one phase to another. For instance, in phase A, let predicate $p(X_2) \equiv equal(X_2, 4.2)$. This is interpreted as follows: if the state variable X_2 "reaches" the value 4.2 (i.e., $X_2 = 4.2$), the phase becomes C.

To discuss abstraction, we first must have an abstraction hierarchy connecting models of different types. Then we take one arc in the graph and consider the mapping between

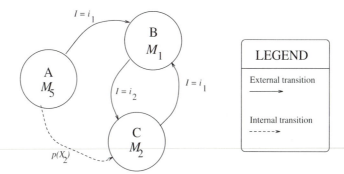

Figure 8.13   Generic phase graph.

the two objects on either end of the arc. In this section, we cover the method of refining a model of one type into another; however, we consider only the FSA (as an example of a declarative model type) and the block model (as an example of functional modeling). For instance, in our first refinement, we break down part of an FSA into a second-level FSA. This is an example of refining as follows: *declarative* $\overset{R}{\rightarrow}$ *declarative*. The form A $\overset{R}{\rightarrow}$ B defines a refinement from a model of type A to a model of type B. States are labeled as s_j, inputs as i_j, outputs as o_j, functions as f_j, and predicates as p_j, where j is an index.

When we use the term "FSA" we are actually referring to an *extended* FSA which can be mapped directly to the normal FSA when the predicates p are based on equalities and inequalities. An FSA accepts only inputs on its arcs. Inequality predicates can be mapped to inputs in the following manner. Consider an arc from state s_1 to s_2 which has the arc label $x > 3$. In such a case, the variable x is defined in a lower-level state space. Moreover, the inequality can be mapped to an "input" by creating a new input variable j and setting $j = 0$ when $x < 3$, $j = 1$ when $x = 3$, and $j = 2$ when $x > 3$. In this way, $x < 3$ serves to label an *internal event* since the input, j, that will cause the state change (to s_2) comes, not externally from the FSA abstraction level but instead, from a lower abstraction level which has a state space containing the variable x.

Consider the FSA in Fig. 8.14. This model, along with the one in Fig. 8.15, involves top-down *homogeneous* decompositions since one model is defined in terms of models of the same type. State s_1 is refined into another FSA at level two, which has two states, s_{11} and s_{12}. The predicate $p1(i)$ means that there is a predicate involving external input variable i such as $i = 0$ [$p2(i)$ is then another predicate, such as $i = 1$ if the input is binary valued]. Figure 8.15 defines a two-level functional hierarchy where function f is defined in terms of a composition of three other functions f_1, f_2, and f_3. In functional decomposition, one sometimes increases the dimension of the input and output sets. For instance, suppose that $i_1 \in \mathcal{Z}$. Then the decimal input can be broken down into a binary input which would involve more inputs to f_1. *Heterogeneous* decomposition of models involves describing the semantics of a model in terms of different model semantics. For instance, in Fig. 8.16 we have an FSA that defines the internal state transitions associated with function f. Indeed, all FSAs should have this kind of functional box around them—even though the functional box is not always explicit in the FSA graphical representation. The predicate $p(i)$ involves

$p_1(i)$
s_1
s_2
$p_2(i)$
s_{11}
s_{12}

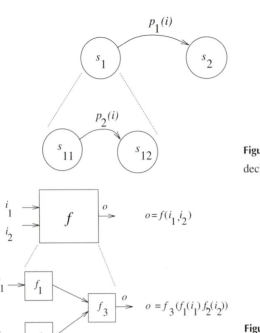

Figure 8.14 Homogeneous refinement: declarative $\stackrel{R}{\rightarrow}$ declarative.

i_1
i_2
f
o
$o=f(i_1,i_2)$
f_1
f_2
f_3
$o = f_3(f_1(i_1)f_2(i_2))$

Figure 8.15 Homogeneous refinement: functional $\stackrel{R}{\rightarrow}$ functional.

f
$p(i)$
i
o
s_1
s_2

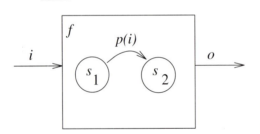

Figure 8.16 Heterogeneous refinement: functional $\stackrel{R}{\rightarrow}$ declarative.

the test of input variable i. Imagine, for instance, that f is the functionality associated with a light switch that "remembers" its current state (ON or OFF). i represents someone flicking the light switch, so $i \in \{TURN_OFF, TURN_ON\}$, s_1 and s_2 mean "light is off" and "light is on," respectively. Then $p(i)$ is defined as $i = TURN_ON$. The arc coming out of a state can be *external*, where the predicate involves the i, or it can be *internal*, where the predicate involves the use of a state variable (o) connected with the internal block model. For example, Fig. 8.17 displays a state s whose semantics are defined in terms of a functional block model with two functions f_1 and f_2. State s is assumed once $p1(i)$ evaluates to true. At this point the value of i into f_1 becomes the value of i, causing $p1(i)$ to become true. Figure 8.17 involves a state-to-state space mapping, which is not immediately apparent from the figure. Specifically, most functional block models that represent some aspect of physical reality involve state transitions. This means that f_1 and f_2 contain internal state transitions as in Fig. 8.16. The state space for the block model within state s has its own state space—call it X. Given the partition method of abstraction, we will associate s with

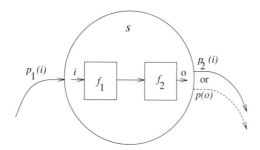

Figure 8.17   Heterogeneous refinement: declarative $\overset{R}{\to}$ functional.

a partition of the state space X by creating a homomorphic mapping from X onto S, where $s \in S$ (i.e., S is the state space of the higher-level FSA model whose states contain the block models).

There is a transition with two possible types of semantics coming out of state s in Fig. 8.17. An external transition would be of the form $p2(i)$ since it is a predicate involving the input i. An internal transition would be based on a variable (o) that is a component of state space X. Most likely, there would be a test (i.e., using a predicate) involving the final output of the block model, $p(o)$, as shown in the figure, although predicate p can have as its argument any variable from the block model.

8.8 PENDULUM PHASE SPACE

Consider the dynamics of an undamped, unforced pendulum of unit length (i.e., damping factor of $\epsilon = 0.0$):

$$\dot{x}_1 = x_2$$
$$\dot{x}_2 = -\epsilon x_2 - \sin x_1$$

Figure 8.18 displays the phase trajectories and phase region A (defined below). Figure 8.19 shows phase regions B and C. Since phase area is conserved during flow, the phase trajectories serve to describe equipotential energy contours. The regions display basins of attraction that partition the state space. The topology of the phase space provides us with three distinct regions (after considering the symmetry by "wrapping" the phase space onto a cylinder):

1. *(Region A) CCW Rotation.* The bob swings counterclockwise in a complete orbit.
2. *(Region B) Vibration.* The bob oscillates back and forth.[6]
3. *(Region C) CW-Rotation.* The bob swings clockwise in a complete orbit.

To control the pendulum requires a forcing function of the general form $f(t)$ so that it is possible to traverse from one basin to another.[7] The geometry of the basins will change depending on parameters such as ϵ and the input trajectory $f(t)$; however, our partitioning of phase space is independent of these modifications since regions A, B, and C are fixed

---

[6]Also known as *libration*.

[7]The general method for basin switching can be chosen from the literature on the control of nonlinear systems.

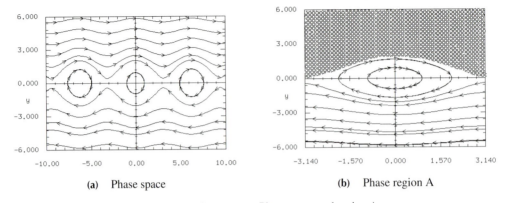

(a)   Phase space   (b)   Phase region A

Figure 8.18   Phase space and region A.

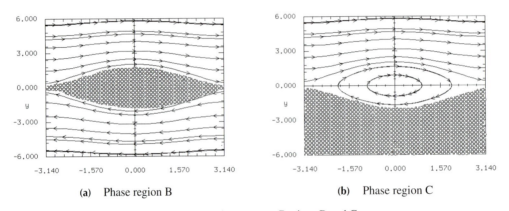

(a)   Phase region B   (b)   Phase region C

Figure 8.19   Regions B and C.

and meaningful in any context regardless of basin geometry. We create an FSA with three states, one for each phase region in Fig. 8.18(a). For our pendulum example, there is one underlying equational model; however, we need not use the same mathematical model when transitioning from one phase to another. Later, we consider other examples that demonstrate that each phase can represent a different model.[8]

We control the pendulum by using $f(t)$, and the boundaries of regions A, B, and C are equal to the separatrices present in the phase graph. These dividing lines provide a method where we can label the FSA appropriately. The Hamiltonian for the pendulum (with $m = 1$) is defined as the sum of the kinetic and potential energies:

$$H(\theta, p) = \frac{p^2}{2} + V(\theta) \tag{8.1}$$

[8]Where each model may have different state spaces.

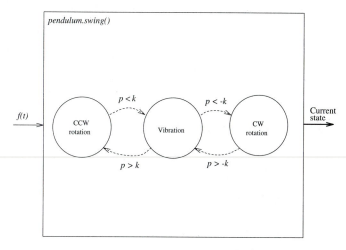

Figure 8.20   FSA pendulum control.

where $V(\theta) = -\alpha^2 \cos\theta$ and $\alpha = (gl)^{1/2}$. Equation (8.1), when expanded, is

$$H(\theta, p) = \frac{p^2}{2} - \alpha^2 \cos\theta \qquad (8.2)$$

This provides a method for measuring the energy level for any point (θ, p) in the phase plane. The separatrix defining the boundary between regions A, B, and C is determined by first noting that the potential energy will be at its maximum when the pendulum bob at located at $\theta = \pi$. So $H(\theta, p) = \alpha^2$ since $\cos\pi = -1$. Therefore, we obtain the following derivation for p:

$$\alpha^2 = \frac{1}{2}p^2 - \alpha^2 \cos\theta$$

$$p^2 = 2\alpha^2(1 + \cos\theta)$$

$$p^2 = 4\alpha^2\left(\frac{1 + \cos\theta}{2}\right)$$

$$p = \pm\sqrt{4\alpha^2\left(\frac{1 + \cos\theta}{2}\right)}$$

$$p = \pm 2\alpha\left(\cos\frac{\theta}{2}\right)$$

This provides the FSA in Fig. 8.20, where $k = 2\alpha\cos(\frac{1}{2}\theta)$.

8.9 THE TWO-JUG SYSTEM

Let's say that we want to represent a model where two jugs of water will undergo various operations, such as emptying, filling, and transferring. We will create a picture model and then build a class hierarchy. Jug A is the 3-gallon jug, and jug B is the 4 gallon jug (Figs.

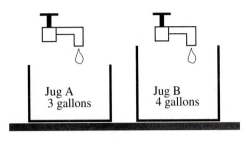

Figure 8.21   Picture model: two jug system.

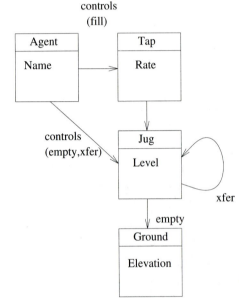

Figure 8.22   Objects in the two jug system.

8.21 and 8.22).   We will take this system and partition state space to design a two-jug multimodel. There is no automatic method for state space partitioning; however, we can use a heuristic to help us: *form new states using participles created from the objects*. That is, the present participle form of "fill" is "filling." A similar approach creates the state "emptying." We can combine both the emptying state and filling state to form a state called *In-between* or *Emptying-or-filling*. Figures 8.23 through 8.25 suggest a partitioning of state space that provides a lumped, more qualitative model. Since these segments overlap and do not form a *minimal* partition, we must combine substates —reflecting levels in jugs A and B—to form nine new states. The greatest "lumping" effect is contained in the state (jug A filling, jug B filling), where the term "filling" means neither empty nor full.   We should not underestimate the lumping effect when considering other possible, but similar, systems. For instance, if the jugs could contain arbitrarily large amounts, the state space for Fig. 8.26 would also be large, but the state space for the lumped model would remain the same (nine states). It is possible to further reduce the number of states by mapping and partitioning state or event space in accordance with natural language terminology; state space could easily be broken into "wet" and "dry," where dry corresponds to both jugs being

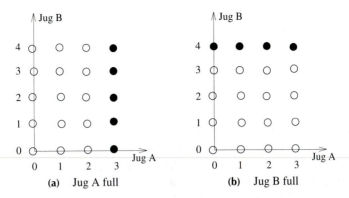

Figure 8.23   Full phase.

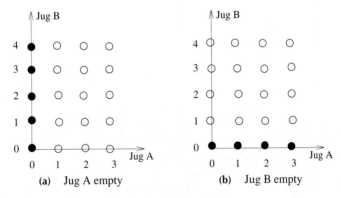

Figure 8.24   Empty phase.

empty, and wet covers the remainder of state space. If we consider each tap and each jug to be a function, we can create the functional block model illustrated in Fig. 8.27. This dynamical model should be simulated as it is represented; that is, messages are transmitted to blocks that *delay* the corresponding outputs to represent the passage of time. Use of a delay, therefore, represents a simple way to create an abstract model. If we want to increase model detail, the delay can be further decomposed into an FSA. For the tap dynamics, we use a delay; however, for the jug dynamics, we choose an FSA. The dynamics for jug A are shown in Fig. 8.28. Empty (A) and Xfer (A, B) on the left side of the jug A object in Fig. 8.28 represent input controlling signals whereas these same identifiers are also used to designate output flow of water from Jug A. The entire functional model shown in Fig. 8.27 has four functional blocks, with two of the blocks (jugs A and B) having an internal FSA to further refine state-to-state behavior. It is worthwhile to contrast this model with the declarative model in Fig. 8.26. In the declarative model, each state represents the state of the entire system, whereas for the functional model, states are "local" to the specific object.

Each level or segment of a multimodel can be represented by a different *type* of model; by permitting different types of models, we create a more flexible modeling environment where each level is represented by its most appropriate form. We create a two-jug

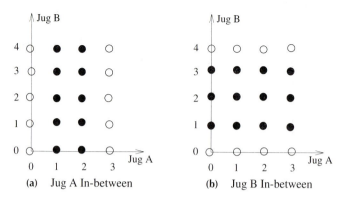

Figure 8.25   In-between phase.

multimodel consisting of two heterogeneous models. Let's summarize a progression of increasingly complex model forms:

- An FSA as depicted in Fig. 4.25 is perhaps one of the simplest forms of models. The FSA is depicted graphically as a set of circles and arcs; however, there is also an implicit function box placed around the entire graph. The input to the function box is the same input referred to on the FSA arc labels; the output from the box is the same as the FSA output represented using either the Moore or Mealy conventions.
- A functional block model captures system dynamics from a functional or procedural perspective rather than the declarative form of the FSA. In most functional models, the only representation of state is the one in the *integrator* and *delay* blocks. For instance, an integrator functional block using a single-step Forward Euler approach keeps track of the last value for a state variable.

Figure 8.29 is a multimodel that contains several layers—or abstraction levels. As we proceed down the figure, we use more powerful refinement techniques. The topmost two levels represent a homogeneous refinement which is common in most model languages. In this case, we have taken a single function and hierarchically decomposed it into four subfunctions. The breakdown is "homogeneous" since each level uses the same type of model—functional block. The dynamics of $JugA$ is performed by an FSA. In such a model, there are external events which occur as a result of changing input and internal events which occur as a result of a time advance specification. A time advance value is created by solving analytically for the underlying equation representing dynamics for that state. If we cannot specify a time advance, we can represent the dynamics for this state by associating yet another level of detail with the state. In this case, the state becomes a "phase" and we heterogeneously decompose the phase into a functional block model. In Fig. 8.29 we represent the act of water filling (or emptying) by a first-order differential equation $\dot{X} = U - kX$, where U is the control input and X is the continuously changing level of the water.

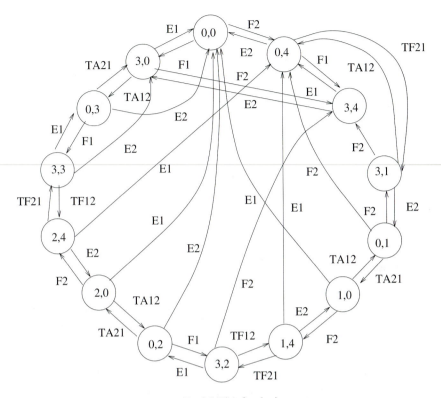

Figure 8.26   Partial FSA for the jug system.

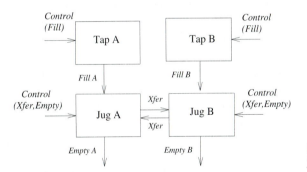

Figure 8.27   Functional model of two jug system.

8.10 THE DINING PHILOSOPHERS REVISITED

Recall the dining philosopher examples given in Ch. 4. For declarative modeling, we approximated the system with an FSA and then used a Petri net as our more detailed model. To illustrate that multimodeling is not restricted to FSAs and block models, we can construct a Markov model of the dining philosophers which operates as follows. Each philosopher will take the same amount of time to eat, and there are five possible pairs. Let each pair [such as (1, 3)] represent a state and transitions be probabilistic based on a uniform random

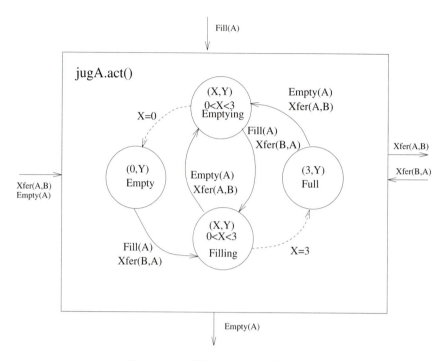

Figure 8.28   FSA for jug A in fig. 8.27.

distribution. This means that after, say, philosophers 1 and 3 eat, there is a 0.2 probability that they will eat again. Figure 8.30 displays the fully connected graph. Now we must decide how we can refine this model to create a more detailed **DP** system. Recalling one of our heuristics for refining models, we can map each state in the Markov model to a lower-level state with more dimensions. Recall the Petri net given in Fig. 4.43. The state space for this net has fifteen dimensions (one for each place). A particular permutation of places in the net conform directly to the state "1 and 3 are eating." For example, the Petri net states below map into the single Markov state $(1, 3)$: $(0, 0, 0, 0, 0, 0, 0, 0, 0, 0, 1, 0, 1, 0, 0, 0)$. This is because when 1 and 3 are eating, places p_{10} and p_{12} each contain a token.

8.11 INDUSTRIAL PLANT CONTROL

Industrial plants (chemical, oil, nuclear) have fairly complicated systems. These systems involve the sensing of temperature and pressure. When either the temperature or pressure increases beyond certain limits, it is important for the plant operator either to correct the problem or shut down part or all of the system. Consider the following aspects of the control procedure for the plant:

- Assume one sensor (each) for pressure and temperature.
- A normal operating range is specified for pressure ($P_{min} < P < P_{max}$) and temperature ($T_{min} < T < T_{max}$). When a sensor detects a value out of this range, an

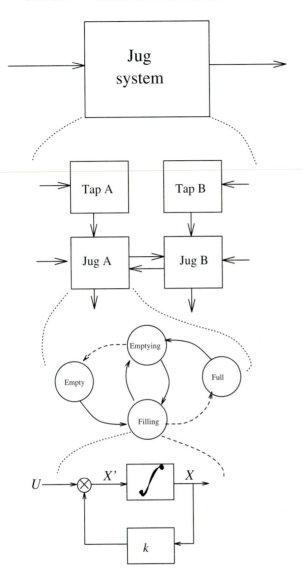

Figure 8.29 Heterogeneous refinement in the jug multimodel.

alarm sounds and a signal is raised. If both sensors are out of range, the plant is shut down.

- If, after a specified period of time K, there is a successful recovery, the plant automatically returns to normal operation. Otherwise, the plant is shut down.

Figure 8.31 displays an FSA which controls the plant. The following symbols are used in Fig. 8.31: T (temperature), P (pressure), T_r (reference[9] temperature), P_r (reference pressure), ΔT (amount of time passed in the current state), K (amount of time before plant

[9]The term "reference" refers to the control value set by the external influence (i.e., the operator).

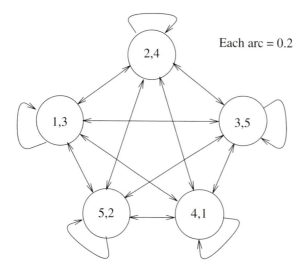

Each arc = 0.2

Figure 8.30   **DP** Markov chain.

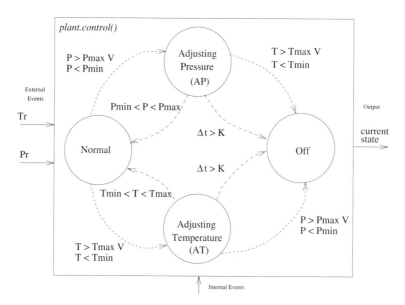

Figure 8.31   Industrial plant control multimodel.

shutdown), and the minimum and maximum values defining the range for temperature (T_{min}, T_{max}) and pressure (P_{min}, P_{max}).

Each of the four basic states has lower-level semantics (i.e., another abstraction level). For instance, consider the state of "adjusting temperature." It can be refined into a block model defining a proportional method of control shown in Fig. 8.32.

The phase space after partitioning is shown in Fig. 8.33 for the space $T \times P$.

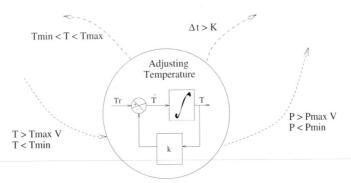

Figure 8.32   Decomposition of "adjusting temperature" state.

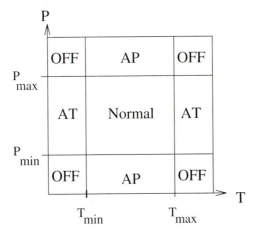

Figure 8.33   Partition of phase space for the plant.

8.12 BOILING LIQUIDS

8.12.1 Overview

Consider a pot of boiling water[10] on a stovetop electric heating element. Initially, the pot is filled to some predetermined level with water. A small amount of detergent is added to simulate the foaming activity that occurs naturally when boiling certain foods. This system has one input or control—the temperature knob. The knob is considered to be in one of two states: on or off (on is $190°C$; off is α, ambient temperature). We make the following assumptions in connection with this physical system:

1. The input (knob turning) can change at any time. The input trajectories are piecewise continuous with two possible values (ON, OFF).

2. The liquid level (height) does not increase until the liquid starts to boil.

---

[10]The liquid used can be any chemical. We will use water for our discussion of the dynamics of a single pot of water. When we extend the multimodel to include a Petri net level, the domain is switched from a pot of boiling water to the coordination of two flasks of boiling chemicals.

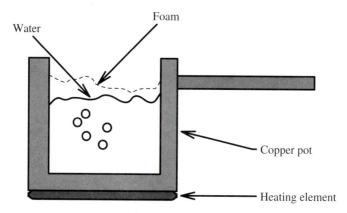

Figure 8.34   A pot of boiling water.

3. When the liquid starts to boil, a layer of foam increases in height until it either overflows the pot or the knob is turned off.

4. The liquid level decreases during the boiling and overflow phases only.

Using the multimodeling approach, we proceed as follows:

1. Define the domain and assumptions.

2. Create a two-state FSA and refine the FSA into two more FSA levels.

3. Create a heterogeneous refinement of each FSA state (i.e., phase) into a block model.

4. Display partitions that illustrate the mapping between levels.

5. Suggest questions that can be asked of the multimodel.

A system of boiling water is shown in Fig. 8.34.

To create a mathematical model, we must start with data and expert knowledge about the domain. If enough data can be gathered in a cost-effective way, then our model engineering process will be simplified since we will not have to rely solely on heuristics to identify the model. By analyzing a pot of boiling water, we may derive simple causal models with individual transitions that may be *knob_on* $\Rightarrow$ *water_getting_hotter* or *water_getting_hotter* $\Rightarrow$ *water_boiling*. An important facet of system modeling is that we choose certain modeling methods that require a categorization of informally specified system components. Key components of any system model are *input*, *output*, *state*, *event*, *time*, and *parameter*. Different modeling methods include these components in different ways. For instance, an FSA focuses on state-to-state transitions, with input being labeled on each arc. A dataflow model, on the other hand, focuses on the transfer function between input and output.

8.12.2 Two Homogeneous Refinements

Homogeneous refinement is the refinement of a model to more detailed models of the same type. For instance, consider a printed circuit board. There are many levels, each of which can be modeled using the same block formalism. The model is defined as having type

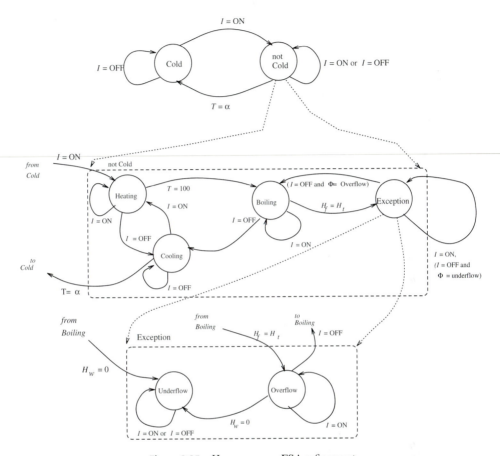

Figure 8.35   Homogeneous FSA refinement.

"block," just as a model might have type "Petri net" or "compartmental model." The chip level of the PC board model will contain function blocks that are decomposed into block networks. In this way, the modeler can build a hierarchy of models without having to represent one model at the lowest abstraction/aggregation level.

Figure 8.35 shows three levels of finite state automata for the boiling water process. The topmost FSA in Fig. 8.35 shows a simple two-state automaton with input. We label this FSA-1. The input has been discretized so that the knob control is either ON or OFF. Input can occur at any time and will facilitate a change in state. A change in input is denoted by $I = i1$ on an arc in Fig. 8.35 defining where the state transition is accomplished when the input becomes $i1$. If the knob is turned on while in state *cold*, the system moves to state *not cold*. When temperature reaches the ambient temperature (denoted by $T = \alpha$), the system returns to *cold*. The second level includes a detailed representation of state "not Cold." By combining this new FSA with FSA-1, we create FSA-2 (a complete model of the boiling process). FSA-3 is constructed similarly.

A transition condition may sometimes refer to a more detailed state specification than

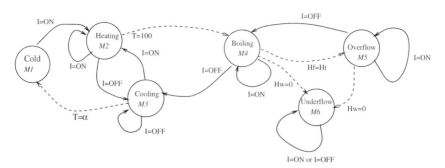

Figure 8.36 Six state automaton controller for boiling water multimodel (FSA-3).

is available at the current level of abstraction. For example, the transition from *Exception* to *Boiling* refers to the phase *Overflow*. In model building, such conditions are evidence that further state refinement is necessary; however, the hierarchy is useful not only for model building but also for facilitating question answering using a variety of abstractions. Figure 8.36 shows a compressed version of the boiling water model by compressing each of the three levels in Fig. 8.35.

8.12.3 A Heterogeneous Refinement

Heterogeneous refinement takes homogeneous refinement a step further by loosening the restriction of equivalent model types. For instance, we might have a Petri net at the high abstraction level and we may choose to decompose each transition into a block graph so that when a transition fires within the Petri net, one may "drop down" into the functional block level. For the last FSA in Fig. 8.35 we choose to represent each state as a continuous model. Specifically, each state will define how three state variables, T (temperature), H_w (height of water), and H_f (height of foam on the top of the water), are updated. Also, the parameter H_t is the height of the pot. In all cases, $H_f \geq H_w$ and $H_w, H_f \leq H_t$. The end result will eventually be a multimodel that will be coordinated by FSA-3.

The continuous models contained within states *heating* and *cooling* require explaining some physical theory before stating them. We model heat conduction since convection and radiation do not play major roles in this system. To derive a good continuous model for *heating* and *cooling*, we first define resistance. Let's define the thermal resistance $R = H/kA$ (H is the height of water, A is the surface area of the pot, and k is the thermal conductivity of water). We will ignore the resistance of the pot since it is not as significant as the resistance of water. The definition for thermal capacitance C is $C\dot{T} = q_h$, with q_h being the flow of heat from the heating element to the water. We will let C_1 be the capacitance of the metal pot, C_2 be the capacitance of water, and C be the total capacitance. Newton's law of cooling states that $Rq_h = \Delta T = T_1 - T_2$, where T_1 is the temperature of the source (heating element) and T_2 is the temperature of the water. Since T_2 is our state variable, we let $T = T_2$ for convenience. By combining Newton's law with the capacitance

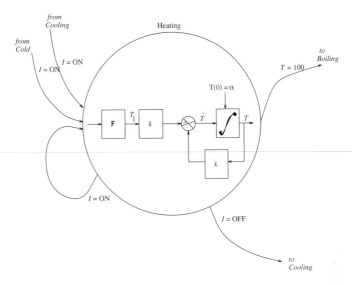

Figure 8.37   Decomposition of *heating* state.

law, and using the law of capacitors in series, we arrive at

$$k = \frac{C_1 + C_2}{RC_1C_2} \tag{8.3}$$

$$\dot{T} = k(T_1 - T) \tag{8.4}$$

Hence, Eq. (8.4) is a first-order lag with a step input representing the sudden change in temperature as affected by a control knob. Figure 8.37 displays a block diagram of *heating* within the *heating* state. Proper coupling is essential in heterogeneous refinements. That is, it must be made clear how components at one level match components at the higher level. Note in Fig. 6.8 the transfer function, which takes the ON/OFF input detected by the FSA and converts these input values to temperature values for the block network. Specifically, the block labeled 'F' performs the mapping from 'ON/OFF' to real-valued temperatures $\alpha \leq T \leq 100$. Due to the latent heat effect, T of water cannot exceed 100 unless all the water has vaporized. After all of the water has turned to steam, the temperature increases beyond 100; however, the system passes to the underflow state since $H_w = 0$.

The low-level continuous models $M_1, \ldots, M_6$ are defined as follows:[11]

1. (M_1) COLD: $T = \alpha$, $\dot{H}_w = 0$, $\dot{H}_f = 0$.
2. (M_2) HEATING: $\dot{T} = k_1(100 - T)$, $\dot{H}_w = 0$, $\dot{H}_f = 0$.
3. (M_3) COOLING: $\dot{T} = k_2(\alpha - T)$, $\dot{H}_w = 0$, $\dot{H}_f = -k_3$.
4. (M_4) BOILING: $T = 100$, $\dot{H}_w = -k_4$, $\dot{H}_f = k_5$.
5. (M_5) OVERFLOW: *same as BOILING with constraint* $H_f = H_t$.

---

[11]Models M_2 and M_3 exhibit first-order exponential behaviors and are, therefore, rough approximations of the actual boiling water system.

6. (M_6) UNDERFLOW: $T = undefined$, $H_w = H_f = 0$.

The system phase is denoted by Φ and the state variables are:

- T: temperature of water
- H_w: height of the water
- H_f: height of the foam

The continuous models share a common set of state variables. However, in general, state variables may be different for each M_i model.

There are also some constants, such as H_t for the height of top of pot, H_s for the starting height of water when poured into the pot; and k_i for rate constants. The initial conditions are: $\Phi = cold$, $T(0) = \alpha$, $H_w(0) = H_f(0) = H_s$, and $knob = OFF$.

8.12.4 Execution Results

We constructed a multilevel simulation in C that permits an execution of an FSA-controlled multimodel. There are four abstraction levels for the boiling water example: three FSAs and one level containing six sets of equations (or block models). The model is encoded as input to the simulator as follows:

1. *Output Type* (1 digit). The type of output: (1) output the state variable time trajectory specified in *Output Value*, (2) output the state trajectory for the abstraction level specified in *Output Value*, (3) output the input state trajectory, and (4) output a phase trajectory.
2. *Output Value* (1 digit). Used in conjunction with the *Object Type*.
3. *Number of States*. The number of states in the lowest FSA level.
4. *External FSA Transition Table* (1 line per state). The lowest-level FSA (level 3 for boiling water) transition function for external events. There are as many columns as there are possible input values.
5. *Internal FSA Transition Table* (1 line per state). The lowest-level FSA transition function for internal events. Each internal event is associated with a condition and there is an integer value assigned to each condition. For instance, condition $T = 100$ is assigned the value 2.
6. *FSA Abstractions*. The number of FSA abstraction levels (3 for boiling water).
7. *Abstraction Mapping Table* (1 line per level). This contains the mapping from lower-level to higher-level FSA states so that state/phase output can be provided at any level.

Phases are labeled (1) cold, (2) heating, (3) cooling, (4) boiling, (5) overflow, (6) underflow, and depending on the abstraction level, there are two other assignments: (2) not cold (for level 1) and (5) exception for level 2. Conditions are assigned: (1) $T = \alpha$, (2) $T = 100$, (3) $H_w = 0$, and (4) $H_f = H_t$. Given these assignments the multimodel simulator is executed with the data file (bwater.in) below under Unix as follows: bwater < bwater.in > bwater.out.

```
2 3
6
1 2
3 2
3 2
3 4
4 5
6 6
0 0 0 0
0 4 0 0
1 0 0 0
0 0 6 5
0 0 6 0
0 0 0 0


3
1 2 2 2 2 2
1 2 3 4 5 5
1 2 3 4 5 6
```

File `bwater.out` contains coordinate values for an individual run based on the input file. Figures 8.38 through 8.41 display the following:

- Fig. 8.38: input trajectory; turning the knob on and off over some time period. This was chosen at random to display phase switching.
- Fig. 8.39: temperature trajectory (T vs. time). Note that the temperature rises to $T = 100$ (in a first-order lag) in response to the step input and then it falls before reaching the ambient temperature of $T = 20$. The temperature levels off at $T = 100$. Note that the temperature is undefined when the phase changes to an overflow condition just prior to $t = 20$.
- Fig. 8.40: water height trajectory (H_w vs. time). The height of the water starts at 5 and then moves down using a constant slope, with steps, until it reaches a zero level, indicating underflow.
- Figure 8.41: foam height trajectory (H_f vs. time).

Figure 8.42 indicates a phase trajectory for abstraction level one (FSA-1). The behavior at this high level is simple: the system moves from state cold to not cold as soon as the temperature increases as a result of the step input from the knob. There is no return to phase cold since the water evaporates completely before the temperature can be reduced. Figure 8.43 displays the phases present in level 3 (FSA-3); from this plot we obtain a qualitative explanation of how the system behaves.

Figure. 8.44 displays the common three-dimensional state space (T, H_w, H_f) for MODELS, and Fig. 8.5 illustrates each phase by a shaded region of state space. Table 8.1 provides the formal correspondence of phases and input values with partition blocks. Let $\pi(u, p)$ be the partition block corresponding to the pair (u, p), where u is an input value

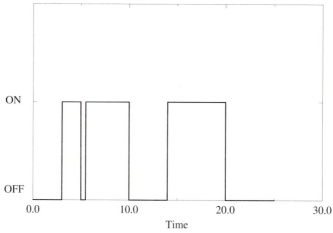

Figure 8.38   Knob input.

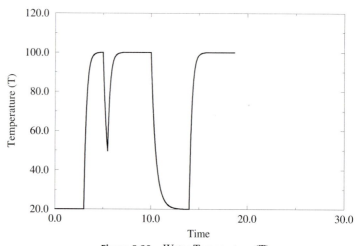

Figure 8.39   Water Temperature (T).

and p is a phase. The mapping π can be viewed as the labeling of partition blocks by input-phase pairs. In this case, the internal event transitions can be expressed in terms of boundary crossings of partition blocks. For example, in the boiling water example, the transition condition when the state is HEATING, under the input $I = ON$, is specified by the entry for $(BOILING, ON)$ in Table 8.1. This transition corresponds to reaching the boundary given by plane $T = 100$ in Fig. 8.45(c).

8.12.5 Question Answering and Choice of Model

We have specified three automata levels and one functional block level, we can use these models to answer questions about the system. We present sample questions and answers that can be practically derived from the presented multimodel. Given an arbitrary question

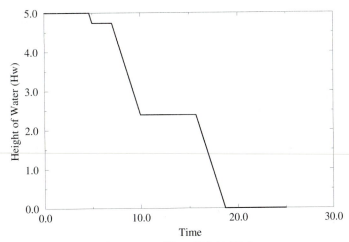

Figure 8.40   Water Height (H_w).

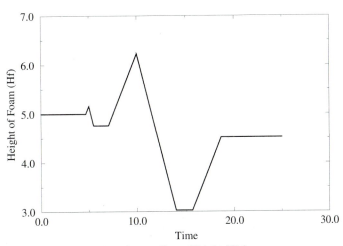

Figure 8.41   Foam Height (H_f).

as shown in Table 8.2, a model is chosen on which to base the answer. An actual natural language processing system does not currently exist, therefore the sort of processing implied by Table 8.2 is idealized.

Note the following acronyms:

- "Q-Type" means category of question. There are four types of questions specified: PRED (prediction), DIAG (diagnostic), EXP (explanation), and REF (reference). A predictive question is one where simulation is required to generate the answer. A diagnostic question yields a short answer specifying the immediate cause, whereas an explanation is more lengthy and may involve a summary of information to be presented. A reference question is one that does not involve analysis or simulation, such as "What is the material in the pot?"

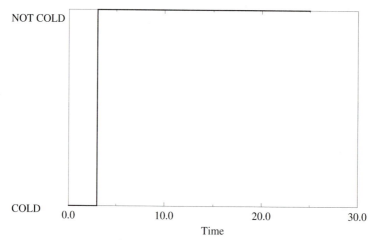

Figure 8.42 FSA-1 phase trajectory.

TABLE 8.1 PARTITIONING OF BOILING WATER STATE SPACE

Phase	I	T	H_w	H_f
COLD	ON	ϕ	ϕ	ϕ
COLD	OFF	$= \alpha$	$> 0 \wedge < H_t$	$> 0 \wedge < H_t$
HEATING	ON	$> \alpha \wedge < 100$	$> 0 \wedge < H_t$	$> 0 \wedge < H_t$
HEATING	OFF	ϕ	ϕ	ϕ
COOLING	ON	ϕ	ϕ	ϕ
COOLING	OFF	$> \alpha \wedge < 100$	$> 0 \wedge < H_t$	$> 0 \wedge < H_t$
BOILING	ON	$= 100$	$> 0 \wedge < H_t$	$> 0 \wedge < H_t$
BOILING	OFF	$= 100$	$> 0 \wedge < H_t$	$> 0 \wedge < H_t$
OVERFLOW	ON	$= 100$	$> 0 \wedge < H_t$	$= H_t$
OVERFLOW	OFF	ϕ	ϕ	ϕ
UNDERFLOW	ON	ϕ	$= 0$	$= 0$
UNDERFLOW	OFF	ϕ	$= 0$	$= 0$

- "Model" refers to the level of abstraction as specified with the models discussed so far. FSA-1 is the most abstract, two-state model. FSA-2 and FSA-3 are the five- and six-state models, respectively. The lowest level—the block model—is labeled COMB (i.e., combined).

- "Knowledge" refers to any additional knowledge that is needed to answer the question. An external knowledge base helps to provide answers to these questions.

Some questions require more information than can possibly be provided by the dynamical model alone. For instance, the "burnt smell" can be attributed to the overflow state.

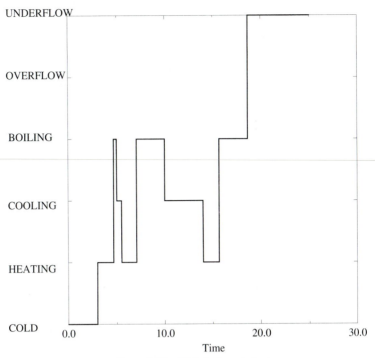

Figure 8.43   FSA-3 phase trajectory.

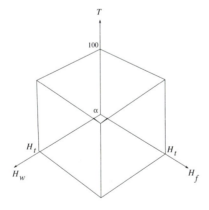

Figure 8.44   Continuous state space for boiling water system.

8.13 REFINING PETRI NETS

8.13.1 Two Flasks of Boiling Water

Even though we have demonstrated an FSA-controlled system with a heterogeneity afforded by blending both FSA models with block control models, we may also incorporate additional levels and model types. For instance, a new boiling water multimodel can be created by starting with FSA-3 as a primitive. Consider a scenario with two flasks of liquid; when both of the liquids are boiling, a human operator takes each flask and mixes the liquids into

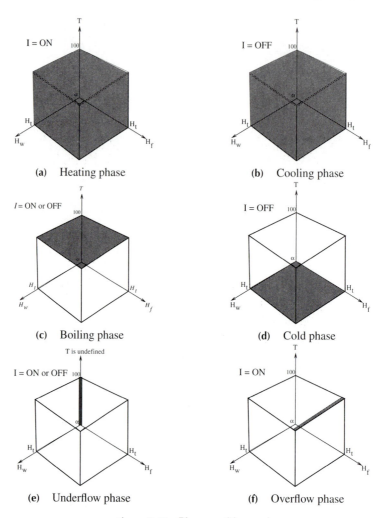

(a)　Heating phase　　　　　　　　(b)　Cooling phase

(c)　Boiling phase　　　　　　　　(d)　Cold phase

(e)　Underflow phase　　　　　　　(f)　Overflow phase

Figure 8.45　Phase partitions.

a separate container. For modeling the flasks, we can use two six-state FSA controllers which "drive" models $M_1, \ldots, M_6$ as before. A five-place Petri net serves to codify the constraint that both liquids must boil before the operator performs the separate function of mixing the liquids into another container. We use an extended Petri net in that (1) tokens can have attributes (i.e., have a *color*), (2) the net can accept a discrete input signal from outside the system (i.e., external events can occur), and (3) transitions can take an arbitrary amount of time (stochastic, timed network). Figure 8.46 shows a Petri net controller for this purpose. The Petri net, PN, is defined as follows:

1. $PNET = \langle P, T, I, O \rangle$

2. $P = \{p_1, \ldots, p_5\}$

TABLE 8.2   QUESTION-ANSWERING USING THE MULTIMODEL

Question	Q-Type	Model	Knowledge	Answer
What happens if I turn the knob (water is cold)?	PRED	FSA-1	None	System moves to "not cold."
If the system is warm, how can it become cold?	DIAG	FSA-1	None	System becomes cold when the water temperature equals the ambient temperature.
The water is now cooling. What happened before?	DIAG	FSA-2	None	The system was either "heating" with knob being turned OFF, or "boiling" with knob being turned OFF.
What is the maximum temperature of the burner?	REF	None	OBJECT	$190°$C
The water is heating and the temperature is now $74°$C. How long will it take to cool to room temperature if the knob is turned off now?	PRED	COMB	None	15.4 minutes
There is a burnt smell! How did that happen?	EXP	FSA-3	ACTION	The water was originally at room temperature. Then it began to heat when the knob was turned on. Eventually, the water boiled until the foam reached the top of the pot. The overflowing foam caused the burnt smell.
Why did the water start to boil?	DIAG	FSA-2	None	The water was being heated and the temperature reached $100°$C, which caused the water to boil.

3. $T = \{t_1, t_2, t_3\}$
4. $I : T \rightarrow P^\infty$
5. $O : T \rightarrow P^\infty$
6. $\mu : P \rightarrow Z_0^+$, for $i \in \{1, \ldots, 5\}$, $\mu(p_i) = 0$
7. $I(t_1) = \{p_1\}, I(t_2) = \{p_2\}, I(t_3) = \{p_3, p_4\}$
8. $O(t_1) = \{p_3\}, O(t_2) = \{p_4\}, O(t_3) = \{p_5\}$

Initially, the Petri net is in state $(0, 0, 0, 0, 0)$, meaning that there are no markers in any of the five places. There are two inputs to the Petri net: I_1 and I_2. These inputs represent discrete signals, with values of either ON or OFF representing the state of the knob controlling the temperature. The two places, p_1 and p_2, accept a discretely sampled input signal from outside the system; tokens will have one of two possible attributes corresponding to ON and OFF. A heterogeneous refinement of the Petri net specifies that transitions t_1 and t_2 refine

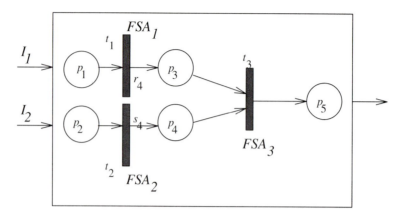

Figure 8.46   Petri net controller for three FSAs.

to FSA_1 and FSA_2, respectively. In this way, the input to the Petri net corresponds directly to the input signal for FSA_1 and FSA_2. Likewise, transition t_3 is refined into another FSA (FSA_3), which defines the states of the human operator: initial state ($INIT$), state of mixing the two liquids (MIX), and a final state (END). We specify an arbitrary amount of time for the second (mixing) state, where the time lapse (Δt) is sampled from a normal probability distribution, for instance. Figure 8.47 displays FSA_1, FSA_2, and FSA_3. Let the state spaces for FSA_1 and FSA_2 be $\{r_1, \ldots, r_6\}$ and $\{s_1, \ldots, s_6\}$, respectively. When the state of FSA_1 is r_4 and the state of FSA_2 is s_4, both liquids are in a boiling state. The input to each FSA is a knob controlling the desired temperature, and the output from each FSA (designating the current FSA state at any given time) is determined as follows: *When an FSA is in the boiling state, output a token; otherwise, do not output anything.* This policy is illustrated, in Fig. 8.46, by attaching the FSA state names (r_4, s_4) next to the relevant transitions (t_1, t_2).

Now, we are in a position to describe the multimodel execution for the Petri net controller. The Petri net is controlled by the knob input, which produces colored tokens into places p_1 (for knob 1) and p_2 (for knob 2). There are two colors: ON and OFF. As this discrete signal is passed to a transition, the transition "fires." Transition firing, for a Petri net controller, means that the systems drops down an abstraction level to both FSA_1 and FSA_2 in parallel. The input signals continue to drive each FSA, moving it from state to state. Each FSA state is further refined into a continuous block model. A combination of internal and external events causes a change in FSA state. Neither FSA produces an output until it enters the boiling state: r_4 for FSA_1 and s_4 for FSA_2. At this time, a token is produced and is put in places p_3 and p_4, depending on which flask is boiling. When both flasks are boiling, there will be tokens in p_3 and p_4, causing t_3 to fire. Since t_3 is further refined into FSA_3, we execute FSA_3 to simulate the action of the human operator—in this case, we model the operator by the amount of time spent mixing the liquids. The token produced by t_3 is realized by FSA_3 as an input value of 11 (representing tokens in both p_3 and p_4), which causes an internal state transition to the mixing state (MIX) of FSA_3. The mixing takes place for Δt and then the multimodel simulation ends.

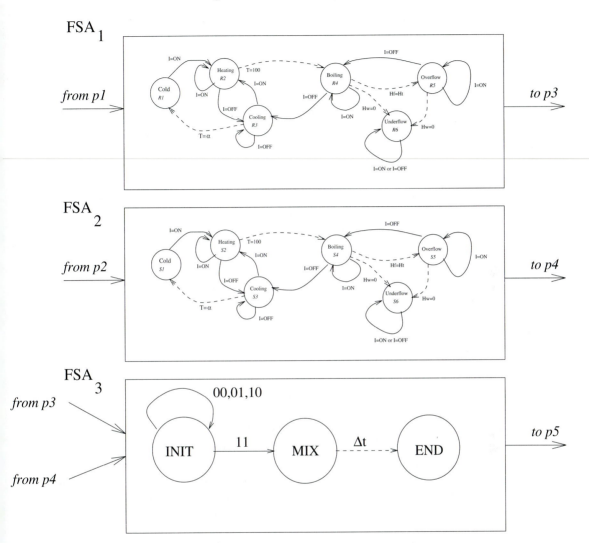

Figure 8.47   Three FSAs: two vessels and the operator.

8.13.2 Robot Operations

Robots perform many kinds of operations, including "pick and place" and tool operation. A robot may use one of many drill bits, such as in a workstation, where the robot must (1) take a part, (2) choose the right size bit, and (3) process the part by applying the bit to the part (i.e., where the part is a piece of sheet metal). Petri nets have been illustrated as models that can encapsulate the concept of *resource contention*. Consider a robot that drills a hole into sheet metal. Before drilling can occur, the robot must first have the correct bit and the part itself—which is traveling along an assembly line. Figure 8.48 displays the robot and its realization in the form of a two-state FSA, and a small Petri net with one

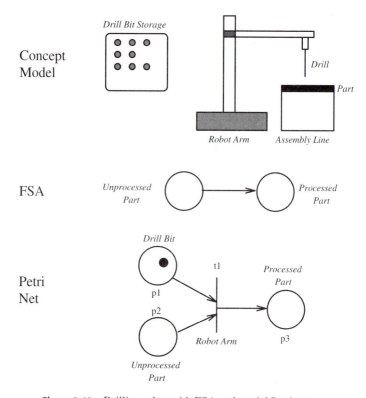

Figure 8.48   Drilling robot with FSA and partial Petri net.

transition, two inputs, and one output. The Petri net for the robot is partial in that it is only a small part of the overall Petri net describing the dynamics of the factory floor, which contains many assembly lines, parts, and robot arms. In this case, the resource is the robot arm, and its operation requires two conditions to be true: the existence of a part to process and the correct tool bit obtained from the bit storage area. Since Petri nets have a discrete state space composed by taking the cross product of all places in the net, we represent a three-place phase space in Fig. 8.49. As before, we can partition the state space to map the two FSA states, *unprocessed part* and *processed part,* to the state space of the Petri net, as illustrated in Fig. 8.50(a) and (b). We also illustrate, in Fig. 8.50, two possible trajectories through the discrete state space. The first trajectory in Fig. 8.50(c) reflects the case where (1) the two input places obtain tokens, (2) transition t_1 fires, (3) the output place obtains the token, and (4) the output place loses its token to a transition in the rest of the network not shown in Fig. 8.48. Figure 8.50(d) displays a second trajectory where two more tokens fill up places p_1 and p_2 after t_1 has fired.   The transition, t_1, of the Petri net in Fig. 8.48 can be further refined into an FSA, where we keep track of precisely which state the robot is in. Figure 8.51 illustrates a three-state refinement. First, we begin with the transition t_1, which we represent using a functional block model. The block model expects discrete values signals as input, so we have represented A' and B' as pulses whose values are stored, summed, and held constant by store-and-hold function (S & H) blocks which model the

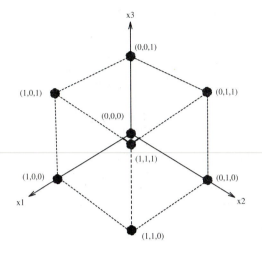

Figure 8.49   Discrete state space for Petri net dynamics.

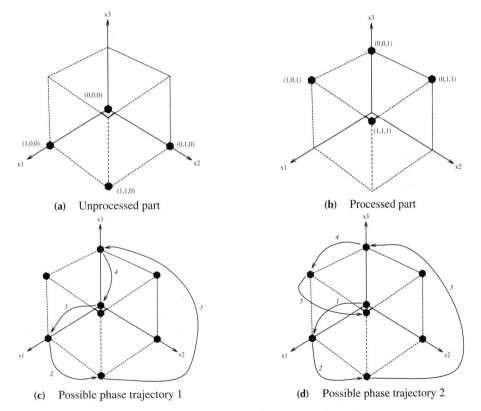

(a)   Unprocessed part   (b)   Processed part

(c)   Possible phase trajectory 1   (d)   Possible phase trajectory 2

Figure 8.50   Phase partitions for robot.

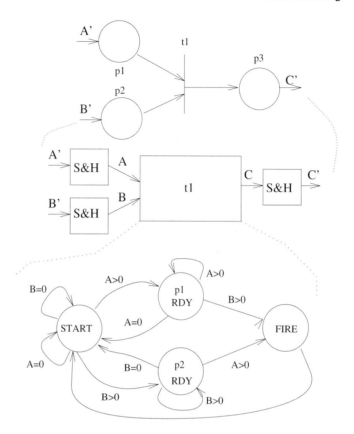

Figure 8.51   Refinement of Petri net transition t_1.

places $\{p_1, p_2, p_3\}$. These constant values are then fed as inputs A and B into the function, which performs the semantics of the Petri net transition. The FSA is the lowest level in Fig. 8.51 which defines how a transition works; nonzero signals, corresponding to a nonzero token count in each input place, will yield a "fired" transition.

8.14 EXERCISES

1. In the description of multimodels, we have focused on well-known model types covered earlier in the book. Clearly, this does not have to be so. Outline advantages and disadvantages to this approach. Should the number of possible model types present in multimodels be decreased, kept the same, or increased?

2. Consider the two FSA models in Fig. 8.52. Model 1 is a simplification (abstraction) of model 2. Demonstrate this by defining a homomorphic mapping which preserves all behavior of model 2 using the mapping.

3. Petri net transitions were refined into functional blocks in examples shown in Figs. 8.47 and 8.51; however, recall from Ch. 4 that the basic unrefined transition acts as an event. What does this say about the relationship between events and functions?

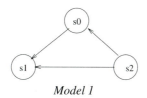

Model 1

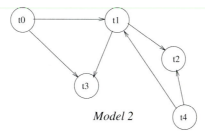

Model 2

Figure 8.52   Two example systems.

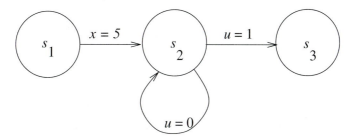

Figure 8.53   FSA multimodel.

4. Construct a multimodel that defines the following scenario. In Mary's bedroom, there are a set of dominoes that are lined up on the floor so that pushing the end dominoe causes a cascade of the other dominoes one after the other. Each domino is either (1) standing, (2) falling, or (3) down.

5. In the boiling water multimodel in Fig. 8.34, our lowest abstraction level represents a state space composed of three dimensions. However, this state space can serve as an input to yet another system if we drop something (modeled as a new system) into the water. We can make the boiling water system (with the newly submersed subsystem) more real world if we consider that instead of simulating boiling water, we are simulating the weather. Then the submersed subsystem can be anything affected by the weather. Discuss this relationship in more depth.

6. Consider the FSA in Fig. 8.53. s_1, s_2, and s_3 are the three states of this automaton, and u is a binary-valued input ($u \in \{0, 1\}$). x is a state variable for a model that refines each state. Let this model be $x(t + 1) = x(t) + u(t)$. Answer the following.
   (a) What is the state space for the lower level?
   (b) How do states s_i partition the lower-level state space?
   (c) Define an input signal that causes the system to go to s_3.
   (d) Can the system be in s_2? If yes, define the signal.

7. [From Ashby [6] (4/7, p. 49)]. Suppose that a machine (transducer), P, is to be joined to another, R. For simplicity, assume that P is going to affect R, without R affecting P, as when a microphone is joined to an amplifier or a motor nerve grows down to supply an embryonic muscle. We must couple P's output to R's input. Evidently, R's behavior, or more precisely the transformation that describes R's changes of state, will depend on, and change with, the state of P. It follows that R must have parameters for input and the values of those parameters must be at each moment some function of the state of P. Suppose that the machine or transducer R has the three transformations

$\downarrow$	a	b	c	d
R_1	c	d	d	b
R_2	b	a	d	c
R_3	d	c	d	b

and that P has the transformation on the three states i, j, and k:

$$P: \downarrow \quad \begin{array}{ccc} i & j & k \\ k & i & i \end{array}$$

P and R are now to be joined by specifying the value of R's parameter, call it α, when P has any one of its states. Suppose that we decide on relation Z as follows: If the state of P is i, j, or k, the value of α is 2, 3, and 2, respectively. Given this information, draw a multimodel representing the coupled machine (i.e., system). Specify all arc labels, blocks, and state transitions.

8. Even though multimodels can be used to define simple two- or three-level systems, they are tailormade for systems that are very large with many abstraction levels. Consider building a model for what happens when a group of people from apartment building B travels to work. We will assume that there are three work shifts and that each person works in one of these shifts. Each person goes through several phases (locations) to get to work and must queue at intersections where the traffic is controlled by lights with phases. Create a multimodel for this scenario.

9. Provide lower-level models for phases "normal," "adjusting pressure," and "off" in Fig.8.31. The model shown in Fig. 8.32 was illustrated for "adjusting temperature."

10. Recall the single function defined in Fig' 8.16. Create additional arcs and arc values if this function is to reflect the operation of a light switch with a memory.

11. In the industrial plant multimodel in Fig. 8.31, we had two state variables: pressure and temperature. If we add a third variable, density, how would our declarative FSA model (inside object *plant*) change?

12. Basins of attraction calculated for nonlinear systems are one way of creating FSA-controlled multimodels and of capturing the qualitative dynamics of the system in a concise form. The reason for this is that basins of attraction can often map directly to linguistic terms as they did for the pendulum phase space. Under what circumstances will there be too many basins—causing a situation where basins no longer have any linguistic correspondence, thereby making it difficult to abstract a higher-level, finite state model?

13. Specify a three-level multimodel for a queuing system by defining each layer as follows: (1) top level: FSA, (2) middle level: functional model (queuing), and (3) spatial model. The queuing system will be composed of a source of entities, two servers in series followed by a sink. The spatial model is necessary to determine the exact position of entities over time.

8.15 PROJECTS

1. Develop a graph-based notation for multimodels which stresses the model types and levels. For instance, a two-level model containing an FSA at the highest level and equations at the lowest level (for all states) would result in a graph with two nodes and one arc. The nodes would be labeled *FSA* and *Constraint*. If just one of the states refined into a block model, your graph would need to be modified. Specify a graph grammar and give examples.

2. Pick out a text or journal article in the area of *nonlinear dynamics* or *complex systems* that contains equational systems along with pictures of their basins of attraction. From this, construct simple two-level multimodels, with level one being an FSA and level two containing the equations of state. Document your research on two such systems.

3. Build a graphical tool that constructs and displays automatically FSAs generated from basins of attraction for nonlinear systems.

4. Provide an expert system tool that helps someone decide what kind of model to use at a particular level in a multimodel. Ask them first to list all attributes of interest to them in the system. Then, group the attributes according to level of abstraction. For instance, if someone says that he or she is interested in *velocity*, *position*, *phase*, and *container car number* (for a set of train cars carrying merchandise), you would need three levels, since velocity and position should be on the same level and are separate from the other two.

5. We have covered many instances of declarative models embedded within functional blocks. Also, we have seen cases where the states of the declarative model have lower-level functional dynamics. How far can this "flipflopping" go? That is, are there any limits to the number of levels that one can create in a multimodel composed of declarative and functional elements? What about for other types of elements (spatial, constraint)?

6. A multimodel containing countless levels can be created when we want to model something like a factory, city, state, or world. Let's consider the factory where something is manufactured, such as clothing or any other industrial product. Pick a specific product, find information on how the product is made, and build a multimodel of the factory using at least six abstraction levels. As an example of the highest abstraction level, consider an FSA demarcating days in a specific month. A low abstraction level would involve phases within a specific factory floor machine or a lower-level characteristic such as change in a low-level state (angle, position of a piece of a machine).

8.16 FURTHER READING

The basics of sets, partitions, and mapping are found in texts on discrete mathematics [210] while automata methods are found in [106]. Further information on handling internal events can be found in any numerical methods text, such as [177]. The text by Hostetter et al. [111] provides a good introduction to multivariable root finding. Pritsker [178] and Cellier [33] present the problem of discontinuities stemming from internal events within the context of simulation. Multimodels are discussed by Fishwick et al. [66, 68, 71, 73]. The term "multimodel" was coined by Oren [162, 163] in his discussion of a taxonomy for modeling systems. The idea of using abstraction in systems is not a new one since the morphisms required for behavior-preserving mappings were explored first in systems theory [165] and later in simulation theory [233, 234]; however, morphisms have gained recent momentum within the simulation community [60, 61] as a way to handle complex models. Sevinc explores the creation of abstract models from model behavior [203]. The use of internal versus external transitions corresponds directly to those methods in the DEVS [237, 238] specification language, which has been shown to have a wide variety of applications [193, 192]. Praehofer has extended DEVS to include nonhomogeneous, combined models of system-theoretic models [174, 176] and has built a combined simulation package around CLOS. Andersson and Mattsson [5, 144, 145] have constructed an object-oriented toolkit called Omola based on hierarchical multimodel principles. Multimodeling is an extension of the more traditionally known *combined modeling* techniques [33, 179, 214] which normally combines discrete event and continuous model formalisms within the same model.

8.17 SOFTWARE

Simpack has the following multimodel implementations:

- *combined:* discrete event and continuous simulation of a cashier with a moving belt for groceries
- *bwater:* simulation of the six-state FSA controlling the boiling water multimodel

An ongoing graphical implementation for multimodeling is underway. For an existing simulator capable of multimodel definitions, consider Omsim and Omola which are available from the University of Lund in Sweden (Sven Erik Mattsson and Mats Andersson). The first file to read is `control.lth.se/pub/cace/README`.

Chapter 9

Parallel and Distributed Simulation

Nature operates in parallel.[1]

9.1 OVERVIEW

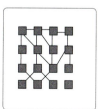

Our model execution methods so far have been serial; that is, we have simulated our model on a sequential, von Neumann architecture characteristically defined as a machine with one central processing unit (CPU) and one memory store. A program is executed by having a counter keep track of the current instruction to execute. The instruction executes which causes register and main memory transfers. We are certainly limited by the hardware that we use for simulating models. Many decades of research into novel architectures have yielded machines that have more than one

---

[1] Anonymous.

memory store and more than one CPU. With multiple CPUs, there are many more issues and considerations: How do we partition programs, and at what level of granularity can we execute the partitions, in order to reduce message passing overhead?

Parallelizing a simulation program requires a partitioning of our model to prepare it for execution on multiple CPUs. Since simulations can take exorbitant amounts of CPU and memory resources, it makes sense that if we parallelize our models, we will get answers sooner and make it possible to build and execute even larger, more complex models. We are concerned with two primary types of concurrent computation. *Parallel simulation* refers to the use of a machine with multiple CPUs and, possibly, multiple data stores. *Distributed simulation* generally refers to parallel computation over a local or wide area network. The two are very similar in concept, and the technical problems associated with each have a strong intersection. There are some differences with parallel simulation because there will be little contention for network resources since the inter-CPU connections will be short and designed specifically for use within an installation. Also, there is an inherent lack of homogeneity with distributed network processing units and more of a potential problem with synchronization. In distributed simulation, we can ostensibly spread our model over a wide area network that has worldwide geographic coverage; the problem with this, though, is that the network channels will be used for other purposes and by other users—causing flooding and contention over the network resource. Moreover, our distributed model may have humans "in the loop." For example, humans may be training within a virtual battlefield or integrated task environment and are physically constrained to operate in a geographic area, thereby forcing the *distributed* approach. The bulk of this chapter covers technical considerations with the parallel "batch" execution of models where humans are not in the loop, although we will also discuss real-time simulation, which characterizes the method of using simulation for training humans in areas such as military combat and nuclear power plant control. We will not separate the two types of simulation (distributed vs. parallel) in our discussion since there are many overlaps, especially considering the variety of machine architectures.

9.2 ARCHITECTURE STYLES

There are four styles of parallel architecture. They are often identified as follows:

1. Single instruction stream, single data stream (SISD)
2. Single instruction stream, multiple data stream (SIMD)
3. Multiple instruction stream, single data stream (MISD)
4. Multiple instruction stream, multiple data stream (MIMD)

The serial, von Neumann architecture is an SISD architecture since it has one instruction executing at any one time, one processing element, and one memory store. The four types of parallel architecture are shown in Fig. 9.1. In SISD, the CPU (P) obtains an instruction and references memory, and then it executes the instruction, after which the program counter increments to the following instruction. In SIMD, there are several pieces of data, all of which apply to an instruction simultaneously. The best example of SIMD execution is

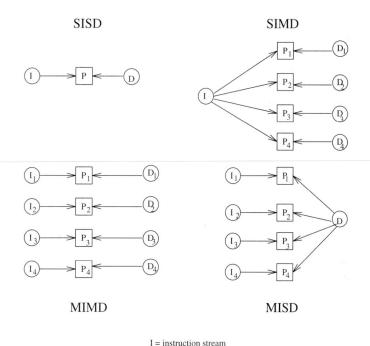

I = instruction stream
D = data stream

Figure 9.1   Four parallel architecture styles.

considering an operation, such as addition, using vectors instead of scalars. Consider adding vector $\langle 1, 3, 10 \rangle$ to $\langle 4, -1, 0 \rangle$. In an SIMD machine the instruction is addition and we can obtain parallel execution of the addition by having multiple streams of data. Processor 1 executes $1 + 4$, 2 executes $3 - 1$, and 3 executes $10 + 0$, all in parallel. The result is placed into the vector $\langle 5, 2, 10 \rangle$. Note that the word "instruction" is more generally defined as a program sequence or fragment; it need not be a single instruction but could, instead, be a function or program (i.e., a stream). The concept of *instruction* in this taxonomy is best envisioned as a *program*. The same holds for data—the data refer to a stream of data and not necessarily one datum.

MISD architectures can be viewed as pipelined machines.[2] A pipeline is one where a single data stream is operated upon by many instructions. The most natural human analog of this is an assembly line operation. In an assembly line for automobile parts, for instance, there is quite a lot of parallelism; each person performs a specialized function while receiving the stream of parts. The first person might inspect the part for defects while the others in the pipeline perform welding and cutting tasks. The original stream is gradually transformed as the parts pass through each of the specialists. In computation, consider functional composition; composition is precisely the same as the assembly process. For instance, consider that our job is to compute $f_1(f_2(x))$; f_2 is a function that operates first on

---

[2]There is controversy about how to characterize pipelined architectures. If the data are assumed to be aggregate in nature (as in a vector), the MISD characterization is reasonable for our discussion.

x and then hands this transformed data to f_1, which operates on it further. Let's build a small parallel MISD machine for evaluating polynomials of the form $x^2 + x$ and assume that we have 1000 values for x that we wish to use. Note that $x^2 + x = x(x + 1) = *(x, +(x, 1))$, where $f_2 = +$ and $f_1 = *$. Our pipeline, composed of the two processing units associated with f_1 and f_2, can now operate in parallel as they receive the 1000 individual items for x. The pipeline looks like: $f_2 \rightarrow f_1$. While f_1 is evaluating the result for one value of x, f_2 can be adding 1 to the next value for x.

MIMD architectures are completely distributed in the sense that each processor has its own memory and communicates with other processors through communication links. Each link has a channel at both ends, with channel buffer to store information being passed in either direction. MIMD machines mimic animals and their interactions with other animals. Each animal has a processor (neural network) and data store (memory); the physical link among animals is the air they breathe through which sound vibrations travel. MIMD machines also are a natural analog to physical networks; one can create a one-to-one mapping between nodes in a phone network, for instance, and nodes in an MIMD machine; the links in the phone network are represented as links among MIMD processors.

9.3 SUITABLE MODEL TYPES FOR PARALLELIZATION

We have covered several key model types in the text and we should first wonder which types are appropriate for parallelization. In general, the answer is *all of them*, although, in practice, functional and spatial models are best suited for parallel and distributed architectures. There are no hard and fast rules to justify this statement, but we can look at current *hotbeds* of activity in parallel simulation and we usually find that our premise generally holds true. Let's discuss each model type and its relation to parallel methods which might be applied to its constituent model subtypes. We'll exclude multimodels from the list by noting that one creates parallel multimodels by parallelizing their functional and spatial components.

1. *Declarative Models.* These models are most useful in describing *sequential* behavior in the form of state-to-state transitions. The states may be defined at different abstraction levels and often describe phases for lower-level model activity. The first obstacle to parallelizing declarative models is the serial execution aspect. There is good reason why the more theoretic term for these models is *sequential machines*. In the case where we consider states in declarative models serving as placeholders for more complex model machinery (as with FSA-controlled multimodels), parallelization may be appropriate, but we can cover this by looking at the model types used during state refinement.

2. *Functional Models.* The *coupling* attribute of functions leads us to parallelization of function execution due to the pipelining effect of this coupling. *Partial order* is the other aspect of functional models that is perfectly adapted to parallelization.

3. *Constraint Models.* Parallel methods exist for difference and ordinary differential equations; however, we have not included them, for the following reasons: Difference models are usually small, and parallelization (although it exists in the form of *cyclic reduction*) can be useful, but many large difference models are better modeled

functionally—as with the case of queuing and digital logic networks. The semantics are the same; however, difference models are not as flexible in modeling large-scale systems as are their functional counterparts, which adhere to connectivity, coupling, and hierarchical design principles. For ordinary differential equations, very large, complex systems are better modeled at higher levels of abstraction. An example of this is use of a digital logic network instead of an analog circuit formulation. There are several layers in an electronic circuit and one moves to the next-higher aggregate level (geometric ⇒ electrical ⇒ gate ⇒ block ⇒ behavioral) for modeling as the number of model components increase. In general, as systems become large and complex, one aggregates and abstracts to form components which form the basis for new functionally or spatially oriented model types.

4. *Spatial Models*. Partial differential equations form the traditional foundation of both classical and statistical physics. These kinds of problems are complex, large scale, and *require* parallelization, where possible to obtain results in a timely manner. In cases where PDEs are not used—as with particle-based approaches—the need for parallelization is clear, and methods of partitioning are based on spatial discretization. Partitioning in space is sometimes termed *domain partitioning*.[3] Spatially based computational methods used in supercomputing applications include finite differencing or finite volume (the computer methods underlying the numerical translation of PDEs), finite element modeling, Monte Carlo methods and particle dynamics. Computational experiments on quantum systems, in particular, cannot practically be carried out without the aid of massive parallelism in one of various forms.

As systems become complex, they often tend toward functional networks or spatial descriptions. A good reason for this is that functional and spatial models lend themselves to *hierarchical decomposition*—that is, as one refines a spatial model component, for instance, it turns into another spatial component and not another type of model. Systems composed of hundreds or thousands of ODEs can be integrated in real time on fast workstations, so it is natural to ask whether parallelism could be used, potentially, for very large scale ODE systems. However, beyond fairly small ODE equation sets, most systems change their model characterizations to represent functional dependencies or spatial composition. A more subtle point relates to the size of constraint models and how this affects potential parallelization of these models. Constraint models are by their nature difficult for humans to work with. This is because of the interaction involved with constraint variables. Constraint relationships can cause systems to behave in nonintuitive ways. For instance, trying to model and follow the behavior of a differential amplifier from first principles (by reasoning about component behavior) is difficult; humans are much more adept at following functional links and flow. If we can, we tend to want to break apart large constraint models by turning them into functional or spatial models, which are easier to understand.

Another issue arises when we are concerned with real-time simulation (Sec. 9.5.2). With human–simulation interaction and real-time constraints, one requires that the simulation run forward in time as opposed to using a relaxation method which may converge to

---

[3]In a recent book on the state of the art in *supercomputing* [120], *all* of the modeling methods described are *spatial*.

a solution, but not necessarily in a temporal sequence. While animating simulation output and coordinating with human controls, we require a method of parallelism that gives us a correct state trajectory at *all times* and in the *correct temporal order*.

9.4 FUNCTIONAL MODELS

A function is defined as an object's method and we associate an object directly with a function in a functional model. In parallel simulation, each object will receive messages as input and produce output messages. The only difference between methods for parallel simulation and those presented in Ch. 5 is that input can come not from the execution of an event routine, but from another computer, which is simulating functions that produce output. For most of our examples in this section, we will map *each* function object in the functional model directly to a physical processor, although we should recognize that a processor could be running its own serial simulation of a subpartition of the entire functional model. The following concepts are central to understanding parallel simulation of functional models:

- Each physical processor (containing a functional object) contains input and output channels, with a first come, first serve queue for each channel. The queue size is finite and filling a queue has an effect on the simulation's progress. The queue has nothing directly to do with data structures that may be associated with the functional object. That is, if the object is a server, which requires a queue data structure, this structure is separate from the channel buffers. Channel buffers are solely for sending messages to and from physical processors. Sending messages to a server queue is accomplished by routing messages from the input channel queue to the server queue.

- We assign a local virtual time (LVT) to each physical processor. Each local virtual time represents the clock time only for that processor. Sometimes it is convenient to associate a time with channels even though there may not be a clock physically stored on a channel unless there is channel hardware. We specify the time for a channel to be the time of the last message which traversed the channel and arrived in the channel buffer at the end of the receiving channel.

9.4.1 Causality

The potential for parallelism is rooted in the concept of *causality*—some instruction sets must be executed before others. More generally, if events A and B modify state variable X, the event with the smaller time stamp must execute first, to avoid a *causality error*. Consider two functions f_1 and f_2, where $f_1(u) = u + 3$ and $f_2(u) = u * 5$. It makes a difference in which order we execute these functions (i.e., instructions). If we let $u = 2$ and calculate $y = f_1(f_2(u))$, then $y = 13$, whereas $y = 25$ if the function order is reversed, as in $y = f_2(f_1(u))$. For functional models, causality is inherently specified in the graph topology of the functional network. If function A's output feeds into function B's input, we can say that events associated with A (such as arrival, departure, or internal state events for state transitions occurring within block A) *cause* events associated with B. For spatial models, causality occurs between *processor boundaries*. So if a large two-dimensional space is partitioned and each partition is delegated to an individual processor,

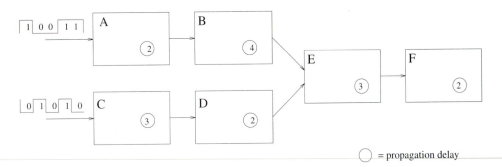

Figure 9.2   Digital logic network with node delays.

events occurring in one partition (*A*) will affect or cause events in an adjacent partition (*B*) if an element from partition *A* moves to partition *B*. Causality and directionality are equivalent concepts. Where a model exhibits a directional flow of entities or information, one can establish a causal dependence.

9.4.2 Synchronous Method

When memory is shared among processors, which read and write to global memory locations, we place certain data structures into the shared memory and then activate multiple parallel processors to perform a task. Two such structures are the most relevant to simulation: the clock and the future event list (FEL). Having a global, shared clock with multiple processors encapsulates the way that serial computer architectures are constructed. In this sense, each serial architecture (composed of many independently executing chips) is a parallel simulation of itself. Having a shared data structure, such as the FEL, means that processors act as schedulers which simultaneously deposit events to be processed in the FEL.

Shared memory methods are associated with tightly controlled, global synchronization of some sort. This is because we cannot let individual processors "run loose" as a result of interdependencies among individual data in shared memory. A shared clock involves small time step synchronization, whereas a shared FEL involves time synchronization when two or more events have the same time stamps. Consider Fig. 9.2, which shows a simple network with nodes having propagation delays (shown in circles). Propagation delays are associated with all real functional elements such as logic gates and represent time taken to process input to produce an output response. The network is generic but can be viewed as a digital logic subnetwork or queuing network which requires simulation when a shared memory approach is desired.

Method 1: shared clock.   With a shared clock, each object is connected to the clock in much the same way as in a digital logic gate network. As time progresses at each tick of the shared clock, *all* objects are updated by executing the function (i.e., method) associated with each object. For the network in Fig. 9.2, consider objects *A* through *D* to be inverters which take a digital signal and invert the bits. Since *A* has a delay of 2, the object will contain an internal array of length 2, which will affect the necessary delay of output. Other objects will contain similar arrays (or circular queues) to enable delaying the signal. If the

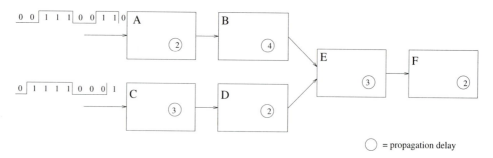

⬭ = propagation delay

Figure 9.3   Digital logic network with node delays.

input trajectories to A and C are 10011 and 01010, respectively, and all outputs are initially reset to zero, the outputs from A and C will be 0001100 and 0001010 for the first seven time steps. The bits enter a block from *left to right* since the bit stream represents a time-based trajectory. At every time step, each block (from A to F) will be executed. So at the first time step, for instance, A and C are executed in parallel, along with all other elements.

At first, this may seem to be an ideal method of parallelism, and in fact, it is quite reasonable on a shared memory architecture with the following reservations:

1. There needs to be a lot of connectivity or message passing so that the global clock synchronizes with each object. In addition to having messages sent for inputs and outputs, it is necessary to install either a global data structure representing the clock, or a separate clock object that sends messages to all other objects. Whether the function blocks read a clock data structure or the clock sends messages, the end effect is the same: increased complexity in message passing and function block-clock connectivity.

2. The synchronization of the system occurs at *every* time step whether or not there is a change in input or output for any particular block. Some signals may not change very often, providing for infrequent messages to blocks; however, the global clock synchronization must nevertheless take place at every time tick.

Method 2: shared FEL.   Method 1 (shared clock) is a kind of parallel time-slicing approach as presented in Ch. 5. To improve it, we can utilize event scheduling instead of time slicing so that, even though we still use a global clock, the time increases by *leaping* forward instead of plodding through every tick of the clock. We let an event occur when the digital signal changes from zero to one, or vice versa. Let's consider the following events to be scheduled in the global future event list (FEL) before any processing starts: Consider Fig. 9.3. Each input will be a time-stamped message denoting a change in signal. The message will be of the form (*timestamp, targetprocessor*). Therefore, a message of the form $(10, C)$ is stamped with time 10 and will be transmitted to processor (or block) C. The initial time ordered FEL is: $(1, C)$, $(2, A)$, $(5, A)$, $(5, C)$, $(7, A)$, $(8, C)$, $(9, A)$. For digital logic simulation, this collection of events represents two input signals to A and C. The input signal to A is 0011100110 and the signal to C is 011110001. The events in the FEL record when the signal changes as opposed to the value of the input signal at every clock step. This is just shorthand, in the form of message events, for input signals to A and C. As usual, the event at the head of the FEL is selected first for execution: $(1, C)$. This means that the global time is changed to 1 and function C executes. C schedules D to occur at time $1 + 3 = 4$, so we need to insert $(4, D)$ in the FEL just prior to $(5, A)$. Next $(2, A)$ moves to the head of the FEL and executes, scheduling $(4, B)$, which is placed in the FEL just behind $(4, D)$. Now $(4, D)$ and $(4, B)$ can be executed in parallel because they have exactly the same time stamp. This is the basic approach for the shared FEL method: events with the same time stamp can execute in parallel. This method can be improved by noting that some messages with different time stamps can also be executed in parallel. For instance, $(4, D)$ and $(5, A)$ can be executed in parallel—the reason being that these two messages are generated on different segments of the functional network which are not causally related.

Given the topology of the functional network, we can more intelligently schedule all events that can occur in parallel without sacrificing causality constraints. Also, we may be able to set up a *sliding time window* that provides for all messages with time stamps in the window (τ_{min}, τ_{max}) to be executed simultaneously without a causality problem. This method is a good one when there are many events with equal (or close to equal) time stamps. In such cases, we can obtain a significant degree of parallelism. If, on the other hand, we have highly interconnected functional networks without much time stamp equivalence, we will have an inefficient method. One way around these problems is to have all functional blocks send messages to one another asynchronously without the need for a global clock or FEL.

9.4.3 Conservative Asynchronous Method

We will now consider the conservative approach to asynchronous parallel simulation described below.

Algorithm 9.1. Conservative Simulation Method
Procedure Main
> *If at least one message in each input channel buffer then*
>> *Select the input message with the* minimal *time stamp*
>> *New time stamp = old time stamp + processing*
>>> *time of block*
>> *If preemption not implemented then*
>>> *Update local clock time to new time stamp*
>> *Else wait to update local clock until a message received*
>>> *with a time stamp $\geq$ new time stamp*
> *Else this object waits/blocks*
End Main

The conservative method picks the lowest time-stamped message from its input channel buffers and then uses this message to create an output message. There is the basic assumption that virtual time stamps on messages could be sent in any order (specifically, they could be sent out of order). In queuing networks, for instance, out-of-order packets arise when certain network resources can be preempted by high-priority messages. When

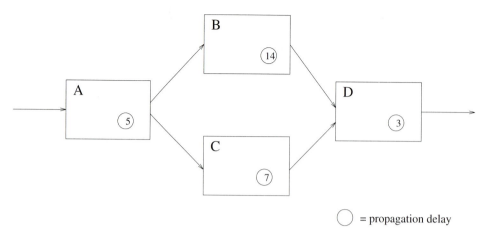

Figure 9.4   Network using asynchronous message passing.

there is no preemption of tokens, the block can immediately update its local clock to the time stamped on the next input message plus the sampled time from the local propagation delay. However, if preemption is permitted, we cannot update the local object clock until we are absolutely sure that there can be no preemptive messages that will arrive *while* the current message is being processed. Consider Fig. 9.4, which represents a preemptive functional model with four functions. Take an arbitrary processor such as C. C will accept messages stamped with time t (t, C) and then might try to send them on to processor D with time $t + 7$ $(t + 7, D)$. Assume that C's LVT is 20. C cannot send message $(27, D)$ until it receives a message from A with a time stamp greater or equal to 27. This is because A could easily send a priority message to C with a time stamp greater than 20 but less than 27. This priority message could then sample a delay less than 7. We have no way of knowing without incorporating special control messages which A might send to C, regardless of whether A is due to send regular messages to C in the future. The processing of a message is usually quite simple and involves (1) adding a sampled propagation delay (in logic networks) or a server delay (in queuing networks) and then (2) making up a new message with this accumulative time stamp. The possibility of a preemption slows message traffic down until functional blocks ensure that enough time has passed for preemption not to have any effect on the message being generated for output.

We will now consider the non-preemptive case. Messages are input to block A in the following time-stamped order: $(0, A)$, $(2, A)$, $(5, A)$. The first and third messages are to be routed from $A \to B$ while the second message is to be routed from $A \to C$. Upon arrival of the first message, channel buffer times are updated and block B is processed yielding the situation in Fig. 9.5. Local clock times are in the squares within each object. If preemption had been allowed, block B would not have been allowed to transmit a message with time stamp 19 until it had received an input message with a time stamp ≥ 19. Figure 9.6 displays the channel buffer times after the second input to block A, routed through C. D now has two messages in its input channel buffers. According to the conservative algorithm, it picks the message with the lowest time stamp 14 and sends a new message by adding its own

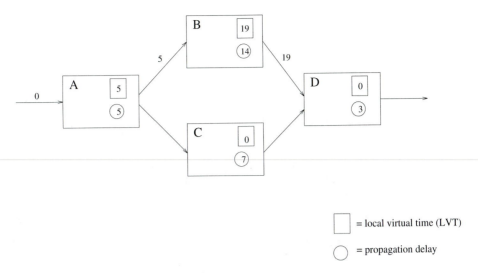

Figure 9.5   Network state after message $(0, A)$.

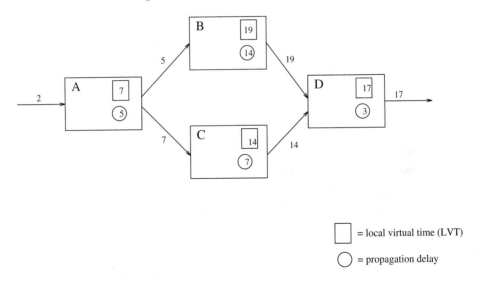

Figure 9.6   Network state after message $(2, A)$.

propagation delay to the input's time stamp. Block D must now wait on channel $C \rightarrow D$ until it is sure that there will be no messages with a time stamp less than 19 (the time value of the channel buffer $B \rightarrow D$). After all, a message could come from block C with a time stamp of 18 for all that D knows about its incoming messages.[4] Thus, with the conservative

---

[4]How much is known about future messages is a central problem in simulating parallel functional models. The ideal situation is if block D can *predict* future message time stamps, thereby allowing it to proceed without blocking.

approach, a functional block is waiting on the lowest time stamp message on each of its input channels and cannot proceed if no messages are forthcoming along these channels.

Supposing D knew, when it looked at each of its two input channels, that the next message to come via C would have a time stamp ≥ 19. This means that D could unblock itself and process the message sent by B, since D would know that no lower time-stamped message would arrive which could create a causality problem. Consider, for instance, a message coming into A which has a time stamp of 8. Further, suppose that this message is routed through C to D. Just by looking at the constant propagation delay times for A and C, we see that any future messages to be routed via C coming after the one with stamp 8 must have a time stamp of at least $8 + 5 + 7 = 20$ when they arrive at D. This means that even though the channel $C \rightarrow D$ time is 14, we can *predict* that no future message will arrive before 20. The capability to use prediction, called *lookahead*, can speed up conservative simulations and help to prevent deadlocks.

The subject of deadlock deserves special consideration for functional models. Deadlock is a condition described in the operating systems literature where a system *hangs* due to a cyclic dependency among processors. Given a functional graph, a deadlock can occur when a cycle exists $A \rightarrow B \rightarrow C \rightarrow A$ (for three example processors): A needs to send something to B, but it must wait on C, which is in turn waiting on B, which waits on A. Consider Fig. 9.4. It is not at all clear that deadlock can occur in this graph. The graph has no apparent cycles. But we note that messages that are transmitted between any two objects can carry information (say, token attributes) as well as control information of the *acknowledgment* variety. The arcs shown in Fig. 9.4 are *information* arcs; they carry token information defining the type and value of the data being transmitted between objects. However, there are usually *control* arcs that reflect the need to synchronize and acknowledge transmitted *information* and these are left out of Fig. 9.4. Thus, we have two things here: (1) the graph representing the functional model (Fig. 9.4), and (2) the graph representing the communications network facilitating the acknowledgment and control-type message passing (which is not shown). Now we see that from a more general communications view, all arrows in Fig. 9.4 are bidirectional even if token messages are transmitted in only one direction. This demonstrates the potential for a cycle. Now, how could deadlock occur in Fig. 9.4, for example? Let's make an assumption first and then discuss it afterward: Each processor will have input and output channel buffers of length equal to 1. This limitation has the effect that processors must wait on other processors until they have *processed* their messages. A deadlock condition is portrayed in Fig. 9.7 along with the cycle of dependency. We will assume no lookahead and analyze the dependency at the time when a new message of time stamp 8 is sent to object A:

1. Object D waits on C to send a message. It must wait without processing the next message from B (when there is one) because channel $C \rightarrow D$ has the lowest time stamp of 14; D cannot assume that the next message along $C \rightarrow D$ will have a time stamp ≥ 19.

2. Object C waits on A to send the next message. C cannot send a message to D unless C first receives one from A. C has already processed the first message that it received from A.

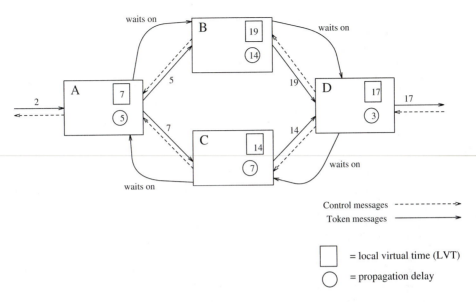

Figure 9.7   Deadlock condition.

3. Object *B* waits on *D*. *B* has already sent a message with a time stamp of 19 and cannot send another until *D* requests *B* to do so. This is because the buffer is full for channel $B \rightarrow D$ as the buffer holds only a single message.

4. Object *A* waits on *B* to clear its input channel buffer (of size 1) so that it can send the next message with time stamp of $5 + 5 = 10$.

The deadlock occurs as a result of two features of the functional model being simulated in parallel on a network: (1) a finite channel buffer size, and (2) an underlying set of control messages preventing channel buffers from overflowing when a message needs to be transmitted. If we have, for all intent and purposes, infinite buffers in an acyclic functional network, deadlock will not occur. The deadlock occurs due to the synchronization problems arising from using buffers that are finite in size and can therefore become saturated. All real processors, of course, have finite buffers; however, the infinite buffer assumption can be made at any time if one knows ahead of time that buffer saturation is not likely to take place. If it is unreliable to make such an assumption, some method for *detecting* or *avoiding* deadlock must be instrumented.

Lookahead allowed one way of avoiding deadlocks by using predictive information associated with propagation delays. Let's see what can be done with lookahead when a processor blocks waiting for input. *Null Messages* are messages sent from one processor to another in a cascading fashion that permits clocks to synchronize. For each feedback cycle (loop) in the functional network, at least one function must have a lookahead value greater than zero. Consider the deadlock condition in Fig. 9.7 with unit-sized channel buffers and null messages:

1. D receives a message from B and also from C. It waits on C since C's message has the lowest time stamp (14).

2. After a period of time Δt, having heard nothing from C, D sends a request for a null message to be sent from C to D.

3. C requests A to send a null message to C.

4. Suppose that A finds that it will have an input message with time stamp 8 and knows that it will take 5 virtual time units to simulate processing this message. It can therefore send a null message to C of the form $(13, null)$. In this manner, null messages incorporate lookahead.

5. This information is propagated back from C to D, which then realizes that it can now process the message from B, $(B, 19)$, without any causality problems.

There are different ways to handle null messages. We have just described a *time-out* or *demand-driven* approach, where D times out and requests local clock time information from other processors. Another method would be to have processors automatically send null messages along other output links whenever a regular message was sent along one of them. The combination of lookahead and null messages is important for breaking deadlock conditions when they occur in systems operating with the finite buffer assumption and associated synchronization overhead.

9.4.4 Optimistic Asynchronous Method

As we have seen with the conservative method, the processor blocks on an input channel, even if it has other pending messages, until it can compare time stamps of incoming messages. The minimal time stamp message gets processed. This procedure is advantageous in that it affords no causality errors. That is, we can run a distributed simulation with the conservative method without fear that the answers we get will be incorrect. Instead of waiting for the lowest time stamp message, suppose that we proceed directly to process the first message that comes to the processor even if the processor has several input channels. This would speed things up considerably, but it relies on the assumption that if message A arrives (in real time) at a processor's input channel buffer before message B, the time stamp of A is less than or equal to B (assuming that the order of equal time-stamped messages is invariant to simulation results). At first, this approach may not appear unreasonable, but the method depends on several factors. The main factor is that the processors utilized may be of different speeds or manufacture. Some processors are slower than others, so they will *physically* take longer to process virtual time-stamped messages. Even with process scheduling, identical processors, and equal-length transmission lines, the optimistic assumption might be able to run without error, but in simulation we cannot often take this chance. If a message arrives at an object with a time stamp less than the LVT of the object, this message is called a *straggler message*.

In general, we need to have a backup mechanism in place to handle errors when time-stamped messages are taken out of turn. This is so we can be assured that the model will be executed correctly regardless of the order in which messages are received by a processor. The algorithm below is a variant of the conservative algorithm. The main difference is that if a causality error is detected, we need to employ a scheme where *anti-messages* are sent out to cancel the effects of messages already sent "ahead of time."

Algorithm 9.2. Optimistic Simulation Method

Procedure Main

   If a message in an input channel buffer then

     Check to see if the message time stamp is less than

      the LVT of the processor object

    If it is not then process the message:

      New time stamp = old time stamp + processing

       time of block

      Update local clock time to new time stamp

    Else this is a straggler message:

      Send an anti-message for all messages whose time stamp is

       greater than the incoming message time stamp

      Roll back the state vector/queue for each affected processor

      Proceed to run the simulation as before by processing

       the straggler message

   Else this object waits/blocks

End Main

If a message occurs with a lower time stamp than the LVT of the processor, the anti-messages must be sent out to cancel the effects of later messages that should not have been sent (under the conservative assumption). There are three key considerations associated with the sending of anti-messages:

1. If the positive message is unprocessed in an input channel buffer, the negative anti-message will be enqueued with the positive message and these two messages will cancel each other.

2. If the anti-message arrives before the positive message, the anti-message is enqueued and cancellation occurs when the positive message reaches the buffer. This may seem somewhat unusual since one would assume that anti-messages, in most implementations, should arrive *after* the positive messages; however, we include this case for completeness.

3. If the positive message has already been processed, rollback must take place so that the states of the processors move back.

Let's use an example to explore what happens during an optimistic approach for functional models. We will have a setup where we have three processors in a row, as shown in Fig. 9.8. We will let our model be composed of functions for A and B. The input to A is $u(t)$ and the output from A (and input to B) is $x(t)$. The output from B (and input to C) is $y(t)$. Functions A and B will operate as follows

$$x(t) = x(t-1) + u(t) \tag{9.1}$$

$$y(t) = \frac{x(t)}{y(t-1)} \tag{9.2}$$

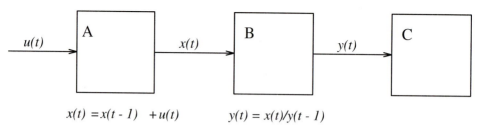

$$x(t) = x(t - 1) + u(t) \qquad y(t) = x(t)/y(t - 1)$$

Figure 9.8   Three function model with sequential coupling.

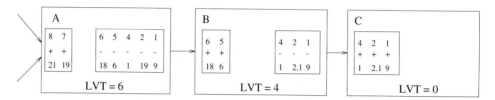

Figure 9.9   State of three processors just *prior* to straggler message (3, 2).

where, for all t, $x(t)$ is defined as zero and $y(t)$ is defined as one. In a distributed simulation, functions are encoded as messages so that $u(5)$ is encoded as $(5, u)$, where u is the value of function $u(t)$ at time 5. Notice from Fig. 9.9 that $u(t)$ may come into A from one of two possible input channels. In this way, it is easier to imagine how time stamps may become reversed. In particular, we will define a set of messages arriving into A as $(1, 9)$, $(2, 10)$, $(4, 1)$, $(5, 5)$, $(6, 12)$, $(7, 19)$, $(8, 21)$. This defines the signal $u(t)$ except that the u value for time 3 is missing. Indeed, it has not yet arrived at A. We will consider its arrival out of order in a moment. First, we will depict the situation for each processor for these given messages in Fig. 9.9. Each processor contain two boxes: the left-most box is the input channel buffer and the right-most box is the output channel buffer. The first processor A receives messages from one of two sources. A processes the first message that it receives without waiting on the other line to see if a lower time stamp message comes along. Messages that have not been processed yet within a processor's input channel are labeled with a plus $(+)$ symbol. Messages that have been processed are labeled with a minus $(-)$ symbol. In some optimistic implementations, negative messages have a higher priority to speed their delivery. A processor keeps track of all messages that have been processed there and uses minus signs for labeling; however, positive messages reflect the true location of messages at any one time. For instance, on processor A, two messages for times 7 and 8 in the input channel buffer have not yet been processed; other processed positive messages have gone on to B and C. The output channel buffer, though, stores all messages processed. The input and output channel buffers within processors A, B, and C are such that:

1. Messages $(1, 9)$, $(2, 2.1)$ and $(4, 1)$, have been processed by A and B.
2. Messages $(5, 6)$ and $(6, 18)$ have been processed by A but not yet by B.
3. Messages $(7, 19)$ and $(8, 21)$ have not been processed by either A or B.

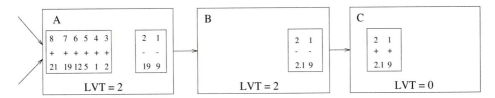

Figure 9.10   State of three processors just *after* the straggler message $(3, 2)$.

We can now study the optimistic method in action when a straggler message $(3, 2)$ arrives at A given the system state in Fig. 9.9.[5] The LVT of A is 6 since it has processed up to time 6. The time 3 in message $(3, 2)$ sent to A is less than 6. This condition causes rollback to occur. The following occurs in rollback:

1. A sends out *anti-messages* whose time is greater than 3 in A's output channel buffer. This means that messages $(4, 1)$, $(5, 6)$, and $(6, 18)$ would be sent out and *propagated* throughout the network.

2. Anti-messages $(5, 6)$ and $(6, 18)$ would cancel the positive messages in B's input channel buffer. In short, B has not yet processed these two messages, so no harm has been done as far as these two messages are concerned, so they are removed from B's input buffer without further action.

3. Anti-message $(4, 1)$, however, does not find its positive counterpart in B's input channel buffer, so it continues to propagate until it finds it in the input buffer of processor C, at which time the two cancel and message $(4, 1)$ is removed from C's input buffer.

Figure 9.10 displays the state of the processor channel buffers after (1) the straggler message arrived and (2) rollback occurs by sending out anti-messages and moving back the state of each processor. We have conveniently stored the state of each processor within the message to be transmitted, but this is not always the case. In a more general case the output of A, for instance, would be a function of state and input and so, in addition to storing input and output buffers for each message, we would have to store the state at past times. When rollback occurs, the system state within each processor is rolled back to a time just before or equal to the time of the straggler message and then the simulation continues as before.

The method we have just described for rollback is called *aggressive cancellation* because as soon as a straggler message was found, all anti-messages with the relevant time stamps were sent out to find their positive counterparts so that they could cancel them and rollback their effects on processor state. It is possible, however, that the introduction of a straggler message will not affect the state of the system composed of all processor states. *Lazy cancellation* represents an approach where we send out anti-messages only when we determine that the state of the system will be different as a result of the straggler message being introduced.[6] Our basic assumption in the last example was that all $x(t)$ were

---

[5]Processor A knew something was wrong when $U(3)$ failed to appear before $U(4)$; however, this example is setup only to demonstrate the effects of rollback.

[6]We assume that processor output messages are a function of the local state of the processor.

instantiated to zero prior to simulation. This meant that the calculation for $x(4)$, which required $x(3)$, used zero for $x(3)$. Suppose instead of the straggler message being $(3, 2)$ we let it be $(3, -19)$. Then $x(3) = x(2) + u(3) = 19 + (-19) = 0$. But this is the same as the default, where $x(3) = 0$, so it should not affect our computation for future time-stamped messages. In such a case, lazy cancellation would result in considerable savings since we would avoid the necessity of having to send anti-messages and perform rollback.

9.5 SPATIAL MODELS

There are two different ways to achieve parallelism with spatial models. All spatial models have two facets: (1) the space and (2) the entities which inhabit the space. The concept of *entity* is very general. Entities are also known as *particles* or *bodies*, depending on which literature is read. Since we have space and entities, we can do one of two things: (1) assign partitions of space to processors, or (2) assign subsets of entities to processors, or both. We will first focus on the former approach since this is the most common, where we have parallelism in the form of a single supercomputer or massively parallel machine.

9.5.1 Space-Based Approach

Spatial models, in general, are very well adapted to implementation on parallel machines. This is because we can assign a geometrical partition of the space onto each processor. Let's look at the one-dimensional equations for heat (or diffusion) and waves (periodic phenomena).

Heat and wave equations.   The one-dimensional (with spatial variable x) equation for heat or diffusion is

$$\frac{\partial^2 \Phi}{\partial x^2} = k \frac{\partial \Phi}{\partial t} \tag{9.3}$$

The wave equation is

$$\frac{\partial^2 \Phi}{\partial x^2} = \frac{1}{c^2} \frac{\partial^2 \Phi}{\partial t^2} \tag{9.4}$$

The first method is to discretize the PDE using a method as discussed in Ch. 7. For instance, the wave equation for an arbitrary processor partition i is

$$\frac{\Phi_{i-1}(t) - 2\Phi_i(t) + \Phi_{i+1}(t)}{\Delta x^2} = \frac{\Phi_i(t - \Delta t) - 2\Phi_i(t) + \Phi_i(t + \Delta t)}{c^2 \Delta t^2} \tag{9.5}$$

Using either equation, our method for parallelization proceeds as follows:

Algorithm 9.3. Partitioning for One-Dimensional PDE Systems
Procedure Main
   Partition space into i subdivisions
   Set initial conditions
   While not end of simulation do:
     Exchange boundary node values with neighbors

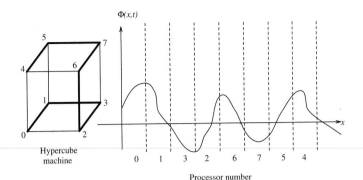

Figure 9.11 Vibrating string exemplifying the wave equation partitioned on a hypercube architecture.

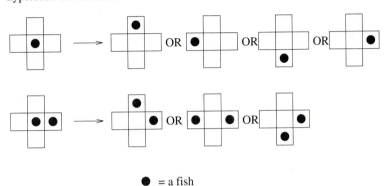

● = a fish

Figure 9.12 CA rules for fish movement.

Perform a state update $\Phi_i(t) \rightarrow \Phi_i(t + \Delta t)$
        Send updated state on adjacent processor boundary strips
    End While
End Main

Figure 9.11 shows the partitioning for some time t during the simulation.

Fish movement on a 2D grid. Let's consider the situation where we have a 2D rectangular space where fish live and move. A fish can move to any of its eight neighbors during a time slice. Its only restriction is that it cannot occupy the same cell as another fish. With toroidal boundary conditions, our fish can move around, performing a random walk over the surface of a doughnut. The cellular automata (CA) rules in Fig. 9.12 demonstrate an implementation of the fish scenario. For a parallel implementation, we let each cell of the CA be handled by a separate processor. Each processor is synchronized so that during one time step, each processor updates its new cell value based on its old value and the old values of the neighbor cells. A SIMD architecture is the most natural for this type of

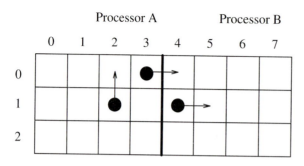

Figure 9.13 Two processor partitioning for a 3 × 8 grid.

parallelism since we have one *instruction* (the application of the rule set) which operates on each cell in the *data* space (grid) in parallel. The more processors on a machine, the larger the grid space that can be simulated and a larger number of fish can move around in their random fashion.

The only problem with the cell per processor partitioning is that it can be expensive; many processors are needed even for relatively small size grids. A grid of 1000 × 1000, which is not that dense, requires 10^6 processors. Instead of being forced to scale down the problem to fit the available number of processors, one can partition the original grid more liberally by having each processor handle the updating of a number of cells instead of just one. Moreover, many parts of the grid will have no fish, so we can use event scheduling to handle the serial simulation on each processor. Event scheduling involves having a linked list data structure to keep track of each fish and its position within any given processor. To update the random movement of fish, we just process the linked list and update the fish positions. Figure 9.13 displays two processors (*A* and *B*) being used to simulate fish on a 3 × 8 grid assuming that our parallel machine contains two processors. From Fig. 9.13 we see that the grid is partitioned into equal-size partitions of size 3 × 4. On each processor, our simulation algorithm, would be the same as described in Sec. 7.2.3. The key item to discuss now is how processors communicate. Processors must communicate with each other when a fish decides that it wants to move on the part of the grid that rests in the other partition. So, on our example grid in Fig. 9.13, one fish at (0, 3) decides to move to (0, 4). How can this fish know that cell (0, 4) is empty? The way we accomplish this is for B to send its *boundary strip*, consisting of cells (0, 4), (1, 4), and (2, 4), to processor *A before A* updates the fish positions, since *A* cannot do a proper update without this information. Likewise, *B* needs *A*'s boundary strip, composed of cells (0, 3), (1, 3), and (2, 3). After processors *A* and *B* exchange their boundary strips, they can each do an update of the fish in *their own partitions only*. So, for instance, after (1) *B* sends its boundary strip to *A*, and (2) *A* decides for the fish in (0, 3) that it can legally move this fish to (0, 4), *A* can now delete this fish from its linked list. But this is not sufficient because it must also tell processor *B* that it now has a new fish in its partition [at cell (0, 4)]. Also, there is yet another problem to worry about. What if the fish in (1, 4) covered by processor *B* decided also to move to (0, 4) instead of (1, 5)? Then we would have two fish in the same location and a conflict would ensue. We, therefore, need some kind of *conflict resolution* to take care of this situation. A conservative method[7] of resolution performs a rollback of fish that were part of the conflict and resimulates those fish motions until the conflict disappears. The basic algorithm for distributed spatial model simulation of fish is:

Algorithm 9.4. 2D Partitioning with Asynchronous Simulation

Procedure Main
> *Exchange boundary fish state information with adjacent processor(s)*
> *Update fish positions on this processor*
> *Send updated fish positions on adjacent processor boundary strips*
> *Resolve conflicts for competing fish*

End Main

Even though we demonstrated a two-processor implementation, the general case for partitioning a 2D grid would involve having four boundary strips around each partition and not one as shown in Fig. 9.13. Also, it is possible to add other species which compete with fish for food resources or eat the fish for food.

N-Body simulation

Overview.   As we saw in Ch. 7, there are several possible methods for improving the execution times for serial architectures. Let's reconsider the naive $O(N^2)$ approach to simulating particle systems and see how we can parallelize this algorithm. There are two nested loops which cause the $O(N^2)$ behavior. The outside loop loops through each body or particle and the inside loop accumulates the effect of the force of other bodies on the body being considered. One may parallelize the outer loop without any problems since the forces acting on a single body are independent of those acting on other bodies. So, with N bodies, we can achieve a parallel situation with N processors—one per body. Let's study the loop:

```
For i = 1 to N
For j = 1 to N
F(i) = F(i) + force(i,j)
End For
End For
```

The loop iterating i can be partitioned one value of i per processor; however, the inner loop does not have the same luxury. To see this, we will break down the force accumulation statement into assembly instructions:

```
1. Load F(i)
2. Add F(i) + force(i,j)
3. Store F(i)
```

This sequence of instructions must be treated as a critical region since the following can

---

[7]Conserving the total number of fish.

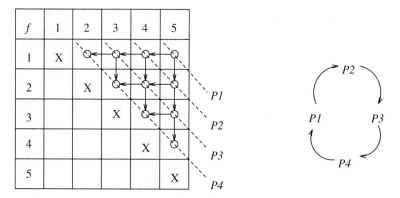

Figure 9.14　Shared memory N body algorithm with barrier synchronization.

occur with two processors P1 and P2 that are trying to accumulate the forces of bodies 2 and 3 on body 1 at the same time:

1. P1: `Load F(1)`
2. P2: `Load F(1)`
3. P2: `Add F(1) + force(1,3)`
4. P2: `Store F(1)`
5. P1: `Add F(1) + force(1,2)`
6. P1: `Store F(1)`

It is necessary for each of these processors to act in serial so that $F(1)$ is calculated correctly for adding the effects of bodies 2 and 3. When P2 has added the effect of body 3 on body 1 in stage 4 above, P1 is still operating on the old value of $F(1)$, so it overwrites what P2 has done.

Shared Memory.　A shared memory approach to the problem is given in the form of a table in Fig. 9.14. Since the forces are symmetric between any two bodies, we need calculate only half the table. The dependencies in the form of arrows in the force calculations form a *barrier synchronization* which specifies what can be done in parallel. Specifically, as shown in the right-hand diagram in Fig. 9.14, processor P1 calculates the force of body 5 on body 1 [i.e., entry $(1, 5)$ in the table]. Then processor P2 calculates the forces $(1, 4)$ and $(2, 5)$ *in parallel*. The workloads for processors P3 and P4 operate accordingly.

Ring Implementation.　The ring topology is useful at solving problems where there are cross-product interactions between elements. Consider the labeled sections of a Rubik's cube[8] as shown in Fig. 9.15. Each section contains nine cubes and these sections can be rotated while keeping the adjacent sections stationary. As section 2, for instance, is turned $360°$ while leaving sections 1 and 3 alone, each small square on section 2 touches each

---

[8]Actually, there are other more suitable *cylindrical*—but less well known—toys that would better serve the purpose of this demonstration.

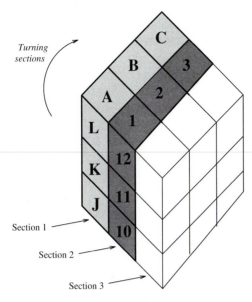

Figure 9.15   Rubik's cube with rotating sections.

small square on section 1 (or section 3). If block X touches block Y, we will label this as (X, Y). Consider the starting position: $\{(A, 1), (B, 2), (C, 3)\}$. Now when we rotate section 2 clockwise one step, we get $\{(A, 12), (B, 1), (C, 2)\}$. A complete revolution causes each cube in section 2 to touch each cube in section 1. This is really a remarkable property due to the nature of rings: one can fill out a cross-product matrix by rotating one matrix dimension around the other one. If our relation between individual elements is symmetric (as the "touching" relation is for Rubik's cubes), using the ring structure is efficient. Instead of rotating rings, let's consider a network of processors connected in a ring topology where messages are circulated or rotated around the ring.

Consider a space of particles or bodies and a regular partitioning of that space where there are approximately the same number of particles per partition. The N-body problem may be simulated on such a topology by having each partition rotate around the ring. The algorithm for this procedure is outlined below.

Algorithm 9.5. Ring Approach to N-Body Simulation
Procedure Main
>   *Receive the partition that is traveling clockwise around the ring*
>   *Compute forces on all bodies in the home partition with bodies*
>   *of the received partition*
>   *Update the current force acting on the home partition*
>   *Send the traveling partition clockwise one position*
End Main

Each processor is in charge of computing all forces on a particular partition of the overall space. It keeps track of the accumulated forces on the *home partition* (i.e., the spatial partition permanently assigned to that processor). Traveling partitions are simply the *fixed*

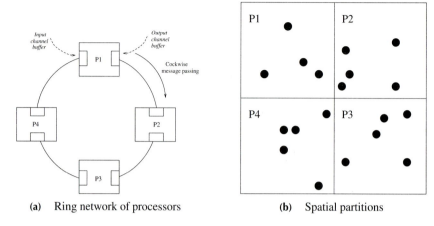

(a)   Ring network of processors   **(b)**   Spatial partitions

Figure 9.16   Ring network with travelling partitions.

partitions (containing all body positions) that rotate around the ring. As traveling partitions reach a given processor, the processor (1) updates the force calculations on the home partition with the bodies present in the traveling partition just received, and (2) sends the received traveling partition on its way around the ring. The direction for the traveling partition can be clockwise or anticlockwise as long as it is consistent throughout the simulation. Figure 9.16 shows a ring network along with a sample space with four geometrical partitions. In each temporal phase, each processor performs a serial simulation of the N-body problem (using a method given in Ch. 7). The phases are shown in Table 9.1 assuming that all processors must be visited.

Our previous discussion has centered around what happens to the traveling partitions. Figure 9.17 shows the state of five processors in a ring where each partition maps directly to a single body (i.e., there are five bodies in the system). The data structure to be gradually filled in as the traveling particles rotate around the ring is a simple matrix. Our goal is to fill it in as we go. As each processor receives the traveler, it is able to accumulate force calculations between bodies i and j. As shown in Fig 9.17, it is only necessary to compute forces between a body and *half* the other bodies. This halves the overall computation time. For five bodies, only two phases are required. Figure 9.17 shows the message passing along with a virtual matrix designating which force is being calculated.

TABLE 9.1   PHASES OF PARALLEL
COMPUTATION FOR FOUR
TRAVELLING PARTITIONS

Time	P1	P2	P3	P4
Phase 1	1	2	3	4
Phase 2	4	1	2	3
Phase 3	3	4	1	2
Phase 4	2	3	4	1

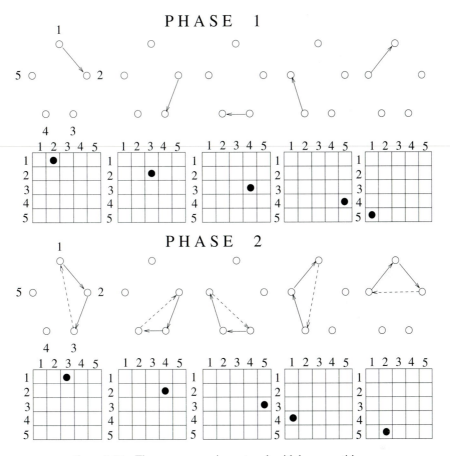

Figure 9.17   Three processor ring network with home partitions.

9.5.2 Entity-Based Approach

If we assign entities or entity groups to processors and proceed with a simulation, we soon realize that we would like to engage in communication transfers with immediate neighbors rather than the whole crowd of entities. This implies that we want a kind of *spatial coherence* as described in the preceding section. So where would the entity-based approach be used? It is used most often in situations involving the human in the loop: distributed interactive simulation (DIS).[9] The emphasis on DIS grew out of the previous work on SIMNET [16, 173], which is a network of homogeneous mockup combat training simulators. SIMNET can be considered a restricted form of DIS where the simulator nodes

---

[9]The term DIS has been coined by the military for the construction of interoperable heterogeneous training machines communicating with one another, although the potential of DIS has far greater implications for basic science and industrial research where teams of people must coordinate with one another to achieve a common, integrated virtual environment.

Figure 9.18   SIMNET M1A1 tank simulator. (Reprinted with permission of the Institute of Simulation and Training, Orlando, FL.)

contain a limited number of simulator types. The following types of simulators are available within SIMNET:

- M1A1 Abrams tank (also configurable as a T72 enemy tank). This is a manned tank which holds room for the Commander, Gunner, Driver, and Loader.
- M2 Bradley fighting vehicle (also configurable as a BMP[10] enemy vehicle). This is a protected vehicle which holds a dismounted infantry fire team.
- Dismounted Infantry Module.
- AH-64 Apache helicopter.

The SIMNET simulators are fiberglass-enclosed boxes containing realistic controls and displays as in the real vehicles. Figure 9.18 displays a simulator. In a pure DIS environment, simulator nodes can be something unsophisticated such as a PC-based platform acting the part of a battlefield entity to something costly such as a fully immersive housing which incorporates physical motion control (a platform) and realistic instrumentation. If an entity is a complex object with a human controlling it, it comes as no surprise that since humans are geographically constrained, we must run a simulation, that is distributed across wide geographical areas. A good example of the use for DIS is battlefield simulation, where entities are things such as M1A1 tanks, Bradley fighting vehicles, or Apache helicopters. Entities can also refer to small objects such as dismounted infantry or communication equipment.

   The primary purpose of DIS is the training of humans in a cooperative task environment such as a battlefield scenario or a task checklist being carried out by NASA space shuttle astronauts. Since training in the field with real equipment is expensive, DIS has the potential to reduce group training costs dramatically. The packets for the DIS protocol are currently oriented toward military applications; however, there are generic aspects of the protocol which lend themselves to any distributed interactive simulation. The simulation community within the military characterizes simulation in three possible ways:

1. *Live* simulation involves people in realistic training environments over terrain which matches the type necessary for the particular exercise (jungle, desert, grassland). This applies to "in the field" training.

---

[10]A BMP is a Soviet vehicle.

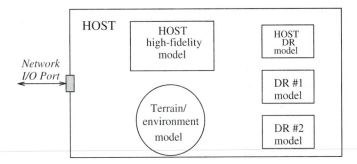

Figure 9.19   Host computer for entity participating in a 3 entity DIS.

2. *Constructive* simulation applies to "board" war games and analytical tools generally not requiring iterative methods. This applies to simulation using basic stochastic processes based on Lanchester equations [219] for combat attrition.

3. *Virtual* simulation applies to computer simulation–based model design and execution as discussed in this book.

Network concepts. In the entity-based approach, entities are associated with a computer. The computer may be attached to a mockup of a tank, a Stewart platform, or an immersive piece of man-machine gear that provides the trainee with the feel of really *being there*. Alternatively, the computer can be a simple personal computer or workstation. In either case, there will be a man-machine interface connecting the human to the computer, and a physical network link connecting the computer, to the outside world. Over the network, it is critical to have some sort of protocol so that machines can communicate with each other. The International Standards Organization (ISO) has set up a seven-layer architecture that is used by most networking schemes [218]. The lowest layer, called the *Physical Layer*, is concerned with transmitting, without error, individual bits from a source to a destination. The highest layer, called the *Application Layer*, is user definable. For computer simulation we are concerned mainly with the application layer since we will have to create packets that can be sent and understood by every machine connected to the network.

Overview of DIS. DIS is composed of a dynamic network[11] of entities where there is an $N{:}1$ map between entity and computer. The current DIS protocol is such that packets sent from one computer are *broadcast* to all other entity computers. Each entity computer reads the information in each packet and may act upon it if necessary. Usually, we consider an entity to have its own host computer ($N = 1$) as shown in Fig. 9.19 (for three entities). In Fig. 9.19, we assume that our DIS is composed of three entities: say, an armored tank, a Bradley vehicle, and an Apache helicopter. The computer HOST has the following information vital to its performance during the simulation:

1. An I/O port, where it communicates with the remaining two entities. The packets used in the communication are called Protocol Data Units (PDUs).

---

[11]The network topology changes over time as new entities enter or leave the simulation.

2. A high-fidelity model of the entity's own behavior. This model changes the state of the entity in real time (see Sec. 9.5.2).

3. A complete model of the environment, including local weather conditions and terrain. For ground vehicles the terrain information is most important, whereas for the helicopter entity, the weather model plays an important role.

4. A host dead reckoning (DR) model. The host dead reckoning model is an aggregate model (of less fidelity and complexity) of the high-fidelity model. The other two entities also have a copy of this model so that they can perform state estimates in between the reception of state PDU packets.

5. Dead reckoning models for the other entities in the DIS. These are used by the HOST entity to calculate the new state of these entities. Usually, the host has models only for entities within a specified range of the host entity.

Real-time synchronization. In both time-slicing and event scheduling approaches, time is incremented according to the *virtual time*. This time is usually different from the *real time* stored on the computer since there is no effort to *synchronize* the virtual time with real time. Sometimes, in computer simulation applications, there is a need to do one of two things:

- Run a simulation in faster than real time.
- Run a simulation in real time.

An example of the first requirement can be found in industrial machines that perform a simulation prior to performing their mechanized task. This is useful to be able to predict possible faults in the system *before* they actually occur. Through prediction, the machine can correct itself to avoid tool or part damage. For the second requirement, whenever humans are in the simulation loop so that simulated entities are driven through human interaction, there is a need to cause entities to behave as they would do in the physical world. The trainee in simulated nuclear power plant rooms or battlefield exercises will want to experience fault conditions (in the case of the power plant) and vehicle movement (in the battlefield domain) in the same way as in real life. Otherwise, the experience will not train them effectively under actual conditions. This need for synchronization occurs in simulation on serial architectures, but is even more acute in distributed simulation since there are multiple models and multiple virtual times (called local virtual times) on separate processors.

The issue of "faster than real time" performance is one that relates to the raw speed of the computer being used to simulate the model. If the simulation runs slower than real time, there are several possible practical options:

1. Optimize the simulation code for greater run-time efficiency. There are a number of techniques to do this, including the optimization of loops to recoding critical parts in assembly language or microcode.

2. Recheck the scheduling policy and priority levels of the operating system software handling the simulation program. Adjust these policies and priorities to see if the real-time constraint can be assured.

3. Buy a faster computer.

The DIS network architecture requires real-time operation of all entities since the changes in state of entities must correspond to what the trainee expects when immersed in the simulation. If a tank moves about on the field in an unrealistically fast or slow manner, the training experience is ineffective since it does not correspond with real-life observation of tank movement. To make a computer simulation operate in real-time, one requires synchronization between the virtual and real time clocks. The following simple algorithm serves to synchronize both clocks. The algorithm should be embedded within the local simulation loop performing time slicing or event scheduling. This is to ensure that the real-time constraint is maintained throughout the simulation. When simulation time exceeds the real time, the simulation time is *reset*. This may cause "jumps" in the entity positions displayed on a computer screen. Sometimes, these jumps may be smoothed out, but one is still playing a game of "catch-up" to resynchronize the simulation with the real-time clock.

Algorithm 9.6. Real-Time Synchronization
Procedure Main
  Update the Local Virtual Simulation Time (LVT)
  Determine the Real Time (RT)
  If LVT > RT then
      Wait until RT ≥ LVT
  Else
      Reset LVT ← RT
  Endif
End Main

The issue of real time is not intrinsic to distributed or parallel simulation since it is also a key element in making single computer interactive simulations realistic. That is, the model being simulated (such as a flight or combat simulator) needs to synchronize with a real-time clock. The effect of real-time synchronization is more critical with distributed simulation since the training experience is not limited to one person, and it is important to have consistency among all changes of entity state. The concept of real-time movement can be greatly relaxed by allowing the human to adjust the simulation to a desired speed. This is a common option in simulation games, where the achieved effect is one of entertainment rather than real-life validation for training.

Entity state estimation: dead reckoning

Overview.   Dead reckoning is the process where one entity tries to predict the movement of other entities in between the state packets it receives from the other entities. The entity performs this operation by having on-line dead reckoning models for all other simulation entities. The term "dead reckoning" comes from the maritime navigational approach [51] for *deducing* (or *ded*) a ship's position in between obtaining *fixed* positions from accurate celestial measurement or via satellite. When a ship is at sea, it is not always possible to fix its position. This may be due to weather conditions or malfunctioning navigation equipment. A DR (dead reckoning) is drawn in between fixes of position.

This is exactly the same problem that we have with networks of distributed entities. If an entity does not frequently receive updated state information from other entities (i.e, fixes of state), it must try to predict the intermediate behavior through some type of dead-reckoning technique. This is required so that the entity's machine can continually update the graphics that represent the other entities in their approximate positions. From a systems perspective, dead reckoning is defined as estimating the future states[12] of another entity given certain information such as current position and velocity. We will treat the two terms (dead reckoning and state estimation) as being equivalent. When an entity receives the next state package from another entity, it has a fix on the other entity's position and can use this to reestablish the correct state for that entity. You may wonder why we just don't increase the frequency with which state packets are transmitted, thereby eliminating much of the need for dead reckoning. There are two good reasons for state estimation on computer networks:

1. We reduce the network traffic by asking other entities to transmit their state packets less frequently.
2. If some entities experience a temporary network shutdown or delay, the other entities should have a way of continuing the simulation.

Of course, the kind of condition in the second reason cannot continue indefinitely; otherwise, the divergence between the true state of an entity and its dead-reckoned state will be too great and will cause unrealistic training scenarios where entities pop on and off the computer image generator (or workstation video monitor).

The HOST entity computer uses a DR model of other entities which is less complex than their high-fidelity models. This is to reduce the time taken to simulate entities. If the HOST is a parallel machine, it may be entirely possible to simulate all other entities at the highest fidelity. To maintain consistency; however, all entity computers use the same level of DR model. The overall DIS network-based simulation will be able to display entities at frame rates of 5, 15, 30, or 60 Hz while receiving PDUs from these entity computers at relatively low rates such as 1 Hz.

State Estimation Methods. Figure 9.20 displays a trajectory of an entity B moving in its environment, according to entity A. Entities A and B both have DR models for B, and so not only can A detect differences between its extrapolation of B's state, but B can send messages to A and all other entities when the difference between the state calculated by the DR model and its high-fidelity model differ by some error bound ϵ. Let's analyze what happens in Fig. 9.20 in some detail. First, the curved line represents the (x, y) state of entity B according to B's own high-fidelity model. Entity A obtains a state PDU packet at times t_1, t_2, t_3, and t_4. Recall that this is known as *fixing* the position of an entity. At these times, A can be assured of the exact state of entity B: B's position and orientation. The tangent lines at the given *fixed* times are oriented in the same direction of the velocity vector for entity B. Let $\mathbf{r}(t)$ represent the position vector with a tip that follows the curved trajectory of Fig. 9.20. Then $\mathbf{v}(t)$ is the tangential velocity vector and $\mathbf{a}(t)$ is the acceleration vector. Consider the fixed time t_1. After simulation time t_1 but before time t_2, entity A calculates the state of B using a state estimation method. One possible method is to expand the Taylor

---

[12]State estimation is a classical approach in the identification and control of systems.

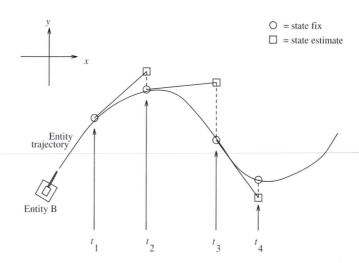

Figure 9.20   Entity state estimation.

series around t_1. $\mathbf{r}(t)$ for $t_1 \leq t < t_2$ can be calculated according to the chosen order, where Δt represents $t_2 - t_1$:

1. Order 0: $\mathbf{r}(t) = \mathbf{r}(t_1)$.
2. Order 1: $\mathbf{r}(t) = \mathbf{r}(t_1) + \mathbf{v}(t_1)\,\Delta t$.
3. Order 2: $\mathbf{r}(t) = \mathbf{r}(t_1) + \mathbf{v}(t_1)\,\Delta t + \frac{1}{2}\mathbf{a}(t_1)(\Delta t)^2$.

The position of entity B is determined by one of the above formulas. We see in Fig. 9.20 that there can be a wide variance in error ϵ, depending on when packets are received by entity A, regardless of how often they are sent by B. At time t_2, ϵ is much less than for t_3, where the difference between the dead-reckoned state and the fixed state is significant. Even though our example uses t_1 and t_2, the previous relationships, in general, hold for any i where we are estimating an entity's state in the interval $[t_i, t_{i+1})$.

   This method for dead reckoning provides for a way to estimate state; however, there is a problem as we shift, for instance, from the estimated state at time t_2 (the circle) and the fixed state at $\mathbf{r}(t_2)$ (the square). Specifically, the entity may appear to *jump* on the display as the estimated state is reset to the fixed state at t_2, as depicted in Fig. 9.21. A method for smoothing the estimated states alleviates this problem and is shown in Fig. 9.22. In this trajectory, the fixed state occurs at times t_{n-3}, t_n and t_{n+3}, and the intermediate times represent DR calculations to update the state associated with the DR model. Smoothing has the advantage of decreasing sudden changes in state that would otherwise cause significant discrete changes for the visual appearance of a simulation entity. The disadvantage is that we must delay the appearance of an entity for some period of time even though we continue to calculate its DR state. This means that the delay prevents the sudden shift in state by allowing an interpolation of past state estimates for purposes of visualization. The time delay used for a simulation depends on the packet transmission delays over the network, and for long-haul networks, the problem will be more acute.

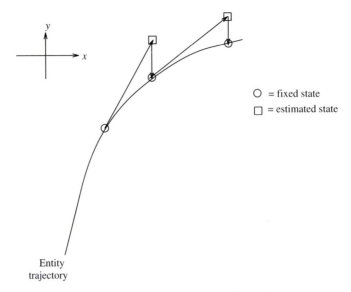

Figure 9.21   Jumping of state when state fix packet arrives.

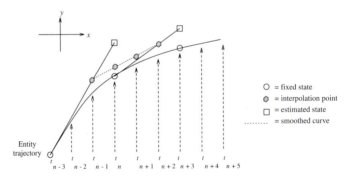

Figure 9.22   Smoothing estimated states.

Let's begin at time t_{n-3}, which we can regard as the beginning of the simulation. Since we have a fix on the entity's position at time t_{n-3} (i.e., we receive a state entity packet at this time), the entity state and DR states are identical. We will let the actual entity state at some time t_i be $f(t_i)$, the DR state be $d(t_i)$ and the interpolated (smoothed) state be $i(t_i)$. At times t_{n-2} and t_{n-1}, the DR states are calculated. Since smoothing has not yet begun, the interpolated states equal the DR states. At time t_n, we receive a fix (actual state). Normally, without smoothing, we would simply update the DR state to reflect the state at the time of the fix—possibly causing a jump in the entity display. With smoothing, we do not update the display immediately, instead continuing to process the DR state. At time t_{n+2}, we linearly interpolate between DR states at times t_{n-1} and t_{n+2}. Moreover, the entity display is updated to reflect a new interpolated position at time t_n. This is depicted in Fig. 9.22. Table 9.2 displays the delay of visual state update as a function of simulation time.

TABLE 9.2   SMOOTHING BY
DELAYED INTERPOLATION OF
VISUAL STATE UPDATE

Time	DR state update	Visual state update
t_{n-3}	$f(t_{n-3})$	$f(t_{n-3})$
t_{n-2}	$d(t_{n-2})$	$d(t_{n-2})$
t_{n-1}	$d(t_{n-1})$	$d(t_{n-1})$
t_n	$f(t_n)$	wait
t_{n+1}	$d(t_{n+1})$	wait
t_{n+2}	$d(t_{n+2})$	$i(t_n)$
t_{n+3}	$f(t_{n+3})$	$i(t_{n+1})$
t_{n+4}	$d(t_{n+4})$	$i(t_{n+2})$
t_{n+5}	$d(t_{n+5})$	$i(t_{n+3})$

Protocol data units.   Protocol data units (PDUs) lie at the heart of DIS since entities send these packets to each other to drive the simulation. Each PDU includes a time stamp[13] and a number of fields applicable to that PDU type. We will overview a subset of PDU types:

1. *Entity State PDU.* This is the most important PDU since it communicates the current state of an entity to all other entities.
2. *Fire PDU.* This is sent out to specify the firing of a weapon. Other entities interpret this visually as a muzzle flash on their displays.
3. *Detonation PDU.* This is sent when a weapon is calculated to strike a target.
4. *Collision PDU.* An entity broadcasts this PDU and specifies which entity will be involved with the collision. The entity must broadcast information that will enable the target entity to calculate its new state (such as velocity and direction).

Nominal times for entity state packet receipt is 5 Hz, while entities update their displays at a rate of 15 Hz. All together there are a large number of PDU types in the Standard. Some general PDU types are clearly applicable to other interactive simulations, such as entity state and collision, whereas the fire PDU is valid only in military simulation. The PDUs can contain varied information. We will look at the precise definition of the entity state PDU as our example:

1. *PDU Header* (32 bits). This includes the protocol version (such as 2.0), exercise identifier, PDU type, and padding.
2. *Entity Information* (48 bits). Identifies the entity issuing the PDU.
3. *Force Identification* (8 bits). Identifies the force to which the entity belongs.
4. *Entity Type* (64 bits). This is category or group information further identifying the entity, such as domain, country, category, and subcategory.

---

[13] In DIS, internal real time synchronization plays a more critical role than does the external synchronization, based on message passing discussed in Secs. 9.4.3 and 9.4.4. Training effectiveness usually requires real time synchronization.

5. *Alternative Entity Type* (64 bits). Alternative information about the entity.

6. *Time Stamp* (32 bits). Time at which the PDU information is valid.

7. *Entity Location* (192 bits). Entity's physical location in the world. The high-fidelity model is used. X, Y, and Z (64 bits each).

8. *Entity Velocity* (96 bits). Entity's linear velocity. The high-fidelity model is used. V_x, V_y, and V_z (32 bits each).

9. *Entity Orientation* (96 bits). Entity's orientation. The high-fidelity model is used. ψ, θ, and ϕ (32 bits each).

10. *Dead-Reckoning Parameters* (320 bits). Identification of the algorithm used for dead reckoning for the entity sending this PDU. Linear acceleration and angular velocity are also encoded according to the dead-reckoning model.

11. *Entity Appearance* (32 bits). Specifies visual aspects of entity as seen by humans associated with other entity displays.

12. *Marking* (96 bits). Specifies unique marking on the entity such as a bumper number or country symbol.

13. *Capabilities* (32 bits). Capabilities of the entity.

14. *Number of Articulation Parameter Records* (8 bits). Number of records to follow.

15. *Articulation Parameters* (variable bits). Parameter values for each articulated part. Examples of articulated entities include a tank (with moving turret), antitank vehicle, or dismounted infantry person (arms, body, legs).

Sample scenario using PDUs.   The previously defined PDUs are necessary to have a *minimal set* for combat purposes. The entity-state PDU is necessary for change in position. This, alone, is sufficient for a simulation involving entities that do not interact with one another. If interaction is taken into account, we can have several kinds. For instance, we have an offensive interaction involving weapon firing and detonation, or we may have one vehicle colliding with another. Consider a two-entity simulation where an A1 tank fires an armor-piercing Sabot round at a BMP. Here are the simulation activities associated with this type of interaction:

1. M1 gunner (who is a human trainee) operates the main tank gun and performs the following interactive sequence:

   (a) Aim gun.
   (b) Detect range with laser finder.
   (c) Fire gun.

2. M1 entity sends a *fire* PDU, which produces a muzzle flash.

3. M1 entity determines if a hit was made on the BMP.

   (a) If the BMP was missed, the M1 entity issues a PDU specifying the piece of terrain hit. All other entities update their terrain databases to reflect the new terrain.
   (b) If the BMP was hit, the M1 entity sends a *detonation* PDU to the BMP.

4. BMP entity receives the detonation packet.

5. BMP assesses the damage to itself given the impact and weapon information provided by the M1 entity.

6. If the BMP state changes (such as partial or complete destruction), it issues an entity state packet to other entities.

Other DIS issues.   The complete DIS battlefield space in which training and simulation occur is virtually shared among all entities. Each entity keeps an updated version of the terrain, along with the position of all entities, and this serves as the digital world in which the trainee operates. The reason for this shared approach is due to the high level of interactivity among entities. If it were the case where an entity had effects only on its nearest neighbors, a more partitioned approach might be more effective. For instance, in a global distributed theatre-level training simulation involving the United States, Canada, and the United Kingdom, there may be very little interaction between any two countries and so it might make more sense to partition the world where separate computers controlled separate partitions and knew nothing of the other partitions unless a computer was sent a PDU.

Another critical issue involves maintaining simulation consistency. It is vital that as training proceeds, (1) each entity has the correct dead-reckoning model for all other entities, and (2) the models are physically realistic and consistent. There is nothing in the DIS architecture itself that enforces the constraint where entities must exhibit rational behavior. One could envision someone entering an ongoing simulation where the person's models have been changed to exhibit faulty, unrealistic, or illegal behavior. Consider a tank (entity A) preparing to fire upon entity B. Suppose that in the case of direct fire, A has no line of sight to B. Likewise, in the case of indirect fire, B may be out of range of A's tank cannon. There is nothing to stop the computer controlling entity A from simply sending detonation or fire PDUs to B, thereby violating the reality associated with intermediate obstacles and effective range of weapons.

9.6 EXERCISES

1. Why is it that spatial models are, in general, more easy to parallelize than functional models? What types of declarative models are suited to parallel execution?

2. Suppose that you have to simulate a network with channels having infinite buffer space (or large enough to avoid overflow). Show an example of a network where deadlock can arise.

3. Comment on the following statement: *A physical system such as a digital circuit, waterfall, or tree is actually a distributed simulation of itself.* Is this statement true? What are the issues?

4. Many model types that we have discussed are based on graphs. If we abstract a model into a generic graph, what graph algorithms can be used to partition the model onto physical processors?

5. With the fish example in Fig. 9.13, is it possible that rollback would have to occur over several steps during a conflict resolution? Explain why or why not.

6. Given an initial randomly populated grid (of density ρ), how often does rollback occur for a given grid size?

7. Provide an example of distributed interactive simulation where entity A obtains entity-state PDUs

from entity B which show that entity B is not moving, but in reality entity B is changing its state. How could something like this happen? Does this problem have a more general name?

8. In current DIS systems, a vehicle that fires a round to another entity determines whether or not that entity was hit. If hit, the entity determines its own damage. Explain how the protocol would work if we let the fired shell be modeled as an entity.

9. Even though dead reckoning reduces the network load, it provides an artificial amount of knowledge about other vehicles. For instance, a tank in an actual battlefield scenario would not have access to the dynamics of many enemy vehicles. Explain the tradeoffs with using dead reckoning.

9.7 PROJECTS

1. Causality in events is intimately related to the algebraic concepts of *group* in pure mathematics. After all, if we don't care in what order something is executed, we do not have to worry about process synchronization. Explore this relationship in a paper and specify examples.

2. Build a simulator for illustrating the principal effects of the conservative and optimistic methods for the parallel simulation of functional models. Essentially, you are required to build a simulation of a simulation since the base-level simulation would have to be executed on an actual parallel machine. Your implementation should be graphical and show the state of each virtual processor as it accepts and processes time-stamped messages. For the conservative method, the user should be able to create a network and a set of time-stamped messages. Local clock times and channel buffers should be visible, and a visual trace facility should be written so that users can step through simulations. For the optimistic method, the user should see the state of each processor along with input and output channel buffers containing messages and anti-messages. When rollback occurs, the trace should show the canceling effect of messages and anti-messages along with the rollback of state.

3. Expand the example in Fig. 9.13 to include sharks as well as fish. Also, add in population dynamics in the form of birth and death. Sharks feed on fish while fish continue to breed based on their age. Once a fish gets to a specific age, a new fish is born on an unoccupied adjacent square (or the nearest one), so you must keep track of the ages of the fish. If a shark does not eat a fish within some period of time, it dies. Create and implement a serial simulation of this new scenario and explain in detail what is necessary to parallelize the simulation.

4. Build a graphical dead-reckoning tool that incorporates different algorithms for (1) numerical integration (i.e., extrapolation) and (2) smoothing. Allow the user to create an arbitrary curve (the actual motion of an entity) in space and the fix the update interval. Display the error along the drawn curve to show the effects of dead reckoning on state estimation.

9.8 FURTHER READING

The characterization of parallel architectures using SISD, SIMD, MISD, and MIMD is attributed to Flynn [76, 77]. There are many good general texts on parallel computing, and some of them contain examples that are relevant to simulation. The book by Fox et al. [82] has a large number of relevant problems. The texts by Almasi and Gottlieb [4] and Bertsekas and Tsitsiklis [21] have some relevant chapter sections. For some of the areas not considered in this chapter, we note that considerable research has been done for parallelizing ODEs. Articles by Worland [233] and Franklin [83] are two early examples. Singh et al. [209] provide a concise collection of articles focusing on continuous-time (ODE) parallelism. Gupta [96] discusses parallelism of non-object-oriented production systems where the issue of time taken to perform matching is critical. Parallel implementations for functional models are found under the heading *parallel discrete event simulation*, or just *distributed simulation*, and there has been considerable activity in this field since the late 1970s. Unfortunately, there is very little in the way of textbook or edited collection coverage of this field, so material was drawn from journal and conference articles. Chandy and Misra [37, 152] have defined the conservative technique useful for simulating functional models in parallel, whereas Jefferson [116] lays out the principles behind the optimistic approach and the Time Warp implementation. The articles by Fujimoto [85], Richter and Walrand [187], and Nicol and Fujimoto [158] contain good overviews of the parallel approach to functional models. The fish population example was inspired by Dewdney [45] and the parallel implementation involving two population types (fish and sharks) and a good discussion of the problems of load balancing for this domain is given by Fox et al. [82]. Many texts on parallel algorithms contain methods applicable only for steady-state solutions, such as those with coverage of relaxation approaches. The text by Fox et al. [82] provides comprehensive material for parallel methods applied to spatial modeling problems in simulation, especially for adaptations to cube-based architectures.

The work in distributed interactive simulation (DIS) is found within the biannual workshops on DIS sponsored by the Advanced Research Projects Agency (ARPA) and the Simulation & Training Research and Instrumentation Command (STRICOM). McDonald [146] provides an overview of DIS. Lin [134] provides a survey of dead-reckoning techniques in use for DIS. Goel and Morris [89] provide a series of performance tests for dead-reckoning algorithms in the context of fast-moving vehicles such as missiles or aircraft. The Institute for Simulation and Training (IST) in Orlando serves as host for the DIS workshops and the emerging IEEE standard for protocol data units (PDUs) [78].

## Chapter 10

SimPack Toolkit

SimPack is dedicated to all programmers who have spent countless hours to design and distribute free software.

10.1 OVERVIEW

SimPack is a collection of C and C++ tools (routines and programs) for computer simulation. The code for SimPack is in two forms: one for the C language and the other for the C++ language. The purpose of SimPack is to provide people with a *starting point* for simulating a system. The SimPack project began with the *queuing* library and grew as we realized how little there was in the way of free simulation software on the net. The intention is that people will view what exists as a template (or "seed") and then "grow a simulation program." SimPack tools have been used for teaching computer simulation at the undergraduate (junior/senior) and graduate levels. The toolkit does not include simulation tools for every model type discussed in the text; however, there are tools for a large number of types, including declarative, functional constraint, and multimodel.

SimPack is obtained freely over the Internet by using *ftp* (file transfer program). Anonymous ftp for SimPack is accomplished by doing the following:

```
ftp ftp.cis.ufl.edu
anonymous
(enter your Internet address for the password)
cd pub/simdigest/tools
binary
ls simpack*
get SimPack software
quit
```

The software is in a Unix compressed tar format, so it will be necessary to apply `uncompress` followed by *tar*. For instance, the following will work:

```
uncompress simpack.tar.Z
tar xvf simpack.tar
```

Some of the SimPack routines and programs output x, y datasets. These data sets can be plotted using a wide variety of programs. Two good ones are *xvgr* and *gnuplot*.

The methods for employing SimPack tools are subject to change, as the package matures. New, online documentation is underway.

10.2 COPYRIGHT AND LICENSE

SimPack is copyrighted by the author, with the exception of those packages built by others to work in conjunction with SimPack. These authors are recognized by identifying footnotes in this chapter. The license agreement is the GNU General Public License as stated below.

This program is free software; you can redistribute it and/or modify it under the terms of the GNU General Public License as published by the Free Software Foundation; either version 2 of the license or (at your option) any later version.

This program is distributed in the hope that it will be useful, but WITHOUT ANY WARRANTY; without even the implied warranty of MERCHANTABILITY or FITNESS FOR A PARTICULAR PURPOSE. See the GNU General Public License for more details.

You should have received a copy of the GNU General Public License along with this program; if not, write to the Free Software Foundation, Inc., 675 Massachusetts Ave, Cambridge, MA 02139.

10.3 EVENT SCHEDULING

Two SimPack routines are used to provide us with a means to simulate basic models involving event scheduling:

- A Next_Event Routine
- A Schedule Routine

Events to occur are stored in the event list sorted by time, and then events are taken off the head of the list with `next_event`. Events are scheduled, or inserted in the list, with `schedule`. Here are the key routines for simulating pseudoparallelism via the discrete event method in C:

1. `schedule(event,delta-time,token)`. Schedule an event by placing it on the future event list. The event is to occur at time: `current-time + delta-time`, where `current-time` is the simulation time when that entity arrives at the facility and `delta-time` is the amount of time that is to transpire before the event occurs. `event` is usually an integer denoting the identification for the event. `token` is a data structure and must be defined as being of type TOKEN.

2. `next_event(&event,&token)`. Take the head node from the event list and update the current simulation time to the time attached to that node. Make `token` the currently active token.

3. `cancel_event(event)`. Cancel the first event specified by *event*.

4. `cancel_token(token)`. Cancel the first event associated with *token*.

In C++, `schedule` and `next_event` are methods for the object called "events." The event list is an object of type `events: event_list`. Therefore, the two methods are invoked as follows:

1. `event_list.schedule(event,delta-time,token)`.

2. `event_list.next_event(&event, &token)`.

Below, we present a general template in C for doing event scheduling with delays. We will see that this template represents the "bare bones" of any event-driven code that we will create.

```
#include "../../queuing/queuing.h"

/* Define event routine constants here */

#define MAX_TIME 1000.0 /* maximum time to run simulation */
#define EVENT1 1 /* identification for event number 1 */
#define EVENT2 2 /* identification for event number 2 */
...

/* Specify global variables and state variables here */

main() {
  int event;
  TOKEN token;

init_simpack(LINKED);
/* You must place something in the event list to start
   or 'bootstrap' the simulation */
schedule(EVENT1,0.0,token);
schedule(EVENT2,0.0,token);
...
/* Main simulation cycle (between 'next_event' and 'schedule')*/
```

```
while (time()< MAX_TIME) {
  next_event(&event,&token);
  switch(event) {
    case EVENT1: /* Event # 1 routine/case */
                 will contain things to do if event1
                 occurs. Often other event routines are scheduled here */
    case EVENT2: /* Similar to Event # 1 routine */
  } /* end switch */
  } /* end while */
} /* end main() */
```

An equivalent template for C++ is shown below.

```
#include "../../queuing/queuing.h"

// Define event routine constants here
enum {EVENT1=1,EVENT2}; // event routine identifications
const MAX_TIME = 1000.0; // maximum time to run simulation
...

// Specify global variables and state variables here

main() {
  int event;
  Token token;

  init_simpack(LINKED);
  event_list.schedule(EVENT1,0.0,token);
  event_list.schedule(EVENT2,0.0,token);
  ...
  /* Main simulation cycle (between 'next_event' and 'schedule')*/
  while (time()< MAX_TIME) {
   event_list.next_event(&event,&token);
   switch(event) {
     case EVENT1: // Event # 1 routine/case
     case EVENT2: // Same as for Event # 1
  } /* end switch */
  } // end while
} // end main()
```

10.4 DECLARATIVE MODEL SIMULATORS

10.4.1 FSA Simulator

The input to FSA is a file formatted as

```
#states
time-state-1 time-state-2 ...
state transition table
```

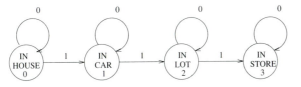

Figure 10.1   Driving to the grocery store.

```
label-for-state-1
label-for-state-2
...
length-of-control-string control-string
```

As an example, consider a simple grocery store simulation where we get into a car and drive to the store, park, and then walk into the store. We will create an FSA for this process shown in Fig. 10.1. We need to assign times to each transition and state. Assigning time durations to states provides a "shorthand" for specifying time-based input trajectories. We will let each state take the following amount of time:

```
State       Time

IN_HOUSE    60 minutes
IN_CAR       5 minutes
IN_LOT       1 minute
IN_STORE    45 minutes
```

Now, we need to provide this automaton with an input (or "control") string to "drive" the simulation: so we give it an input of "111," which should take it from *IN_HOUSE* to *IN_STORE* and all states in between.

```
%fsa < groc.dat
```

For examples of a simulation of a five-state automaton representing a simulation of the dining philosophers, try

```
%fsa < dp.dat
```

or

```
%fsa < dp2.dat
```

The output from the first example is

```
%fsa < groc.dat
FSA Simulator Output

Time: 0.000000, State: IN_HOUSE
Time: 60.000000, State: IN_CAR
Time: 65.000000, State: IN_LOT
Time: 66.000000, State: IN_STORE
```

```
FSA Model Statistics

State Statistics
- - - - - - - - - - - - -
State     Name           Freq     Freq %          Time        Time %
0         IN_HOUSE         1       25.0%           60.0        54.1%
1         IN_CAR           1       25.0%            5.0         4.5%
2         IN_LOT           1       25.0%            1.0         0.9%
3         IN_STORE         1       25.0%           45.0        40.5%


Link Statistics
- - - - - - - - - - - - -
Link      Id            Freq      Freq %
1         0 — > 1         1        33.3%
6         1 — > 2         1        33.3%
11        2 — > 3         1        33.3%
```

Statistics are set for both states and transitions (or links). The columns `Freq%` and `Time%` mean the frequency- and time-based utilization, respectively.

The input data for the dining philosophers (**DP**) simulation are

```
5
3.0 3.0 3.0 3.0 3.0
0   1
1   2
2   3
3   4
4   0
(1,3)-are-eating
(2,4)-are-eating
(3,5)-are-eating
(4,1)-are-eating
(5,2)-are-eating
9 111111111
```

The first line represents the number of states. The second line specifies the length of time taken for each state. This provides a convenient way of coding FSAs. Another, more uniform approach would be to specify an accurate time base and extend the input trajectories to account for the time taken in a state. The next lines (one for each state) specify the transition function which has as many columns as possible inputs. For this model we have two possible inputs (0 or 1), and each cell in the transition array has two values: the next state and the time taken for that transition arc. For instance, when at state 3 and given an input of 1, the FSA model specifies a transition to state 4. State 3 takes time 3.0 and the transition just defined takes time 1.0. Simulation time, therefore, is updated by 4.0.

Here is the output after running the simulation for the "time" indirectly specified by the control string length (9). Note that since the model reflects a ring structure, we can easily continue the simulation for an arbitrary length of time by increasing the control string length.

```
% fsa < dp.dat
FSA Simulator Output

Time: 0.000000, State: (1,3)-are-eating
Time: 3.000000, State: (2,4)-are-eating
Time: 6.000000, State: (3,5)-are-eating
Time: 9.000000, State: (4,1)-are-eating
Time: 12.000000, State: (5,2)-are-eating
Time: 15.000000, State: (1,3)-are-eating
Time: 18.000000, State: (2,4)-are-eating
Time: 21.000000, State: (3,5)-are-eating
Time: 24.000000, State: (4,1)-are-eating
Time: 27.000000, State: (5,2)-are-eating


FSA Model Statistics

State Statistics
- - - - - - - - - - - - -
State    Name              Freq    Freq %    Time     Time %
0        (1,3)-are-eating    2      20.0%     6.0      20.0%
1        (2,4)-are-eating    2      20.0%     6.0      20.0%
2        (3,5)-are-eating    2      20.0%     6.0      20.0%
3        (4,1)-are-eating    2      20.0%     6.0      20.0%
4        (5,2)-are-eating    2      20.0%     6.0      20.0%

Link Statistics
- - - - - - - - - - - - -
Link     Id           Freq      Freq %
1        0 - > 1        2        22.2%
7        1 - > 2        2        22.2%
13       2 - > 3        2        22.2%
19       3 - > 4        2        22.2%
20       4 - > 0        1        11.1%
```

Given this same model, for a control string of 100100, we obtain the following response:

```
% fsa < dp2.dat
FSA Simulator Output

Time: 0.000000, State: (1,3)-are-eating
Time: 3.000000, State: (2,4)-are-eating
Time: 6.000000, State: (2,4)-are-eating
Time: 9.000000, State: (2,4)-are-eating
Time: 12.000000, State: (3,5)-are-eating
Time: 15.000000, State: (3,5)-are-eating
Time: 18.000000, State: (3,5)-are-eating
```

```
FSA Model Statistics

State Statistics
- - - - - - - - - - - -
State     Name                  Freq    Freq %     Time     Time %
0         (1,3)-are-eating        1     14.3%       3.0     14.3%
1         (2,4)-are-eating        3     42.9%       9.0     42.9%
2         (3,5)-are-eating        3     42.9%       9.0     42.9%
3         (4,1)-are-eating        0      0.0%       0.0      0.0%
4         (5,2)-are-eating        0      0.0%       0.0      0.0%

Link Statistics
- - - - - - - - - - - -
Link      Id            Freq       Freq %
1         0 -> 1          1         16.7%
6         1 -> 1          2         33.3%
7         2 -> 2          1         16.7%
12        3 -> 2          2         33.3%
```

10.4.2 Markov Simulator

The input to markov is a file formatted as

```
#states time-for-simulation-run
adjacency matrix
label-for-state-1
label-for-state-2
...
```

Consider a simple cycle composed of five states, as shown in Fig. 10.2. This is a Markov implementation of the dining philosophers. We will simulate a set of dining philosophers (numbered 1 through 5) by letting philosophers 1 and 3 eat simultaneously, and then 2 and 4, and so on. We will let each eating action (the time associated with a state) be 3 minutes. A state, then, is specified as a pair [such as (1,3)]. We will specify each arc to be 1 minute. The arc represents a probabilistic change in state. Probabilities are shown in the following five state Markov process. We can see, for instance, that philosophers 2 and 4 (the second state) have a greater chance of eating again (probability = 0.4) right after their 3-minute eating activity. To try out this simulation, do

```
% markov < dpm.dat
```

The input for the example model that we discussed is as follows:

```
5 5000.0
3.0 3.0 3.0 3.0 3.0
0.1 0.9 0.0 0.0 0.0
0.0 0.4 0.6 0.0 0.0
0.0 0.0 0.1 0.9 0.0
0.0 0.0 0.0 0.2 0.8
```

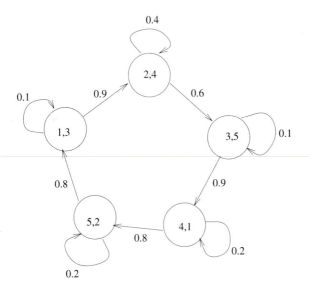

Figure 10.2 Markov dining philosophers model.

```
0.8 0.0 0.0 0.0 0.2
(1,3)
(2,4)
(3,5)
(4,1)
(5,2)
```

The first row specifies a Markov model of five states and a total simulation time of 5000 time units. The second row specifies the time associated with each Markov state. The next set of rows is the adjacency matrix that defines the topology of the graph. Each matrix cell contains two items: the probability of moving from state i to state j in the matrix $adj[i, j]$ and the time associated with that transition. The last rows are the state descriptors (as in the finite state automaton program). Here is the output for this simulation. As expected, state 1 has a high utilization due to its higher probability of being active.

```
% markov < dpm.dat
Markov Simulator Output
/* we commented out this portion to save space */
Markov Model Statistics


State Statistics
- - - - - - - - - - - - -
State     Name       Freq       Freq %       Time       Time %
0         (1,3)      302        18.1%        906.0      18.1%
1         (2,4)      445        26.7%        1335.0     26.7%
2         (3,5)      280        16.8%        840.0      16.8%
3         (4,1)      309        18.5%        927.0      18.5%
3         (5,2)      331        19.9%        993.0      19.9%
```

```
Link Statistics
- - - - - - - - - - - - -

Link      Id             Freq        Freq %
0         0 - > 0         40          2.4%
1         0 - > 1         262         15.7%
6         1 - > 1         183         11.0%
7         1 - > 2         262         15.7%
12        2 - > 2         18          1.1%
13        2 - > 3         262         15.7%
18        3 - > 3         47          2.8%
19        3 - > 4         262         15.7%
20        4 - > 0         262         15.7%
24        4 - > 4         69          4.1%
```

10.4.3 Temporal Logic Simulator

The following Tempura code represents the logistic beetle growth example in Ch. 4:

```
/* Simple Logistic Growth, Limit = population of 100 */

/* Look for a subinterval of length n on which N is stable */
define find_stable_n(n,N) = find_stable_n_aux(n,n,0,N).

define find_stable_n_aux(n,i,pN,N) = {
  if i=0 then empty
  else next{
    if N=pN then find_stable_n_aux(n,i-1,N,N)
    else find_stable_n_aux(n,n,N,N)
  }
}.

define grow3(n) =
  exists N: { N=1 and
              N gets N + N*(100-N)/50 and
  find_stable_n(n,N) and
              always output(N)
            }.
```

The program halts when it finds two successively identical values—corresponding to a weak integer definition of dynamic stability. In this way, we can ask the temporal logic program "Does the beetle growth stabilize?" or "Does the population ever reach 110?"

S-R latch simulation with unit delay.   As an example of simulation with temporal logic models, we consider an SR latch in digital logic using the Tempura language. Below we list the actual SR simulation program followed by the execution trace (i.e., simulation). The representative functional model has the behavior shown in the table below and is depicted graphically in Fig. 10.3. Note that there is a unit time delay associated with both

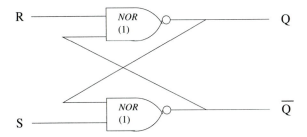

Figure 10.3 SR-Latch functional model.

NOR gates which is accomplished in Tempura using the next $\bigcirc$ operator.

S	R	Q	$\overline{Q}$
1	0	1	0
0	1	0	1
0	0	Q	$\overline{Q}$
1	1	—	—

```
define latch_test() = exists S,R,Q,Qbar: {
    S=0 and R=0 and Q=0 and Qbar=0
    and for l in [[1,0],[0,0],[0,1],[1,0],[0,0]] do {
        len(5)
        and S gets l[0] and R gets l[1]
    }
    and Q gets not(R or Qbar)
    and Qbar gets not(S or Q)
    and format("%4S %3S %3S %3S\n","Qbar"," Q "," R "," S ")
    and always format("%3S %3S %3S %3S\n",wave(Qbar),wave(Q),wave(R),wave(S))}.
define wave(A) = {
    if A=0 then " | " else " | "
}.

% tempura
C-Tempura (Version 2.7)

Tempura 1> load "latch".
Tempura 2> run latch_test().
State    0: Qbar    Q    R    S
State    0: |      |    |    |
State    1:  |     |  |      |
State    2: |     |    |      |
State    3: |       |  |      |
State    4: |       |  |      |
State    5: |       |  |      |
State    6: |       |  |    |
State    7: |       |  |    |
State    8: |       |  |    |
State    9: |       |  |    |
```

```
State  10: |         |  |      |
State  11: |         |  |      |
State  12: |      |     |      |
State  13:   |    |     |      |
State  14:   |    |     |      |
State  15:   |    |     |      |
State  16:   |    |  |         |
State  17: |    |  |         |
State  18: |       |  |      |
State  19: |       |  |      |
State  20: |       |  |      |
State  21: |       |  |    |
State  22: |       |  |    |
State  23: |       |  |    |
State  24: |       |  |    |
State  25: |       |  |    |
```

10.4.4 Timed Petri Net Simulator

The program `petri` permits Petri net simulations for nets with timed transitions. We will
run the program by specifying a data file for the dining philosopher example. The file
`phil.dat` includes the topology for the Petri net:

```
10 15 100
1.0 3 0 1 2 1 10
1.0 3 2 3 4 1 11
1.0 3 4 5 6 1 12
1.0 3 6 7 8 1 13
1.0 3 8 9 0 1 14
1.0 1 10 3 0 1 2
1.0 1 11 3 2 3 4
1.0 1 12 3 4 5 6
1.0 1 13 3 6 7 8
1.0 1 14 3 8 9 0
1 1 1 1 1 1 1 1 1 1 0 0 0 0 0
```

The first line in `phil.dat` specifies (1) the number of transitions (#T), (2) number of
places (#P), and (3) number of states to produce from the simulation run. The next #T data
lines have the information for each transition in the network: (1) the time attached to the
transition (i.e., how long it takes), (2) the number of input places, (3) the id of each input
place, (4) the number of output place, and (5) the id of each output place. The last line in
the data file specifies the initial state for the Petri net. There are #P integers; one for each
place. The value specifies how many tokens are initially located in each place.

Here is the output from the run performed:

```
% petri < phil.dat > states.out
Petri net contains 10 transitions and 15 places.
Simulate for 100 firings.
```

TIME	TRS	0	1	2	3	4	5	6	7	8	9	10	11	12	13	14
0.00	0:	1	1	1	1	1	1	1	1	1	1	0	0	0	0	0
0.00	2:	0	0	0	1	1	1	1	1	1	1	0	0	0	0	0
1.00	7:	0	0	0	1	0	0	0	1	1	1	1	0	1	0	0
1.00	5:	0	0	0	1	0	0	0	1	1	1	1	0	0	0	0
2.00	3:	1	1	1	1	1	1	1	1	1	1	0	0	0	0	0
2.00	1:	1	1	1	1	1	0	0	0	1	0	0	0	0	0	0
3.00	6:	1	1	0	0	0	1	0	0	0	1	0	1	0	1	0
3.00	8:	1	1	0	0	0	1	0	0	0	1	0	0	0	1	0
4.00	1:	1	1	1	1	1	1	1	1	1	1	0	0	0	0	0
4.00	4:	1	1	0	0	0	1	1	1	1	1	0	0	0	0	0
5.00	6:	0	1	0	0	0	1	1	1	0	0	0	1	0	0	1
5.00	9:	0	1	0	0	0	1	1	1	0	0	0	0	0	0	1
6.00	4:	1	1	1	1	1	1	1	1	1	1	0	0	0	0	0
6.00	2:	0	1	1	1	1	1	1	1	1	1	0	0	0	0	0

.....................output removed.............................

TIME	TRS	0	1	2	3	4	5	6	7	8	9	10	11	12	13	14
46.00	2:	1	1	1	1	1	1	1	1	1	1	0	0	0	0	0
46.00	4:	1	1	1	1	0	0	0	1	1	1	0	0	0	0	0
47.00	7:	0	1	1	1	0	0	0	1	0	0	0	0	1	0	1
47.00	9:	0	1	1	1	0	0	0	1	0	0	0	0	0	0	1
48.00	2:	1	1	1	1	1	1	1	1	1	1	0	0	0	0	0
48.00	0:	1	1	1	1	0	0	0	1	1	1	0	0	0	0	0
49.00	7:	0	0	0	1	0	0	0	1	1	1	1	0	1	0	0
49.00	5:	0	0	0	1	0	0	0	1	1	1	1	0	0	0	0

10.5 FUNCTIONAL MODEL SIMULATORS

10.5.1 Time-Slicing Block Simulator

`slice` is a simple example of a block diagram simulator—not unlike a CSMP program for continuous simulation. The input file format is

```
#-of-blocks output-block time-for-simulation
block0-num block-type block-input(s) parameters(if any)
block1-num ............
    . . . .
```

Consider the block model shown in Fig. 5.6. This is coded within `slice` using a data file `block.dat`:

```
7 6 50.0
0 const 0.8
1 const 2.0
2 mult 0 5
```

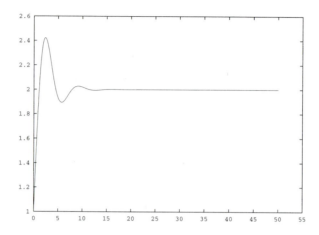

Figure 10.4   Sample block model execution.

```
3 sub 1 2
4 sub 3 6
5 int 4 1.0
6 int 5 1.0
```

The output of block 6 is shown in Fig. 10.4.

10.5.2 Clock

Two routines in the queuing library `queuing.o` provide a good introduction to updating time through event scheduling. The two routines are `schedule` and `next_event`. The output from the clock simulator performs schedules every second, minute, and hour. The output begins as follows and continues while updating the current minute and hour:

```
Time: 0: 0: 0
Time: 0: 0: 1
Time: 0: 0: 2
Time: 0: 0: 3
Time: 0: 0: 4
Time: 0: 0: 5
Time: 0: 0: 6
Time: 0: 0: 7
Time: 0: 0: 8
Time: 0: 0: 9
Time: 0: 0:10
Time: 0: 0:11
Time: 0: 0:12
Time: 0: 0:13
Time: 0: 0:14
Time: 0: 0:15
Time: 0: 0:16
Time: 0: 0:17
```

```
Time:  0:  0:18
Time:  0:  0:19
Time:  0:  0:20
. . . . . . . . . . . . . . . . .
```

10.5.3 Digital Logic Network Simulator

This is a simple digital logic simulator with nominal gate delays. For available gate types, note the program `logic.c`. It is easy to add new gates after seeing this template. The form of the input file for 'logic' is

```
#-of-gates output-gate time-for-simulation
#-of-gate1 gate-type delay input-1 input-2 ... #-of-outputs out-1 out-2 ...
#-of-gate2 gate-type delay input-1 input-2 ... #-of-outputs out-1 out-2 ...
...
```

If there are zero inputs, leave the input part of the input record blank. The program `logic.c` specifies the # of inputs for each type of gate, whereas the number of outputs is arbitrary. Let's consider a logic network that consists of two gates: a generator and an invertor in file `inv.dat`:

```
2 2 18.0
0 gen 4.0 1 1
1 inv 0.0 0 0
```

`logic < inv.dat` generates output:

```
0.000000 1 0
4.000000 0 1
8.000000 1 0
12.000000 0 1
16.000000 1 0
```

A larger network with AND and OR gates is shown in file `andor.dat`:

```
5 4 18.0
0 gen 4.0 1 3
1 gen 4.0 1 3
2 gen 4.0 1 4
3 and 2.0 0 1 1 4
4 or 1.0 3 2 0
```

The output (from block 4) is

```
0.000000 1 1 1 0 0
1.000000 1 1 1 0 1
2.000000 1 1 1 0 1
3.000000 1 1 1 0 1
4.000000 0 0 0 0 1
5.000000 0 0 0 0 0
```

```
 6.000000 0 0 0 0 0
 7.000000 0 0 0 0 0
 8.000000 1 1 1 0 0
 9.000000 1 1 1 0 1
10.000000 1 1 1 0 1
11.000000 1 1 1 0 1
12.000000 0 0 0 0 1
13.000000 0 0 0 0 0
14.000000 0 0 0 0 0
15.000000 0 0 0 0 0
16.000000 1 1 1 0 0
17.000000 1 1 1 0 1
```

10.5.4 Queuing Model Library

Here are the routines that support queuing in C:

1. `facility_id = create_facility(name,num_servers)`. A facility is a system resource that acts as a collection of servers. An example of a facility is the front desk. This facility will have several servers (bank tellers). Often, we will associate one server with each facility. This is known as a single-server queue. Note that there is only one queue per facility regardless of the number of servers.

2. `status = request(facility_id,token,priority)`. An incoming token requests the use of the facility—or the facility's servers to be more precise. If there is at least one server available, then the facility status is FREE; otherwise, it is BUSY. If the facility is busy, the token is placed onto a facility queue in *priority sorted* order.

3. `status = preempt(facility_id,token,priority)`. An incoming token requests use of the facility and will usually preempt (or replace) the token already being serviced. The difference between `preempt` and `request` is that with preemption, one does not normally wait in the queue at all—the idea is to bypass the queue completely to obtain immediate service. If the incoming token's priority is less than or equal to all tokens currently being served in a facility, the incoming token will actually be queued. However, if the incoming token's priority is greater than the lowest-priority server token for the facility, the server token will be preempted and placed back on the facility queue tagged with its remaining time for service.

4. `release(facility_id,token)`. After a token has been served, release this facility to the token at the head of the facility queue.

For C++, the calls are similar but oriented toward an object perspective.

1. `Facility facility(id,num_servers)`, where `facility` is the name of a facility object to create and `id` is an integer identifying the facility.
2. `status = facility.request(token,priority)`.
3. `status = facility.preempt(token,priority)`.
4. `facility.release(token)`.

Here are some key auxiliary routines in SimPack for queuing model executions:

1. `update_arrivals()`. This should be placed at the entry point to the system. That is, this helps SimPack determine when to count arrivals into the system.
2. `update_completions()`. This should be placed at the exit point from the system. That is, this helps SimPack determine when to count departures from the system.
3. `report_stats()`. This function provides statistics for the individual facilities and for the system in general.
4. `trace_facility(facility_id)`. Show each facility queue for facility facility_id. In C++, this is called as: `facility.trace()`.
5. `trace_eventlist()`. Show contents of the event queue. In C++, this is called as: `event_list.trace()`.
6. `trace_eventlist_size()`. Print the time and current event list size to std-out. If output is directed to a file, this file can be plotted to view the event list size fluctuations over time.
7. `trace_visual(mode)`. Display the contents of both the event list and each facility queue in "graphical" form. Link lists show nodes as rectangular boxes. The parameter mode can be either INTERACTIVE or BATCH. With interative mode you can step through the simulation, and with batch mode you can have the graphical output sent to a file for subsequent visual viewing. This trace routine is an excellent way to learn how the simulation works internally, and it can be used to debug your simulation programs. You will find that the following is an effective way to perform an animated, dynamic playback of the simulation procedure. Output the visual trace to a text file: `%queue1 > output`, then edit the file `output` using the Emacs Text Editor. By using control-S on the string `## TIME` to search through the output file, you will find that you can watch an animation of the simulation control mechanism in action: items will be added and deleted from lists and queues dynamically as time increments.

10.5.5 Grocery Store I

Sample simulation output for this system could be a list with three columns (`C#`, `Arrival Time`, and `Departure Time`). Here is sample output when service time is 2 minutes. Note that the kind of simulation output depends on the type of information that is important to the analyst—there is not a unique type of output.

```
------------------------------------------------
SAMPLE OUTPUT FOR GROCERY STORE CHECKOUT SIMULATION
------------------------------------------------


CUSTOMER #    ARRIVAL TIME    DEPARTURE TIME
--------      ----------      ------------
1             0               2
2             2               4
3             4               6
4             6               8
5             8               10
```

The table shows that there are many times when customer departure time is specified by the arrival time plus service time. For instance, when service is 2 minutes, customer 4 arrives at time 6 (6 minutes after the simulation begins) and then leaves at time $6 + 2 = 8$. This activity reflects that a "queue" has not formed for any of the customers—the server (or cashier) is fast enough to handle all of the customers without forcing any customer to wait in line. This type of situation is true when service times are less than or equal to interarrival times. Now, consider what happens when service time is 3 minutes. We see now that the service time dominates system behavior. That is, when service time is greater than all inter-arrival times, *all* customers must wait in line except for the first customer, who is served immediately. Look at how long different customers are in the system: customer 1 is in the system for only 3 minutes, whereas customer 5 is waiting for a total of 7 minutes $(15 - 8)$. Suppose, now that service time is sometimes less than interarrival time and sometimes is more. This is the situation in the last column of the above table. The last column reflects more natural circumstances where the server and interarrival times vary, causing the queue length to fluctuate over time.

10.5.6 Grocery Store II

Sample output for this simulation will look like the output for the previous model; however, we will include some statistics also. Statistics are very common in the simulation output of systems using event scheduling because you are using random variables as part of the model. Note that we have arbitrarily chosen to obtain output for only the first five customers and only a subset of the variables. This demonstrates that we need not output all state values for all time for a simulation. We may want only a few pieces of information and some summary statistics.

```
--------------------------------------------
SAMPLE OUTPUT FOR GROCERY STORE CHECKOUT SIMULATION #2
--------------------------------------------


CUSTOMER #    ARRIVAL TIME    SERVICE BEGINS    SERVICE ENDS
--------      ----------      -------------     ----------
1             0               0                 1
2             3               3                 7
3             10              10                14
4             13              14                16
5             22              22                23


Summary Statistics
---------------


THE AVERAGE CUSTOMER WAITS 4.5 MINUTES
THE SERVER IS IDLE 39% OF TOTAL TIME
```

10.5.7 Single-Server System

The single-server queue is the simplest model that uses the queuing library. For a simulation time of 1000.0, the summary statistics produced by report_stats() are

```
+--------------------+
| SimPack SIMULATION REPORT |
+--------------------+

Total Simulation Time: 1000.000000
Total System Arrivals: 501
Total System Completions: 492


System Wide Statistics
-----------------
System Utilization: 92.4%
Arrival Rate: 0.501000, Throughput: 0.492000
Mean Service Time per Token: 1.877115
Mean # of Tokens in System: 6.682195
Mean Residence Time for each Token: 13.581697


Facility Statistics
---------------
F 1 (queue): Idle: 7.6%, Util: 92.4%, Preemptions: 0
```

Summary statitistics can be augmented with a step by step trace of the internal data structures. Program q2.c demonstrates output for the first few time periods when the visual trace mode is turned on. The trace can be seen interactively or in batch mode.

```
## TIME: 0.0
## BEFORE SCHEDULE Event 1 Inter-Time 0.0 Token 1

## EVENT LIST

Token
Time
Event

## FACILITY 1: (queue). 1 Server(s), 0 Busy.
Server(s): ( 1) TK 0 PR 0

Token
Time
Event
Priority
```

```
## TIME: 0.0
## AFTER SCHEDULE Event 1 Inter-Time 0.0 Token 1

## EVENT LIST
          +----+
Token     |  1 |
Time   <= | 0.0 |
Event     |  1 |
          +----+
## FACILITY 1: (queue). 1 Server(s), 0 Busy.
Server(s): ( 1) TK 0 PR 0

Token
Time
Event
Priority



## TIME: 0.0
## BEFORE NEXT_EVENT

## EVENT LIST
          +----+
Token     |  1 |
Time   <= | 0.0 |
Event     |  1 |
          +----+

## FACILITY 1: (queue). 1 Server(s), 0 Busy.
Server(s): ( 1) TK 0 PR 0

Token
Time
Event
Priority



## TIME: 0.0
## AFTER NEXT_EVENT Event 1 Token 1

## EVENT LIST

Token
Time
Event

## FACILITY 1: (queue). 1 Server(s), 0 Busy.
Server(s): ( 1) TK 0 PR 0

Token
Time
Event
Priority
```

```
## TIME: 0.0
## BEFORE SCHEDULE Event 2 Inter-Time 0.0 Token 1

## EVENT LIST

Token
Time
Event

## FACILITY 1: (queue). 1 Server(s), 0 Busy.
Server(s): ( 1) TK 0 PR 0

Token
Time
Event
Priority


## TIME: 0.0
## AFTER SCHEDULE Event 2 Inter-Time 0.0 Token 1

## EVENT LIST

          +----+
Token   |    1 |
Time  <=|  0.0 |
Event   |    2 |
          +----+
## FACILITY 1: (queue). 1 Server(s), 0 Busy.
Server(s): ( 1) TK 0 PR 0

Token
Time
Event
Priority


## TIME: 0.0
## BEFORE SCHEDULE Event 1 Inter-Time 2.0 Token 2

## EVENT LIST
          +----+
Token   |    1 |
Time  <=|  0.0 |
Event   |    2 |
          +----+
## FACILITY 1: (queue). 1 Server(s), 0 Busy.
Server(s): ( 1) TK 0 PR 0

Token
Time
Event
Priority
```

```
## TIME: 0.0
## AFTER SCHEDULE Event 1 Inter-Time 2.0 Token 2

## EVENT LIST
          +----+  +----+
Token  |   1 |  |   2 |
Time  <=|  0.0 |<=|  2.0 |
Event  |   2 |  |   1 |
          +----+  +----+
## FACILITY 1: (queue). 1 Server(s), 0 Busy.
Server(s): ( 1) TK 0 PR 0

Token
Time
Event
Priority



## TIME: 0.0
## BEFORE NEXT_EVENT

## EVENT LIST
          +----+  +----+
Token  |   1 |  |   2 |
Time  <=|  0.0 |<=|  2.0 |
Event  |   2 |  |   1 |
          +----+  +----+
## FACILITY 1: (queue). 1 Server(s), 0 Busy.
Server(s): ( 1) TK 0 PR 0

Token
Time
Event
Priority



## TIME: 0.0
## AFTER NEXT_EVENT Event 2 Token 1

## EVENT LIST
          +----+
Token  |   2 |
Time  <=|  2.0 |
Event  |   1 |
          +----+
## FACILITY 1: (queue). 1 Server(s), 0 Busy.
Server(s): ( 1) TK 0 PR 0

Token
Time
Event
Priority
```

```
## TIME: 0.0
## BEFORE REQUEST Facility 1 Token 1 Priority 0

## EVENT LIST
        +----+
Token   |   2 |
Time  <=|  2.0 |
Event   |   1 |
        +----+

## FACILITY 1: (queue). 1 Server(s), 0 Busy.
Server(s): ( 1) TK 0 PR 0

Token
Time
Event
Priority


## TIME: 0.0
## AFTER REQUEST Facility 1 Token 1 Priority 0

## EVENT LIST
        +----+
Token   |   2 |
Time  <=|  2.0 |
Event   |   1 |
        +----+
## FACILITY 1: (queue). 1 Server(s), 1 Busy.
Server(s): ( 1) TK 1 PR 0

Token
Time
Event
Priority


## TIME: 0.0
## BEFORE SCHEDULE Event 3 Inter-Time 3.5 Token 1

## EVENT LIST
        +----+
Token   |   2 |
Time  <=|  2.0 |
Event   |   1 |
        +----+
## FACILITY 1: (queue). 1 Server(s), 1 Busy.
Server(s): ( 1) TK 1 PR 0

Token
Time
Event
Priority
```

```
## TIME: 0.0
## AFTER SCHEDULE Event 3 Inter-Time 3.5 Token 1
## EVENT LIST
          +----+  +----+
Token  |   2 |  |   1 |
Time  <=|  2.0 |<=|  3.5 |
Event  |   1 |  |   3 |
          +----+  +----+
## FACILITY 1: (queue). 1 Server(s), 1 Busy.
Server(s): ( 1) TK 1 PR 0


Token
Time
Event
Priority



## TIME: 0.0
## BEFORE NEXT_EVENT

## EVENT LIST
          +----+  +----+
Token  |   2 |  |   1 |
Time  <=|  2.0 |<=|  3.5 |
Event  |   1 |  |   3 |
          +----+  +----+
## FACILITY 1: (queue). 1 Server(s), 1 Busy.
Server(s): ( 1) TK 1 PR 0


Token
Time
Event
Priority



## TIME: 2.0
## AFTER NEXT_EVENT Event 1 Token 2

## EVENT LIST
          +----+
Token  |   1 |
Time  <=|  3.5 |
Event  |   3 |
          +----+
## FACILITY 1: (queue). 1 Server(s), 1 Busy.
Server(s): ( 1) TK 1 PR 0


Token
Time
Event
Priority
```

```
## TIME: 2.0
## BEFORE SCHEDULE Event 2 Inter-Time 0.0 Token 2

## EVENT LIST
            +----+
Token   |    1 |
Time   <=|  3.5 |
Event   |    3 |
            +----+
## FACILITY 1: (queue). 1 Server(s), 1 Busy.
Server(s): ( 1) TK 1 PR 0

Token
Time
Event
Priority


## TIME: 2.0
## AFTER SCHEDULE Event 2 Inter-Time 0.0 Token 2

## EVENT LIST
            +----+  +----+
Token   |    2 |  |    1 |
Time   <=|  2.0 |<=|  3.5 |
Event   |    2 |  |    3 |
            +----+  +----+
## FACILITY 1: (queue). 1 Server(s), 1 Busy.
Server(s): ( 1) TK 1 PR 0

Token
Time
Event
Priority

## TIME: 2.0
## BEFORE SCHEDULE Event 1 Inter-Time 2.0 Token 3

## EVENT LIST
            +----+  +----+
Token   |    2 |  |    1 |
Time   <=|  2.0 |<=|  3.5 |
Event   |    2 |  |    3 |
            +----+  +----+
## FACILITY 1: (queue). 1 Server(s), 1 Busy.

Server(s): ( 1) TK 1 PR 0
Token
Time
Event
Priority
```

```
## TIME: 2.0
## AFTER SCHEDULE Event 1 Inter-Time 2.0 Token 3

## EVENT LIST
          +----+  +----+  +----+
Token  |    2 |  |    1|  |    3 |
Time   <=|  2.0 |<=|  3.5|<=|  4.0 |
Event  |    2 |  |    3|  |    1 |
          +----+  +----+  +----+
## FACILITY 1: (queue). 1 Server(s), 1 Busy.
Server(s): ( 1) TK 1 PR 0


Token
Time
Event
Priority




## TIME: 2.0
## BEFORE NEXT_EVENT

## EVENT LIST
          +----+  +----+  +----+
Token  |    2 |  |    1|  |    3 |
Time   <=|  2.0 |<=|  3.5|<=|  4.0 |
Event  |    2 |  |    3|  |    1 |
          +----+  +----+  +----+
## FACILITY 1: (queue). 1 Server(s), 1 Busy.
Server(s): ( 1) TK 1 PR 0


Token
Time
Event
Priority




## TIME: 2.0
## AFTER NEXT_EVENT Event 2 Token 2

## EVENT LIST
          +----+  +----+
Token  |    1 |  |    3 |
Time   <=|  3.5 |<=|  4.0 |
Event  |    3 |  |    1 |
          +----+  +----+
## FACILITY 1: (queue). 1 Server(s), 1 Busy.
Server(s): ( 1) TK 1 PR 0


Token
Time
Event
Priority
```

```
## TIME: 2.0
## BEFORE REQUEST Facility 1 Token 2 Priority 0

## EVENT LIST
             +----+  +----+
Token    |    1 |  |    3 |
Time   <=|  3.5 |<=|  4.0 |
Event    |    3 |  |    1 |
             +----+  +----+
## FACILITY 1: (queue). 1 Server(s), 1 Busy.
Server(s): ( 1) TK 1 PR 0

Token
Time
Event
Priority


## TIME: 2.0
## AFTER REQUEST Facility 1 Token 2 Priority 0

## EVENT LIST
             +----+  +----+
Token    |    1 |  |    3 |
Time   <=|  3.5 |<=|  4.0 |
Event    |    3 |  |    1 |
             +----+  +----+

## FACILITY 1: (queue). 1 Server(s), 1 Busy.
Server(s): ( 1) TK 1 PR 0
                 +----+
Token        |    2 |
Time       <=|  2.0 |
Event        |    2 |
Priority     |    0 |
                 +----+


## TIME: 2.0
## BEFORE NEXT_EVENT

## EVENT LIST
             +----+  +----+
Token    |    1 |  |    3 |
Time   <=|  3.5 |<=|  4.0 |
Event    |    3 |  |    1 |
             +----+  +----+
## FACILITY 1: (queue). 1 Server(s), 1 Busy.
Server(s): ( 1) TK 1 PR 0
                 +----+
Token        |    2 |
Time       <=|  2.0 |
Event        |    2 |
Priority     |    0 |
                 +----+
```

```
## TIME: 3.5
## AFTER NEXT_EVENT Event 3 Token 1

## EVENT LIST
            +----+
Token   |    3 |
Time  <=|  4.0 |
Event   |    1 |
            +----+
## FACILITY 1: (queue). 1 Server(s), 1 Busy.
Server(s): ( 1) TK 1 PR 0
               +----+
Token      |    2 |
Time     <=|  2.0 |
Event      |    2 |
Priority   |    0 |
               +----+


## TIME: 3.5 ##
BEFORE RELEASE Facility 1 Token 1

## EVENT LIST
            +----+
Token   |    3 |
Time  <=|  4.0 |
Event   |    1 |
            +----+
## FACILITY 1: (queue). 1 Server(s), 1 Busy.
Server(s): ( 1) TK 1 PR 0
               +----+
Token      |    2 |
Time     <=|  2.0 |
Event      |    2 |
Priority   |    0 |
               +----+


## TIME: 3.5
## AFTER RELEASE Facility 1 Token 1

## EVENT LIST
            +----+  +----+
Token   |    2 |  |    3 |
Time  <=|  3.5 |<=|  4.0 |
Event   |    2 |  |    1 |
            +----+  +----+
## FACILITY 1: (queue). 1 Server(s), 0 Busy.
Server(s): ( 1) TK 0 PR 0

Token
Time
Event
Priority
```

(job cycle)

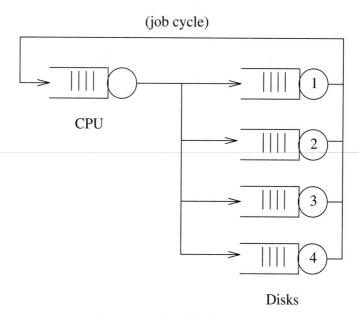

Figure 10.5   CPU with four disk units.

10.5.8 CPU/Disk System

Consider a combined CPU/Disk system in Fig. 10.5. There is a single CPU (central processing unit) that will execute a set number of tasks. Each task will have disk I/O requirements. The cyclical nature of this model reflects the fact that a single task will be simulated as follows:

```
Use CPU
Access a disk (1-4)
Use CPU
Access a disk (1-4)
Use CPU
Access a disk (1-4)
Use CPU
Access a disk (1-4)
... etc ...
```

This occurs until a set number of *tours* have been completed. One tour for a task is the use of the CPU and then the disk before cycling back. Let's now consider the CPU/Disk system primary code (within the `switch` statement). This will be shown using SimPack. The code for the CPU/Disk system is now specified:

The first thing to notice is the selection of events that have been defined after constructing the graph. Events have no associated time interval; they occur at an *instant* of time.

Therefore, `disk-operating` would not make a good event since it involves an inherent time duration. `start-disk` would be fine since it "brackets" (along with `end-disk`) the disk operation procedure. In this model we have events that bracket the (1) operation on the CPU, and (2) the operation on the Disk. Let's now discuss the `switch` statement code.

The purpose of `schedule` is to allow you to place time durations in the simulation. For instance—the CPU will take a certain amount of time per task. This time can be obtaining by sampling a random distribution. By scheduling the "end of CPU usage" for a token (i.e., task), we are simulating the use of the CPU for that time period. Note that the CPU resource is a *preemptive resource*, whereas the disks are not. This means that the CPU can be preempted but the disks cannot. Let's consider the piece of code where we request the use of the CPU:

```
case 2: /* request cpu */
  if (preempt(cpu,i,j) == FREE)
  schedule(3,expntl(tc[j]),i);
  break;
```

This states that token i with priority j wants to use the facility. If the facility is free, there is no need to preempt anybody. It is only when the facility is busy that preemption can take place. `preempt` returns FREE in one of two cases: (1) the facility is not busy and can therefore be obtained by the incoming token, or (2) the facility is busy but can be preempted due to the incoming token's high priority—that is, the incoming token's priority is higher than the minimum priority of token(s) being presently being served. The `schedule` simply tells the simulation how long the CPU is to "operate" for this token.

We will study the effects of each of the key routines by seeing what the main internal system data structures look like *before* and then *after* each routine is executed. The visual trace output can, therefore, be coordinated with the C source by comparing the current command being executed (schedule, release, preempt, request, etc.) against the corresponding C routine in the main program. Figures 10.6 through 10.10 define the following operations:

- `request`. Request a facility. If the facility is free, the requesting token obtains one of the facility's servers. If the facility is busy, this token is placed behind all queued tokens with the *same or higher priority*.
- `preempt`. Preempt a facility. If the facility is free, the preempting token obtains one of the facility's servers. If the facility is busy, is a check is made to compare the server token's priority with the priority of the preempting token. If the preempting token has a higher priority, preemption takes place. This means that (1) the preempting token is placed in the server, and (2) the token that was just displaced from service is placed in the front of the queue awaiting service (again). The preempted token only need wait the *remaining* length of time necessary until its next event. On the other hand, if preempting token has a lower or the same priority as the server token, the preempting token does not preempt and is placed behind all queued tokens with the *same or higher priority*. In the latter case, `preempt` behaves like *request*.
- `release`. Release a server of a facility so that the token waiting on that server (if one exists) can now occupy it.

```
## TIME: 0.0
## BEFORE NEXT_EVENT

## EVENT LIST
            +----+  +----+  +----+  +----+  +----+  +----+
Token   |     5|  |    6|  |    7|  |    8|  |    9|  |    1|
Time   <=|   0.0|<=|  0.0|<=|  0.0|<=|  0.0|<=|  0.0|<=|  5.8|
Event   |     2|  |    2|  |    2|  |    2|  |    2|  |    3|
            +----+  +----+  +----+  +----+  +----+  +----+
```

```
## TIME: 0.0
## AFTER NEXT_EVENT Event 2 Token 5

## EVENT LIST
            +----+  +----+  +----+  +----+  +----+
Token   |     6|  |    7|  |    8|  |    9|  |    1|
Time   <=|   0.0|<=|  0.0|<=|  0.0|<=|  0.0|  5.8|
Event   |     2|  |    2|  |    2|  |    2|  |    3|
            +----+  +----+  +----+  +----+  +----+
```

Figure 10.6   Cause next event to occur.

10.5.9 Sample Traces for the CPU/Disk Problem

We will enclose selective trace snapshots from the CPU/Disk simulation: Figs. 10.6 and 10.7 provide a visual trace of what occurs during the schedule and next event routines. These traces are initiated using the routine trace_visual(mode), where mode is equal to INTERACTIVE or BATCH, depending on how the output is to be viewed. In the INTERACTIVE mode, the user is able to step through the simulation process one step at a time. This is especially useful in debugging simulation code and learning exactly how the event scheduling mechanism occurs.

10.5.10 Blocking in Queuing Networks

A program blocking implements the serial queuing network described in Sec. 5.2.6. The example is where we have a generator of raw parts, an accumulator, a lathe, and a "finished part" inspector. A cost value is assigned to this operation by assigning subcosts for the part generation and labor associated with each server.

10.5.11 General Network Simulator

qnet is a simulator that accepts as input a file specifying the topology of a queuing network. The purpose of qnet is to demonstrate how easy it is to "encode network structure in data" instead of having to embed the topology within the code itself. To run qnet, we need to

```
## TIME: 0.0
## BEFORE SCHEDULE Event 3 Inter-Time 7.0 Token 7

## EVENT LIST
              +----+  +----+
Token    |     8|  |     9|
Time    <=|   0.0|<=|   0.0|
Event    |     2|  |     2|
              +----+  +----+


## TIME: 0.0
## AFTER SCHEDULE Event 3 Inter-Time 7.0 Token 7

## EVENT LIST
              +----+  +----+  +----+
Token    |     8|  |     9|  |     7|
Time    <=|   0.0|<=|   0.0|<=|   7.0|
Event    |     2|  |     2|  |     3|
              +----+  +----+  +----+
```

Figure 10.7   Schedule Event to Occur.

create a data file. For a single-server queue, for instance, we create the following file:

```
4 1000.0
0 gen 2.0
1 request queue-1 1 expon 5.0 0.0
2 release
3 sink
0 0 1 1
1 1 0 1 2
2 1 1 1 3
3 1 2 0
```

The format of the file is as follows:

1. Number of blocks, end time to stop simulation.
2. A list of the blocks with their specifications. For GEN blocks, we need the interarrival time. For REQUEST blocks, we need a queue name, number of servers, type of distribution for service, and the distribution parameters. For a FORK block we need the number of fork tines along with a probability value that each tine will be taken (probabilities must add to 1.0).
3. A list of the blocks with their inputs and outputs (net structure). Each line contains the following: block number, number of inputs, inputs, number of outputs, outputs.

There are six kinds of blocks available for modeling using the queuing network simulator: gen, request, release, join, fork, and sink. A combination request/release connection is used

```
## TIME: 43.2
## BEFORE REQUEST Facility 2 Token 9 Priority 0

## EVENT LIST
              +----+  +----+
Token         |   7|  |   8|
Time       <=| 45.4|<=| 78.4|
Event         |   3|  |   5|
              +----+  +----+
## FACILITY 1: (CPU). 1 Server(s), 1 Busy.
Server(s): ( 1) TK 7 PR 1
              +----+  +----+  + ----+  +----+  +----+  +----+
Token         |   1|  |   2|  |   3|  |   4|  |   5|  |   6|
Time       <=| -5.8|<=|  0.0|<=|  0.0|<=|  0.0|<=|  0.0|<=|  0.0|
Event         |   3|  |   2|  |   2|  |   2|  |   2|  |   2|
Priority      |   0|  |   0|  |   0|  |   0|  |   0|  |   0|
              +----+  +----+  +----+  +----+  +----+  +----+
## FACILITY 2: (disk). 1 Server(s), 1 Busy.
Server(s): ( 1) TK 8 PR 0

Token
Time
Event
Priority


## TIME: 43.2
## AFTER REQUEST Facility 2 Token 9 Priority 0

## EVENT LIST
              +----+  +----+
Token         |   7|  |   8|
Time       <=| 45.4|<=| 78.4|
Event         |   3|  |   5|
              +----+  +----+
## FACILITY 1: (CPU). 1 Server(s), 1 Busy.
Server(s): ( 1) TK 7 PR 1
              +----+  +----+  + ----+  +----+  +----+  +----+
Token         |   1|  |   2|  |   3|  |   4|  |   5|  |   6|
Time       <=| -5.8|<=|  0.0|<=|  0.0|<=|  0.0|<=|  0.0|<=|  0.0|
Event         |   3|  |   2|  |   2|  |   2|  |   2|  |   2|
Priority      |   0|  |   0|  |   0|  |   0|  |   0|  |   0|
              +----+  +----+  + ----+  +----+  +----+  +----+
## FACILITY 2: (disk). 1 Server(s), 1 Busy.
Server(s): ( 1) TK 8 PR 0

              +----+
Token         |   9|
Time       <=| 43.2|
Event         |   4|
Priority      |   0|
              +----+
```

Figure 10.8   Request Facility

```
## TIME: 0.0
## BEFORE PREEMPT Facility 1 Token 7 Priority 1


## EVENT LIST
              +----+  +----+  +----+
Token         |   8|  |   9|  |   1|
Time       <=|  0.0|<=|  0.0|<=|  5.8|
Event         |   2|  |   2|  |   3|
              +----+  +----+  +----+
## FACILITY 1: (CPU). 1 Server(s), 1 Busy.
Server(s): ( 1) TK 1 PR 0
              +----+  +----+  +----+  +----+  +----+
Time       <=|  0.0|<=|  0.0|<=|  0.0|<=|  0.0|<=|  0.0|
Event         |   2|  |   2|  |   2|  |   2|  |   2|
Priority      |   0|  |   0|  |   0|  |   0|  |   0|
              +----+  +----+  +----+  +----+  +----+



## TIME: 0.0
## AFTER PREEMPT Facility 1 Token 7 Priority 1


## EVENT LIST
              +----+  +----+
Token         |   8|  |   9|
Time       <=|  0.0|<=|  0.0|
Event         |   2|  |   2|
              +----+  +----+
## FACILITY 1: (CPU). 1 Server(s), 1 Busy.
Server(s): ( 1) TK 7 PR 1
              +----+  +----+  +----+  +----+  +----+  +----+
Token         |   1|  |   2|  |   3|  |   4|  |   5|  |   6|
Time       <=| -5.8|<=|  0.0|<=|  0.0|<=|  0.0|<=|  0.0|<=|  0.0|
Event         |   3|  |   2|  |   2|  |   2|  |   2|  |   2|
Priority      |   0|  |   0|  |   0|  |   0|  |   0|  |   0|
              +----+  +----+  +----+  +----+  +----+  +----+
```

Figure 10.9   Preempt Facility.

to simulate the behavior of a queue/facility. In the input file above, there are four blocks as shown in Fig. 10.11.

If we have a CPU/Disk system where there is a single CPU and two disks we can represent the system in the graph shown in Fig. 10.12. Each entity is a job that arrives at the CPU, takes a certain amount of time, and then requests access to either disk1 or disk2 using the distribution specified by the probability attached to each arc. In this case, we have specified a distribution of 50% for each disk.

```
## TIME: 45.4
## BEFORE RELEASE Facility 1 Token 7

## EVENT LIST
              +----+
Token         |    8|
Time      <=|  78.4|
Event         |    5|
              +----+
## FACILITY 1: (CPU). 1 Server(s), 1 Busy.
Server(s): ( 1) TK 7 PR 1
              +----+ +----+ +----+ +----+ +----+ +----+
Token         |    1| |    2| |    3| |    4| |    5| |    6|
Time      <=|  -5.8|<=|  0.0|<=|  0.0|<=|  0.0|<=|  0.0|<=|  0.0|
Event         |    3| |    2| |    2| |    2| |    2| |    2|
Priority      |    0| |    0| |    0| |    0| |    0| |    0|
              +----+ +----+ +----+ +----+ +----+ +----+
## FACILITY 2: (disk). 1 Server(s), 1 Busy.
Server(s): ( 1) TK 8 PR 0

              +----+
Token         |    9|
Time      <=|  43.2|
Event         |    4|
Priority      |    0|
              +----+

## TIME: 45.4
## AFTER RELEASE Facility 1 Token 7

## EVENT LIST
              +----+ +----+
Token         |    1| |    8|
Time      <=|  51.1|<=|  78.4|
Event         |    3| |    5|
              +----+ +----+
## FACILITY 1: (CPU). 1 Server(s), 1 Busy.
Server(s): ( 1) TK 1 PR 0
              +----+ +----+ +----+ +----+ +----+
Token         |    2| |    3| |    4| |    5| |    6|
Time      <=|  0.0|<=|  0.0|<=|  0.0|<=|  0.0|<=|  0.0|
Event         |    2| |    2| |    2| |    2| |    2|
              +----+ +----+ +----+ +----+ +----+
## FACILITY 2: (disk). 1 Server(s), 1 Busy.
Server(s): ( 1) TK 8 PR 0
              +----+
Token         |    9|
Time      <=|  43.2|
Event         |    4|
Priority      |    0|
              +----+
```

Figure 10.10   Release Facility.

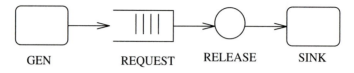

GEN   REQUEST   RELEASE   SINK

Figure 10.11   Single server queue.

The graph in Fig. 10.12 is encoded into data form in the same way as in the single-server queue example:

```
9 1000.0
0 gen 2.0
1 request cpu 1 expon 2.0 0.0
2 release
3 fork 2 0.5 0.5
4 request disk1 1 expon 30.0 0.0
5 release
6 request disk2 1 expon 40.5 0.0
7 release
8 sink
0 0      1 1
1 1 0    1 2
2 1 1    1 3
3 1 2    2 4 6
4 1 3    1 5
5 1 4    1 8
6 1 3    1 7
7 1 6    1 8
8 2 4 6 0
```

When executed, the simulator produces the following output:

```
+---------------------+
| SimPack SIMULATION REPORT |
+---------------------+

Total Simulation Time: 1000.000000
Total System Arrivals: 501
Total System Completions: 53

System Wide Statistics
------------------
System Utilization: 92.0%
Arrival Rate: 0.501000, Throughput: 0.053000
Mean Service Time per Token: 17.350454
Mean # of Tokens in System: 224.908951
Mean Residence Time for each Token: 4243.564941

Facility Statistics
----------------
F 1 (cpu): Idle: 3.0%, Util: 97.0%, Preemptions: 0
F 2 (disk1): Idle: 12.0%, Util: 88.0%, Preemptions: 0
F 3 (disk2): Idle: 9.2%, Util: 90.8%, Preemptions: 0
```

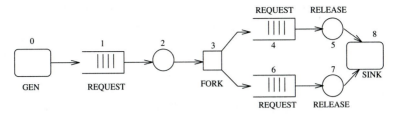

Figure 10.12   One CPU, two disk network.

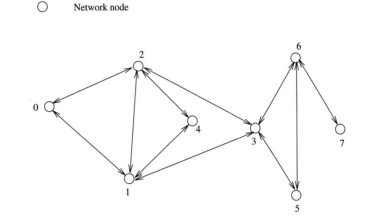

Figure 10.13   Sample Communications Network.

10.5.12 Communications Network Simulator

Let's consider the communications network shown in Fig. 10.13. This network is encoded as data to the network program as follows:

```
8 1.0
0.0 1.0 1.0 0.0 0.0 0.0 0.0 0.0
1.0 0.0 1.0 1.0 1.0 0.0 0.0 0.0
1.0 1.0 0.0 1.0 1.0 0.0 0.0 0.0
0.0 1.0 1.0 0.0 0.0 1.0 1.0 0.0
0.0 1.0 1.0 0.0 0.0 0.0 0.0 0.0
0.0 0.0 0.0 1.0 0.0 0.0 0.0 1.0 0.0
0.0 0.0 0.0 1.0 0.0 1.0 0.0 1.0
0.0 0.0 0.0 0.0 0.0 0.0 1.0 0.0
6
9 0 1 3 5 6 7 6 5 3
3 3 1 0
3 6 3 5
3 3 5 6
3 5 3 6
2 3 1
```

The first two numbers in the file are the number of nodes in the network and the time taken for a node to process a message. The next set of data is the adjacency matrix, similar to the matrices defined for the automata simulation programs. Each floating-point number in this matrix defines the link traversal time. After the matrix comes the number of message paths followed by the path information for each message. Path information starts with a path length followed by the actual path. For instance, the second message path is of length 3 and the path defines a message that originates at node 3 and then is routed through 1 and 0 in that order. The message terminates at node 0. Here is the output from the following command:

```
%network < net.dat

+---------------------+
| SimPack SIMULATION REPORT |
+---------------------+

Total Simulation Time: 16.000000
Total System Arrivals: 0
Total System Completions: 1

System Wide Statistics
------------------
System Utilization: 4.8%
Arrival Rate: 0.000000, Throughput: 0.062500
Mean Service Time per Token: 0.772727
Mean # of Tokens in System: 0.000000
Mean Residence Time for each Token: 0.000000

Facility Statistics
----------------
F 1 (tr): Idle: 93.8%, Util: 6.2%, Preemptions:    0
F 2 (tr): Idle: 100.0%, Util: 0.0%, Preemptions:   0
F 3 (tr): Idle: 93.8%, Util: 6.2%, Preemptions:    0
F 4 (tr): Idle: 100.0%, Util: 0.0%, Preemptions:   0
F 5 (tr): Idle: 93.8%, Util: 6.2%, Preemptions:    0
F 6 (tr): Idle: 100.0%, Util: 0.0%, Preemptions:   0
F 7 (tr): Idle: 100.0%, Util: 0.0%, Preemptions:   0
F 8 (tr): Idle: 100.0%, Util: 0.0%, Preemptions:   0
F 9 (tr): Idle: 100.0%, Util: 0.0%, Preemptions:   0
F 10 (tr): Idle: 100.0%, Util: 0.0%, Preemptions:  0
F 11 (tr): Idle: 87.5%, Util: 12.5%, Preemptions:  0
F 12 (tr): Idle: 100.0%, Util: 0.0%, Preemptions: 0
F 13 (tr): Idle: 81.2%, Util: 18.8%, Preemptions: 0
F 14 (tr): Idle: 93.8%, Util: 6.2%, Preemptions:   0
F 15 (tr): Idle: 100.0%, Util: 0.0%, Preemptions: 0
F 16 (tr): Idle: 100.0%, Util: 0.0%, Preemptions: 0
F 17 (tr): Idle: 87.5%, Util: 12.5%, Preemptions: 0
F 18 (tr): Idle: 87.5%, Util: 12.5%, Preemptions: 0
F 19 (tr): Idle: 93.8%, Util: 6.2%, Preemptions:   0
F 20 (tr): Idle: 93.8%, Util: 6.2%, Preemptions:   0
F 21 (tr): Idle: 93.8%, Util: 6.2%, Preemptions:   0
F 22 (tr): Idle: 93.8%, Util: 6.2%, Preemptions:   0
```

```
- - - - - - - - - - - - - -
NODE Utilization
- - - - - - - - - - - - - -


Node 0 Utilization is 6.2%
Node 1 Utilization is 18.8%
Node 2 Utilization is 0.0% Node 3 Utilization is 25.0%
Node 4 Utilization is 0.0%
Node 5 Utilization is 25.0%
Node 6 Utilization is 25.0%
Node 7 Utilization is 6.2%
```

We see that nodes 3, 5, and 6 are quite busy since many messages are routed through these nodes. No messages encounter node 4 during their routing, so node 4 is never utilized. Individual transceiver statistics are obtained from the standard SimPack simulation report at the start of the output.

10.5.13 Pulse Processes

The data file `herb.dat` is

```
3 50 0
−1 1 0
−1 0 1
0 −1 0
Plant
Herbivore
Carnivore
10 10 10
1 0 0
```

The file is arranged as follows:

- #vertices, #time units, #output vertex
- Adjacency matrix
- Names for each vertex
- Initial value vector
- Initial pulse vector

The simulation will be run for a total of 9 time units. The values in the adjacency matrix are $\{-1, 0, +1\}$ for the standard pulse process; however, they can be assigned any real value for the above implementation. The #output vertex stands for the integer number of the value vector variable requested for the point output file. In this way, the output file `pulse.pnts` will contain a two-column list of values (time, value), where value is the referenced state variable. Since zero is specified for our example file, we will obtain a time series of the plant population. The execution is run on this data file and the program produces the transient behavior of the system. Time is indexed on the left, and the values of the pulse and value

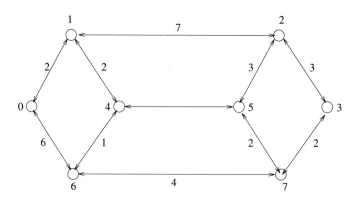

Figure 10.14 Sample network for optimal routing.

vectors are presented for each time stamp.

```
fish% pulse < herb.dat
Pulse Process Results
- - - - - - - - - - - - - - - - -
0: P 1.00 0.00 0.00 V 10.00 10.00 10.00
1: P -1.00 1.00 0.00 V 9.00 11.00 10.00
2: P 0.00 -1.00 1.00 V 9.00 10.00 11.00
3: P 1.00 -1.00 -1.00 V 10.00 9.00 10.00
4: P 0.00 2.00 -1.00 V 10.00 11.00 9.00
5: P -2.00 1.00 2.00 V 8.00 12.00 11.00
6: P 1.00 -4.00 1.00 V 9.00 8.00 12.00
7: P 3.00 0.00 -4.00 V 12.00 8.00 8.00
8: P -3.00 7.00 0.00 V 9.00 15.00 8.00
9: P -4.00 -3.00 7.00 V 5.00 12.00 15.00
```

10.5.14 Optimal Path Route Simulator

The optimal path route simulator[1] is used to simulate a subset of Routing Management Protocol (RMP), which is used in Inter-Switching System Interface (ISSI) of Switched Multimegabit Data Service (SMDS).

Shortest path routing.   The shortest path routing algorithm is widely used in many forms because it is simple and easy to understand. The concept shortest path deserves some explanation. One way of measuring path length is the number of hops. Using this metric, the paths 012 and 014 in Fig. 10.14 are equally long. Another metric is the geographic distance in kilometers, in which case 012 is clearly much longer than 014 (assuming that the figure is drawn to scale).

However, many other metrics are also possible. For example, each could be labeled with the mean queuing and transmission delay for a standard test packet as determined by

---

[1]Written by Hyun Park as a University of Florida simulation class project.

hourly or daily test runs. With this graph labeling, the shortest path is the fastest path, rather than the path with the fewest arcs or kilometers.

Several algorithms for computing the shortest path between two nodes of a graph are known. This one is from Dijkstra [48]. Each node is labeled with its distance from the source node along the best known path. Initially, no paths are known, so all nodes are labeled with infinity. As the algorithm proceeds and paths are found, the labels may change, reflecting better paths. A label may be either tentative or permanent. Initially, all labels are tentative. When it is discovered that a label represents the shortest possible path from the source to that node, it is made permanent and never changed thereafter.

Adjacency.   The following matrix shows the connectivity between nodes and metrics of all the links connecting nodes. Any nonzero number indicates the existence of direct link between two nodes and the value represents the metric of the link. The following is the topology of the example network shown in Fig. 10.14:

```
0.0 2.0 0.0 0.0 0.0 0.0 6.0 0.0
2.0 0.0 7.0 0.0 2.0 0.0 0.0 0.0
0.0 7.0 0.0 3.0 0.0 3.0 0.0 0.0
0.0 0.0 3.0 0.0 0.0 0.0 0.0 2.0
0.0 2.0 0.0 0.0 0.0 2.0 1.0 0.0
0.0 0.0 3.0 0.0 2.0 0.0 0.0 2.0
6.0 0.0 0.0 0.0 1.0 0.0 0.0 4.0
0.0 0.0 0.0 2.0 0.0 2.0 4.0 0.0
```

Packet generator.   Each message is generated from the message generator according to the input data regarding the nature of the packets. This will include the interarrival time distribution of packets, the source address distribution, destination address distribution, and parameters such as mean or minimum and maximum values.

The source and destination address distribution is by default the uniform integer distribution ranging from the smallest node number to the largest node number. The packet interarrival time distribution can be given as an input such as exponential, uniform, Erlang, or normal, and the packets are generated according to the given distribution type and mean and/or range values.

Input and output.   The following is the format of input file:

```
simulation-time #nodes node-processing-time
adjacency matrix specifying connectivity of network
packet-interarrival-time-distribution mean-value(or low range) (high range)
```

The simulation run is initiated by `route < inputfile` and the following is the sample input file for the example network in Fig. 10.14:

```
100.0 8 2.0
0.0 2.0 0.0 0.0 0.0 0.0 6.0 0.0
```

```
2.0 0.0 7.0 0.0 2.0 0.0 0.0 0.0
0.0 7.0 0.0 3.0 0.0 3.0 0.0 0.0
0.0 0.0 3.0 0.0 0.0 0.0 0.0 2.0
0.0 2.0 0.0 0.0 0.0 2.0 1.0 0.0
0.0 0.0 3.0 0.0 2.0 0.0 0.0 2.0
6.0 0.0 0.0 0.0 1.0 0.0 0.0 4.0
0.0 0.0 0.0 2.0 0.0 2.0 4.0 0.0
```

The following is a sample output when the input shown above is used:

```
+----------------------+
| SimPack SIMULATION REPORT |
+----------------------+


Total Simulation Time: 100.061264
Total System Arrivals: 18
Total System Completions: 16

System Wide Statistics
------------------
System Utilization: 2.4%
Arrival Rate: 0.179890, Throughput: 0.159902
Mean Service Time per Token: 0.149668
Mean # of Tokens in System: 1.329356
Mean Residence Time for each Token: 8.313566


Facility Statistics
----------------
F 1 (tr): Idle: 91.6%, Util: 8.4%, Preemptions: 0
F 2 (tr): Idle: 100.0%, Util: 0.0%, Preemptions: 0
F 3 (tr): Idle: 100.0%, Util: 0.0%, Preemptions: 0
F 4 (tr): Idle: 93.7%, Util: 6.3%, Preemptions: 0
F 5 (tr): Idle: 98.0%, Util: 2.0%, Preemptions: 0
F 6 (tr): Idle: 98.0%, Util: 2.0%, Preemptions: 0
F 7 (tr): Idle: 100.0%, Util: 0.0%, Preemptions: 0
F 8 (tr): Idle: 100.0%, Util: 0.0%, Preemptions: 0
F 9 (tr): Idle: 96.0%, Util: 4.0%, Preemptions: 0
F 10 (tr): Idle: 98.0%, Util: 2.0%, Preemptions: 0
F 11 (tr): Idle: 94.0%, Util: 6.0%, Preemptions: 0
F 12 (tr): Idle: 98.0%, Util: 2.0%, Preemptions: 0
F 13 (tr): Idle: 96.0%, Util: 4.0%, Preemptions: 0
F 14 (tr): Idle: 100.0%, Util: 0.0%, Preemptions: 0
F 15 (tr): Idle: 92.0%, Util: 8.0%, Preemptions: 0
F 16 (tr): Idle: 100.0%, Util: 0.0%, Preemptions: 0
F 17 (tr): Idle: 100.0%, Util: 0.0%, Preemptions: 0
F 18 (tr): Idle: 98.0%, Util: 2.0%, Preemptions: 0
F 19 (tr): Idle: 100.0%, Util: 0.0%, Preemptions: 0
F 20 (tr): Idle: 98.0%, Util: 2.0%, Preemptions: 0
F 21 (tr): Idle: 96.0%, Util: 4.0%, Preemptions: 0
F 22 (tr): Idle: 100.0%, Util: 0.0%, Preemptions: 0
```

```
- - - - - - - - - - - - - -
NODE Utilization
- - - - - - - - - - - - - -


Node 0 Utilization is 2.8%
Node 1 Utilization is 2.8%
Node 2 Utilization is 0.7%
Node 3 Utilization is 2.0%
Node 4 Utilization is 4.0%
Node 5 Utilization is 2.7%
Node 6 Utilization is 0.7%
Node 7 Utilization is 2.0%
```

Notice that there are some transceivers which show zero utilization due to the high link metrics.

10.5.15 Mini GPSS Compiler

Mini-GPSS[2] is a classified as a discrete event system simulation language and it fits within the scope of functional models. The language Mini-GPSS is a subset of the GPSS (which stands for General Purpose Simulation System) programming language. There are four phases involved in Mini-GPSS:

1. *Scanning Phase.* The scanner reads the input file, which is a Mini-GPSS program. The scanner recognizes all program constructs, identifiers, and parameters. These are called tokens.
2. *Parsing Phase.* The parser uses information from the scanning phase to check the Mini-GPSS program for correct syntax. As the syntax check is underway, the parser executes semantics which build intermediate data structures to be used in the next phase. These data structures will hold, for instance, the topology of the *flow network* defined by the code, along with information on facilities and server information.
3. *Code Creation Phase.* Now that data structures contain all relevant information provided by the Mini-GPSS program, a C code source file is generated which, when executed, will provide program output.
4. *Code Execution Phase.* The code is executed.

Syntax.   Below, we specify the syntax for Mini-GPSS programs.

```
Mini-GPSS Program Syntax


==============================================================
{}      Enclosed items which may be repeated zero or more times
()      Enclosed items occur one time
^()     Means "any item but enclosed item(s) "
```

[2]Mini-GPSS was written by David Bloom for his Senior Thesis at the University of Florida. The information in this section was drawn from his on-line documentation and written thesis.

```
.        Means "end of syntax definition "
::=      Means "is defined as "
|        Means "or "
[$1-$2]  Means "any character >= $1 and <= $2 "
================================================================
        program    ::= remarks sim-st remarks sto-blk.
        sto-blk    ::= (sto-st remarks sto-blk) | prog-blk.
        prog-blk   ::= gen-st tra-blk remarks opt-prog.
        prog-blk2  ::= serv-blk remarks opt-prog.
        tra-blk    ::= fac-blk | (remarks tra-blk fac-blk).
        opt-prog   ::= prog-blk | prog-blk2 | opt-sta.
        rel-blk    ::= (rel-st remarks new-f) | ter-st.
        lev-blk    ::= (lev-st remarks new-f) | ter-st.
        opt-sta    ::= (sta-st remarks opt-sta) | end-st.
        serv-blk   ::= (que-st remarks sei-st remarks dep-st adv-serv1) |
                       (que-st remarks ent-st remarks dep-st adv-serv2).
        fac-pt     ::= (que-st remarks sei-st remarks dep-st adv-serv1) |
                       (que-st remarks ent-st remarks dep-st adv-serv2) |
                       rel-blk.
        fac-blk    ::= (remarks adv-st fac-blk) | (remarks fac-pt).
        new-f      ::= ter-st | new-f-blk.
        new-f-blk ::= serv-blk | (adv-st remarks new-f-blk).
        adv-serv1 ::= (remarks adv-st adv-serv1) | (remarks rel-blk).
        adv-serv2 ::= (remarks adv-st adv-serv2) | (remarks lev-blk).
        sim-st     ::= simulate end-rmks.
        sto-st     ::= storage ident blanks comma blanks num end-rmks.
        tra-st     ::= transfer blank blanks number dot number blanks
                       comma blanks comma blanks ident end-rmks.
        que-st     ::= (queue blank blanks ident end-rmks) | (ident
                       blank blanks queue blank blanks ident end-rmks).
        gen-st     ::= generate blank opt-pms newline.
        sei-st     ::= seize blank blanks ident end-rmks.
        ent-st     ::= enter blank blanks ident end-rmks.
        lev-st     ::= leave blank blanks ident end-rmks.
        dep-st     ::= depart blank blanks ident end-rmks.
        rel-st     ::= release blank blanks ident end-rmks.
        sta-st     ::= start blank blanks digit end-rmks.
        ter-st     ::= (terminate blank opt-num newline) | (terminate
                       newline).
        end-st     ::= end end-rmks.
        adv-st     ::= (advance blank opt-pms newline)| (advance
                       newline).
ident      ::= letter{letter|num}.
enter      ::= ("E"|"e")("N"|"n")("T"|"t")("E"|"e")("R"|"r").
storage    ::= ("S"|"s")("T"|"t")("O"|"o")("R"|"r")("A"|"a")
               ("G"|"g")("E"|"e")(blank blanks)("S"|"s")("$").
transfer   ::= ("T"|"t")("R"|"r")("A"|"a")("N"|"n")("S"|"s")
               ("F"|"f")("E"|"e")("R"|"r").
leave      ::= ("L"|"l")("E"|"e")("A"|"a")("V"|"v")("E"|"e").
generate   ::= ("G"|"g")("E"|"e")("N"|"n")("E"|"e")("R"|"r")
               ("A"|"a")("T"|"t")("E"|"e").
simulate   ::= ("S"|"s")("I"|"i")("M"|"m")("U"|"u")("L"|"l")
               ("A"|"a")("T"|"t")("E"|"e").
```

```
release   ::= ("R"|"r")("E"|"e")("L"|"l")("E"|"e")("A"|"a")
                ("S"|"s")("E"|"e").
start     ::= ("S"|"s")("T"|"t")("A"|"a")("R"|"r")("T"|"t").
end       ::= ("E"|"e")("N"|"n")("D"|"d").
queue     ::= ("Q"|"q")("U"|"u")("E"|"e")("U"|"u")("E"|"e").
sieze     ::= ("S"|"s")("E"|"e")("I"|"i")("Z"|"z")("E"|"e").
depart    ::= ("D"|"d")("E"|"e")("P"|"p")("A"|"a")("R"|"r")
                ("T"|"t").
advance   ::= ("A"|"a")("D"|"d")("V"|"v")("A"|"a")("N"|"n")
                ("C"|"c")("E"|"e").
terminate ::= ("T"|"t")("E"|"e")("R"|"r")("M"|"m")("I"|"i")
                ("N"|"n")("A"|"a")("T"|"t")("E"|"e").
num       ::= ["0"-"9"].
letter    ::= (["A"-"Z"]|["a"-"z"]).
digit     ::= num{num}.
comma     ::= ",".
comm      ::= "*".
dot       ::= ".".
newline   ::= "\n".
      anytext  ::= {^(newline)}.
      note     ::= blank anytext
      end-rmks ::= note newline
      opt-num  ::= (blank opt-num) | (^(newline|blank|digit) note) |
                     digit note | null.
      opt-pms  ::= (blank opt-pms) | (^(newline|blank|digit) note) |
                     params | null.
      params   ::= digit opt-pm2.
      opt-pm2  ::= (blank opt-pm2) | (^(newline|blank|digit) note) |
                     (comma blanks digit note)| null.
      remark   ::= comm anytext | null.
      remarks  ::= (remark newline remarks) | (space remarks) | null.
blank     ::= " ".
blanks    ::= {blank}.
      null     ::= .
```

Mini-GPSS flowchart.   Figure 10.15 gives the Mini-GPSS architecture starting with lexical analysis and parsing phases and ending with the construction of the C simulation source code.

```
token.l   : input to lex (scanner specification)
parse.y   : input to yacc (parser specification)
lex.yy.c  : "C" generated code for scanner
y.tab.c   : "C" generated code for parser
queuing.c : simulation routines (simpack)
main.c    : simulation event handler
error.c   : error handler for parser
simgen.c  : "C" generated code for simulation
gp        : Mini-GPSS executable
program.g : Mini-GPSS source file
```

We now define this procedure in more detail. The first subprogram is the scanner, which

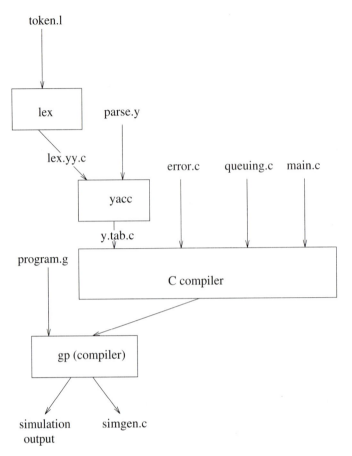

Figure 10.15   Mini-GPSS processing flow chart.

reads in the file token by token and returns to the parser each token it encounters. The scanner is implemented using a tool called `lex`. `lex` has its own programming language in which one specifies how tokens are recognized. Actions are associated with each token as it is recognized. The actions to be performed are written in C and are also in the scanner specification. File `token.l` contains the scanner specification. `lex` generates C code for the scanner by invoking the command `lex token.l`. The C-generated code is in the file `lex.yy.c`. The function which returns the current token is called `yylex`.

     The second subprogram is the parser, which interfaces with the scanner (by calling the function `yylex`) to check a program for proper syntax. The parser was implemented in `yacc`. `yacc` has its own programming language for specifying actions associated with syntax. The file `parse.y` contains the parser specification. `yacc` generates C code for the parser by invoking the command `yacc parse.y`. The generated C code for the parser is in the file `y.tab.c`. The program is parsed by calling the function `yyparse`. The error handler for syntax errors is located in the file `error.c`. If there are no syntax errors,

0 is returned from yyparse. Otherwise, a 1 is returned from yyparse, yyerror is called, and the program terminates. The error handler gives a description of the error and the line where the error is located. The parser, when checking syntax, creates two data structures with all pertinent information that the simulation event handler needs. The third subprogram is the simulation event handler which gets its input from the data structures produced by the parser.

Running Mini-GPSS.   The first step is to write a Mini-GPSS program according to the defined syntax. A simple example is a double-server queue, say a *barbershop system* with two barbers. Here is file barbershop.g:

```
* Barbershop Model with two barbers

        Simulate
        Storage S$BARBERSHOP,2     Declare 2 barbers

* Model Definition

        Generate   18,6               Customers arrive uniformly (12,24)
        Queue      BARBERSHOP         Customer queued
        Enter      BARBERSHOP         Enter Barbershop
        Depart     BARBERSHOP         When free server, release from queue
        Advance    16,4               Customer is served
        Leave      BARBERSHOP         Leave Barbershop
        Terminate 1                   Decrement termination count
        Start      25                 # of terminations
        end
    *
    * End Simulation
```

In another model, we permit branching based on probability (branching.g):

```
* model with two independent facilities branching

    Simulate

    Generate 1
    transfer 0.30,,Fac2 thirty percent of arrivals branch to fac2

        Queue      Line1
        Seize      Line1
        Depart     Line1
        Advance    1
        Release    Line1
        Terminate 1

Fac2 Queue      Line2
        Seize      Line2
```

```
      Depart    Line2
      Advance   1
      Release   Line2
      Terminate 1


 Start 50
 end
```

To run the compiler, do: `gp < source-file`, where `source-file` could be `bar-bershop.g` or `branching.g`, for example. If there are syntax errors the simulator will specify the error and the line where the error is located. If there are no syntax errors, the simulator produces the simulation output and will generate C equivalent code to the simulation. The C code is simply the *object* code, which can be rerun independently without retranslation. The object code file name is `simgen.c`.

10.5.16 Xsimcode

User interface. Xsimcode[3] is an X windows–based interface which builds C code which is then executed to produce simulation results. In many ways, Xsimcode is similar to Mini-GPSS. They both are compilers. In the case of Xsimcode, the user interface is graphical, whereas the Mini-GPSS interface is textual. Xsimcode's interface provides four buttons at the top of the screen. These are the `file`, `options`, `draw`, and `translate` menus. Below the buttons is a status line. The status line gives the user feedback on each operation. Check the status line after each operation. Below the status line is a large blank area, called the "pad." The pad is the user's drawing pad. The model is displayed on the pad. Clicking on an empty spot on the pad gives the ADD menu. Here, you can add a facility, generic block, or a routing point. Clicking on an occupied spot on the pad gives the EDIT menu. You can change the settings of an element, and delete or get information about the element.

To create a model, add elements to the screen. Elements must be linked together to show the flow of tokens through the system. There are five types of elements: arrivals, processes, fan-in routing points, fan-out routing points, and sinks. Each model has only one arrival point and one sink. Processes may be grouped together with a pair of fan-in and fan-out routing points. A fan-out routing point sends its token to one of several possible facilities. The varied destinations of the tokens are brought into a single stream with a fan-in routing point. Elements must be linked together to show the flow of tokens through the system.

Figure 10.16 shows the X window layout for the airport passport example model described in [99]. We summarize the structure of the model:

- The left-hand icon is the source of entities (people).

---

[3]Xsimcode was written by Brian E. Harrington as partial fulfillment of his Master's thesis [98] at the University of Florida. The material for this section was drawn from his on-line documentation and from Harrington and Fishwick [99], who describe the relationship of Xsimcode to the literature on simulation programming languages in more detail.

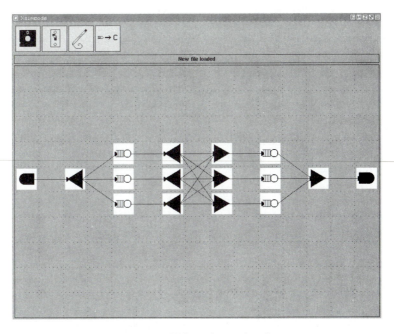

Figure 10.16   Xsimcode user interface.

- The left-hand triangle is a fork where people split to go to passport checking stations.
- The first three facilities represent the passport checkers.
- From here, passengers can choose one of three baggage checking stations. The ability to move to any station is represented by the fork/join network in the center of the model.
- Passengers continue to wait for the baggage checker and then proceed out of the terminal.
- The right-hand icons are join and sink icons.

Menus.   Xsimcode's user interface provides six menus. Four of these menus—the file menu, the options menu, the draw menu, and the translation menu—have pull-down lists. You should click the mouse and continue to hold down the mouse button until the appropriate option is highlighted. Releasing the mouse button selects a menu item. Each pull-down menu item generates a dialog box. A dialog box may ask you to type in input. Move the mouse within the dialog box. Use control-D or backspace to delete any unnecessary characters and type the input in the box. Do not press return. Just select the appropriate button in the box. Most dialog boxes with input have an "OK" and "CANCEL" button. Some dialog boxes do not ask for input. Simply select the appropriate button.

The add and modify menus lead to a dialog box with several buttons. After you have selected the appropriate option, another dialog box will appear. This box, like those used by the pull-down menu, may or may not ask for input. Use it the same way as the dialog boxes called by the pull-down menu. The following are menu options:

```
File              Options            Draw

New               File Name          Connect
Load              Run Time           Disconnect
Save              Departure          Grid
Save As           Mode               Refresh
Quit              Report
                  Trace Data Struct
                  Trace Event List
                  Trace Facility


        Modify                       Add

        Delete                       Facility
        Edit                         Fan-In
              Name                   Fan-Out
              Number of Servers      Generic
              Priority
              Initial Count
              Distribution
              Parameter 1
              Parameter 2
              Parameter 3
```

The parameters of individual elements can be changed using the edit menu. The model parameters can be changed from the options menu. See Simpack's documentation for information on Simpack options. However, all options have defaults, so a user does need to change anything to create a working model.

Translation.   You can save your model and all its parameters at any time. "Save" saves the model to the current directory under the current name. "Save As" allows the name to be changed prior to saving the file. The filename will end in `.sim`. When the model is complete, it can be translated to C by pressing the translate button. The default filename is `model.c`. The file is written to the current directory. To run the simulation, the user would exit Xsimcode, perform a `make model`, and run the model executable.

10.6 CONSTRAINT MODELS

The three types of constraint models supported in SimPack are:

1. Difference equations
2. Ordinary differential equations
3. Delay-differential equations

10.6.1 Difference Equations

Difference equations are solved using one of two methods: (1) converting the difference equations to first-order form, or (2) using a circular queue. We will present a simulator for each type for the logistics equation: $x(t) = ax(t-1)(1.0 - x(t-1))$. The first method employing a first-order canonical form is

```
#include <math.h>

int time,delta_time;
float a,in[2],out[2];

main()
{
  init_conditions();
  while (time <= 100) {
   state();
   /* output time,x */
   printf("%d %f\n",time,in[1]);
   update();
  } /* end while */
} /* end main */

init_conditions()
{
  out[1] = 0.8;
  time = 0;
  delta_time = 1;
  a = 1.2;
}

state()
{
/* Calculate state: x(t) = a*x(t-1)(1.0 - x(t-1)) */
in[1] = a*out[1]*(1.0 - out[1]);
}

update()
{
  int i;
  out[1] = in[1];
  time += delta_time;
}
```

Here is a program that will solve the second-order equation $x(t) = ax(t-1)(1-x(t-2))$ with the specified initial conditions.

```
#include <math.h>

int time,delta_time;
float a,out[3],in[3];
```

```
main()
{
  init_conditions();
  while (time <= 100) {
   state();
   update();
   /* output time,x(t) */
   printf("%d %f\n",time,in[2]);
  } /* end while */
} /* end main */

init_conditions()
{
  out[1] = 0.8;
  out[2] = 0.8;
  time = 0;
  delta_time = 1;
   a = 1.5;
}

state()
{
/* Calculate state: x(t) = a*x(t-1)(1.0 - x(t-2)) */
in[1] = out[2];
in[2] = a*out[2]*(1.0 - out[1]);
}

update()
{
 int i;
 for(i=1;i<=2;i++)
   out[i] = in[i];
 time += delta_time;
}
```

The second, circular queue, method is encoded as follows:

```
#define X_BUFFER 2
int pointer;

main()
{ float x[X_BUFFER],a;
  int time,t();

  pointer = X_BUFFER - 1;
  x[t(-1)] = 0.8; a = 1.2; /* set initial conditions and parameters */
  time = 0;
  while (time < 100) {
   x[t(0)] = a*x[t(-1)]*(1.0 - x[t(-1)]); /* difference equation */
   printf("%d %f\n",time,x[t(0)]);
   pointer = (pointer+1) % X_BUFFER;
   time++;
```

```
     } /* end while */
} /* end main */

int t(index)
int index;
{int new_pointer;
 new_pointer = pointer + index;
 if (new_pointer < 0)
    return(X_BUFFER + new_pointer);
  else
    return(new_pointer % X_BUFFER);
}
```

10.6.2 Ordinary Differential Equations

Basic method. The basic method of integrating ordinary differential equations
(ODEs) involves (1) converting the equation set to first-order form, and (2) integrating
the left-hand side `in[]` array to produce the right-hand-side `out[]` array. There are
numerous methods for solving differential equations [177] and we provide code for only
forward Euler and fourth-order Runge-Kutta. Examples are included for integrating the
Lorenz equations, Lotka-Volterra equations, and some generic equations. Here is a program that will solve $x''' + x' = 2$ with the specified initial conditions using Euler's method
of integration.

```
float time,delta_time,in[4],out[4];

main()
{
  init_conditions();
  while (time < 5.0) {
   state();
   integrate();
   /* output time,x */
   printf("%f %f\n",time,out[1]);
  } /* end while */
} /* end main */

init_conditions()
{
   out[1] = 1.0;
   out[2] = 1.0;
   out[3] = 1.0;
   time = 0.0;
   delta_time = 0.01;
}

state()
{
/* Calculate state: x''' + x' = 2 */
   in[1] = out[2];
   in[2] = out[3];
```

```
              in[3] = 2 - out[2];
          }

          integrate()
          {
            int i;
            for(i=1;i<=3;i++)
             out[i] += in[i]*delta_time;
            time += delta_time;
          }
```

To use RK4, we simply substitute different code for the `integrate()` routine.

```
float in[4],out[4],f[4],savevar[4],time,delta_time;
int num_equations;

main()
{
  init_conditions();
  while (time < 5.0) {
   state();
   integrate();
   /* output time,x */
   printf("%f %f\n",time,out[1]);
  } /* end while */
} /* end main */

init_conditions()
{
  num_equations = 3;
  out[1] = 1.0;
  out[2] = 1.0;
  out[3] = 1.0;
  time = 0.0;
  delta_time = 0.01;
}

state()
{
/* Calculate state: x"' + x' = 2 */
  in[1] = out[2];
  in[2] = out[3];
  in[3] = 2 - out[2];
}

integrate()
{
  int i;

  for (i=1;i<=num_equations;i++) {
   f[i] = delta_time*in[i];
   savevar[i] = out[i];
```

```
      out[i]  += f[i]/2;
    }
    time += delta_time/2;
    state();

    for (i=1;i<=num_equations;i++) {
     f[i] += 2*delta_time*in[i];
     out[i] = savevar[i] + delta_time*in[i]/2;
    }
    state();
    for (i=1;i<=num_equations;i++) {
     f[i] += 2*delta_time*in[i];
     out[i] = savevar[i] + delta_time*in[i];
    }
    time += delta_time/2;
    state();

    for (i=1;i<=num_equations;i++)
     out[i] = savevar[i] + (f[i] + delta_time*in[i])/6;
  }
```

DEQ. Since converting to first-order form can be automated, DEQ[4] parses input formulas, produces the required first-order equations, and creates a plot of the desired state variables. To make DEQ, do a `make all`. Then you can run `deq` from the UNIX command line as: `deq`. You will have the choice of entering one of two types of systems:

1. A single equation with high-order terms (derivatives). The system may be homogeneous or nonhomogeneous, and that the left-hand side of the equation must be the higher-order term. Note the following notation: Use x[1] for the first derivative of x with respect to time (i.e., x' or dx/dt), x[2] is the second derivative, and so on. x[0] represents x.

2. A set of first-order equations for an n-dimensional system. You will be prompted for each equation in turn. Here x[1] means the first state variable, x[2] means the second state variable, and so on.

An example of a single equation is to enter

$$x[2] = -16*x[0] + 10*\cos(3.7*t)$$

This equation is $x'' = -16x + 10\cos(3.7t)$ using standard notation. Try plotting this over time (always denoted as t) by specifying t as the x-axis variable, and x[0] as the y-axis variable to be plotted by gnuplot.

An example of entering multiple first-order equations is to enter (for the Lorenz system)

---

[4]DEQ was written by James Bowman for his Senior Thesis at the University of Florida. The information in this section was drawn from his on-line documentation.

```
x[1]' = 10*(-x[1] + x[2])
x[2]' = 28*x[1] - x[2] - x[1]*x[3]
x[3]' = -2.666*x[3] + x[1]*x[2]
```

Try plotting x[1] vs. x[2] (for a phase plane view) when prompted for this information. As far as the internals are concerned, each equation is parsed and transformed into a system of first-order equations if it is not already in that form. Then this system of equations is piped into another parser which generates the C source code. This code is compiled and run, generating the raw data, which is given to a graphing program for visual feedback.

10.7 SPATIAL MODELS

10.7.1 Gas Dynamics

There are two programs that perform basic molecular gas simulations within a box. The first program (gas1) uses a time-slicing technique and the second program (gas2) uses a discrete event scheduling simulation method. Gas molecules are traveling at constant velocity, therefore, it is easy ahead of time to calculate the next collision time for any number of gas molecules (thereby making the discrete event approach possible).

The inspiration for these programs came from one of A. K. Dewdney's books on computer recreations (chapters originally appeared in *Scientific American*).

Each program prints records of the form

```
#-of-molecule X-position Y-position
```

Any number of initial molecules may be specified. One can change the NUM_BALLS value to be something other than 4 (the current value).

10.7.2 Diffusion

The following code was used to produce the diffusion behavior shown in Sec. 7.2.3.

```
# Copyright (C) 1992 J Dana Eckart
#
# Simulate gas diffusion by 'randomly' rotating the cell values in a
# Margolus-neighborhood either clockwise or counter-clockwise. The
# groupings are alternations of the two overlapping square neighborhoods:
#
# 0 1
# 2 3 2
# 1 0
#
2 dimensions of
which of 0..3
rand  of 0..1
gas   of 0..1
end
```

```
# Calculate the sum of the random fields in the neighborhood.
#
random := cell.rand + [1, 0].rand + [1, -1].rand + [0, -1].rand
when (cell.which = 0 & time%2 = 1) | (cell.which = 3 & time%2 = 0)

      := [-1, 0].rand + cell.rand + [0, -1].rand + [-1, -1].r and
when (cell.which = 1 & time%2 = 1) | (cell.which = 2 & time%2 = 0)

      := [-1, 1].rand + [0, 1].rand + cell.rand + [-1, 0].rand
when (cell.which = 3 & time%2 = 1) | (cell.which = 0 & time%2 = 0)

      := [0, 1].rand + [1, 1].rand + [1, 0].rand + cell.rand
otherwise

# Alternate between the overlapping neighborhoods
#
if time(%2 = 1 then)
if random%2 = 1 then
# Clockwise rotation
cell.gas := [0, -1].gas when cell.which = 0
         := [-1, 0].gas when cell.which = 1
         := [0, 1].gas when cell.which = 3
         := [1, 0].gas when cell.which = 2
else
# Counter-clockwise rotation
cell.gas := [1, 0].gas when cell.which = 0
         := [0, -1].gas when cell.which = 1
         := [-1, 0].gas when cell.which = 3
         := [0, 1].gas when cell.which = 2
end
else
if random%2 = 1 then
# Clockwise rotation
cell.gas := [0, -1].gas when cell.which = 3
         := [-1, 0].gas when cell.which = 2
         := [0, 1].gas when cell.which = 0
         := [1, 0].gas when cell.which = 1
else
# Counter-clockwise rotation
cell.gas := [1, 0].gas when cell.which = 3
         := [0, -1].gas when cell.which = 2
         := [-1, 0].gas when cell.which = 0
         := [0, 1].gas when cell.which = 1
end
end

# Stir the random values, so that they keep changing.
#
cell.rand := ([1, 0].rand + [0, 1].rand + [-1, 0].rand + [0, -1].rand
+ cell.rand)%2
```

Bibliography

[1] Harold Abelson and Andrea diSessa. *Turtle Geometry*. MIT Press, Cambridge, MA, 1980.

[2] Amir Wadi Al-Khafaji and John R. Tooley. *Numerical Methods in Engineering Practice*. Holt, Rinehart and Winston, New York, 1986.

[3] M. P. Allen and D. J. Tildesley. *Computer Simulation of Liquids*. Oxford University Press, New York, 1987.

[4] George S. Almasi and Allan Gottlieb. *Highly Parallel Computing*. Benjamin-Cummings, Redwood City, CA, 1989.

[5] Mats Andersson. Modelling of Combined Discrete Event and Continuous Time Dynamical Systems. In *Preprints of 12th IFAC World Congress*, volume 9, pp. 69–72, Sydney, Australia, July 1993.

[6] W. Ross Ashby. *An Introduction to Cybernetics*. John Wiley & Sons, New York, 1963.

[7] Wanda Austin and Behrokh Khoshnevis. Intelligent Simulation Environments for System Modeling. In *Institute of Industrial Engineering Conference*, May 1988.

[8] Norman I. Badler, Brian A. Barsky, and David Zeltzer. *Making Them Move: Mechanics, Control and Animation of Articulated Figures*. Morgan Kaufmann, San Mateo, CA, 1991.

[9] Norman I. Badler, Kamran Manoochehri, and Graham Walters. Articulated Figure Positioning by Multiple Constraints. *IEEE Computer Graphics and Applications*, 7(6):28–38, 1987.

[10] Norman I. Badler and Stephen W. Smoliar. Digital Representations of Human Movement. *ACM Computing Surveys*, 11(1):19–38, 1979.

[11] John E. Baerwald. *Transportation and Traffic Engineering*. Prentice Hall, Englewood Cliffs, NJ, 1976.

[12] Lorna Balian. *An Elephant?* Abingdon Press, New York, 1964 (pictures by Lorna Balian adapted from a poem by John Godfrey Saxe).

[13] Lech Banachowski, Antoni Kreczmar, and Wojciech Rytter. *Analysis of Algorithms and Data Structures*. Addison-Wesley, Reading, MA, 1991.

[14] Jerry Banks and John S. Carson. *Discrete Event System Simulation*. Prentice Hall, Englewood Cliffs, NJ, 1984.

[15] Joshua Barnes and Piet Hut. A Hierarchical $O(N \log N)$ Force Algorithm. *Nature*, 324(4):446–449, December 1986.

[16] BBN. SIMNET: Advanced Technology for the Mastery of Warfighting. Technical Report 6787, BBN Laboratories Inc., Cambridge, MA, 1988.

[17] Howard W. Beck and Paul A. Fishwick. Incorporating Natural Language Descriptions into Modeling and Simulation. *Simulation Journal*, 52, (3) 102–109, March 1989.

[18] Howard W. Beck and Paul A. Fishwick. Natural Language, Cognitive Models and Simulation. In Paul A. Fishwick and Paul A. Luker, editors, *Qualitative Simulation Modeling and Analysis*. Springer-Verlag, New York, 1990.

[19] Edward Beltrami. *Mathematics for Dynamic Modeling*. Academic Press, New York, 1987.

[20] Ludwig von Bertalanffy. *General System Theory*. George Braziller, New York, 1968.

[21] Dimitri P. Bertsekas and John N. Tsitsiklis. *Parallel and Distributed Computation: Numerical Methods*. Prentice Hall, Englewood Cliffs, NJ, 1989.

[22] G. M. Birtwistle. *Discrete Event Modelling on SIMULA*. Macmillan, New York, 1979.

[23] H. M. Blalock. *Theory Construction: From Verbal to Mathematical Formulations*. Prentice Hall, Englewood Cliffs, NJ, 1969.

[24] H. M. Blalock. *Causal Models in the Social Sciences*. Aldine Publishing Co., Chicago, 1971.

[25] H. M. Blalock. *Causal Models in Panel and Experimental Designs*. Aldine Publishing Co., Chicago, 1985.

[26] Grady Booch. On the Concepts of Object-Oriented Design. In Peter A. Ng and Raymond T. Yeh, editors, *Modern Software Engineering*, chapter 6, pp. 165–204. Van Nostrand Reinhold, New York, 1990.

[27] Grady Booch. *Object Oriented Design*. Benjamin-Cummings, Redwood City, CA, 1991.

[28] Alan H. Borning. THINGLAB: A Constraint-Oriented Simulation Laboratory. Technical report, Xerox PARC, 1979.

[29] Ronald Brachman and Hector Levesque, editors. *Readings in Knowledge Representation*. Morgan Kaufmann, San Mateo, CA, 1985.

[30] Peter C. Breedveld. A Systematic Method to Derive Bond Graph Models. In *Second European Simulation Congress*, Antwerp, Belgium, September 1986.

[31] P. W. Bridgman. *The Logic of Modern Physics*. Macmillan, New York, 1927.

[32] Randy Brown. Calendar Queues: A Fast O(1) Priority Queue Implementation for the Simulation Event Set Problem. *Communications of the ACM*, 31(10):1220–1227, October 1988.

[33] François E. Cellier. Combined Continuous System Simulation by Use of Digital Computers: Techniques and Tools. Ph.D. thesis, Swiss Federal Institute of Technology Zurich, 1979.

[34] François E. Cellier. *Continuous System Modeling*. Springer-Verlag, New York, 1991.

[35] François E. Cellier. Qualitative Simulation of Technical Systems using the General System Problem Solving Framework. *International Journal of General Systems*, 13(4):333–344, 1987.

[36] François E. Cellier and David W. Yandell. SAPS-II: A New Implementation of the Systems Approach Problem Solver. *International Journal of General Systems*, 13(4):307–322, 1987.

[37] K. M. Chandy and J. Misra. Distributed Simulation. *IEEE Transactions on Software Engineering*, 5(5):440–452, September 1979.

[38] P. B. Checkland. *Systems Thinking, Systems Science*. John Wiley & Sons, New York, 1981.

[39] L. O. Chua and P.-M. Lin. *Computer Aided Analysis of Electronic Circuits: Algorithms and Computational Techniques*. Prentice Hall, Englewood Cliffs, NJ, 1975.

[40] Necia Grant Cooper. *From Cardinals to Chaos: Reflections on the Life and Legacy of Stanislaw Ulam*. Cambridge University Press, Cambridge, 1989.

[41] Shamus Culhane. *Animation: From Script to Screen*. St. Martin's Press, New York, 1990.

[42] Alan M. Davis. A Comparison of Techniques for the Specification of External System Behavior. *Communications of the ACM*, 31(9):1098–1115, September 1988.

[43] Morton M. Denn. *Process Fluid Mechanics*. Prentice Hall, Englewood Cliffs, NJ, 1980.

[44] Alan A. Desrochers. *Computer Integrated Manufacturing*. IEEE Computer Society Press, Troy, NY, May 1990.

[45] A. K. Dewdney. Computer Recreations. *Scientific American*, December 1984.

[46] Stephen C. Dewhurst and Kathy T. Stark. *Programming in C++*. Prentice Hall, Englewood Cliffs, 1989.

[47] Dominique D'Humières, Pierre Lallemand, and T. Shimorua. Lattice Gas Cellular Automata: A New Experimental Tool for Hydrodynamics. Technical report, Los Alamos National Laboratory, 1985. Preprint LA-UR-85-4051.

[48] E. W. Dijkstra. A Note on Two Problems in Connexion with Graphs. *Numerische Mathematik*, 1:269–271, 1959.

[49] Richard C. Dorf. *Modern Control Systems*. Addison-Wesley, Reading, MA, 1986.

[50] Richard Durret. *Lecture Notes on Particle Systems and Percolation*. Wadsworth and Brooks/Cole, Belmont, CA, 1988.

[51] Benjamin Dutton. *Navigation and Piloting*. Naval Institution Press, Annapolis, MD, 1978.

[52] Leah Edelstein-Keshet. *Mathematical Models in Biology*. Random House, New York, 1988.

[53] Maurice S. Elzas, Tuncer I. Ören, and Bernard P. Zeigler. *Modelling and Simulation Methodology in the Artificial Intelligence Era*. North-Holland, Amsterdam, 1986.

[54] Maurice S. Elzas, Tuncer I. Ören, and Bernard P. Zeigler. *Modelling and Simulation Methodology: Knowledge Systems' Paradigms*. North-Holland, Amsterdam, 1989.

[55] Herbert Enderton. *A Mathematical Introduction to Logic*. Academic Press, New York, 1972.

[56] Stanley J. Farlow. *Partial Differential Equations for Scientists and Engineers*. Dover Publications, Mineola, NY, 1982.

[57] Doyne Farmer, Tommaso Toffoli, and Stephen Wolfram, editors. *Cellular Automata: Proceedings of an Interdisciplinary Workshop*. North-Holland, Amsterdam, 1983.

[58] Richard Feynman. *The Character of Physical Law*. MIT Press, Cambridge, MA, 1965.

[59] Nicholas V. Findler, editor. *Associate Networks: Representation and Use of Knowledge by Computers*. Academic Press, New York, 1979.

[60] Paul A. Fishwick. The Role of Process Abstraction in Simulation. *IEEE Transactions on Systems, Man and Cybernetics*, 18(1):18–39, January/February 1988.

[61] Paul A. Fishwick. Abstraction Level Traversal in Hierarchical Modeling. In Bernard P. Zeigler, Maurice Elzas, and Tuncer Ören, editors, *Modelling and Simulation Methodology: Knowledge Systems Paradigms*, pp. 393–429. Elsevier/ North-Holland, New York, 1989.

[62] Paul A. Fishwick. Qualitative Methodology in Simulation Model Engineering. *Simulation Journal*, 52(3):95–101, March 1989.

[63] Paul A. Fishwick. Toward an Integrated Approach to Simulation Model Engineering. *International Journal of General Systems*, 17(1):1–19, May 1990.

[64] Paul A. Fishwick. Extracting Rules from Fuzzy Simulation. *Expert Systems with Applications*, 3(3):317–327, 1991.

[65] Paul A. Fishwick. Fuzzy Simulation: Specifying and Identifying Qualitative Models. *International Journal of General Systems*, 9(3):295–316, 1991.

[66] Paul A. Fishwick. A Functional/Declarative Dichotomy for Characterizing Simulation Models. In *1992 Artificial Intelligence, Simulation and Planning in High Autonomy Systems*, pp. 102–109, Perth, Australia, July 1992. IEEE Computer Society Press, New York, 1993.

[67] Paul A. Fishwick. An Integrated Approach to System Modelling Using a Synthesis of Artificial Intelligence, Software Engineering and Simulation Methodologies. *ACM Transactions on Modeling and Computer Simulation*, 2(4):307–330, 1992.

[68] Paul A. Fishwick. A Simulation Environment for Multimodeling. *Discrete Event Dynamic Systems: Theory and Applications*, 3:151–171, 1993.

[69] Paul A. Fishwick and Paul A. Luker, editors. *Qualitative Simulation Modeling and Analysis*. Springer-Verlag, New York, 1991.

[70] Paul A. Fishwick and Richard B. Modjeski, editors. *Knowledge Based Simulation: Methodology and Application*. Springer-Verlag, New York, 1991.

[71] Paul A. Fishwick, Hari Narayanan, Jon Sticklen, and Andrea Bonarini. Multimodel Approaches for System Reasoning and Simulation. *IEEE Transactions on Systems, Man and Cybernetics*, 24(10) October 1994.

[72] Paul A. Fishwick and Bernard P. Zeigler. Qualitative Physics: Towards the Automation of Systems Problem Solving. *Journal of Theoretical and Experimental Artificial Intelligence*, 3:219–246, 1991.

[73] Paul A. Fishwick and Bernard P. Zeigler. A Multimodel Methodology for Qualitative Model Engineering. *ACM Transactions on Modeling and Computer Simulation*, 2(1):52–81, 1992.

[74] William I. Fletcher. *An Engineering Approach to Digital Design*. Prentice Hall, Englewood Cliffs, NJ, Englewood Cliffs, NJ, 1980.

[75] Robert L. Flood and Ewart R. Carson. *Dealing with Complexity: An Introduction to the Theory and Application of Systems Science*. Plenum Press, New York, 1988.

[76] Michael J. Flynn. Very High Speed Computing Systems. *Proceedings of the IEEE*, 54(12):1901–1909, December 1966.

[77] Michael J. Flynn. Some Computer Organizations and Their Effectiveness. *IEEE Transactions on Computers*, 21:948–960, 1972.

[78] Institute for Simulation and Training. Miltary Standard Version 2.0 (Draft) Protocol Data Units for Entity Information and Entity Interaction in a Distributed Interactive Simulation. Technical report, UCF Institute for Simulation and Training, September 1992.

[79] J. W. Forrester. *Industrial Dynamics*. MIT Press, Cambridge, MA, 1961.

[80] J. W. Forrester. *Urban Dynamics*. MIT Press, Cambridge, MA, 1969.

[81] J. W. Forrester. *World Dynamics*. Wright-Allen Press, Cambridge, MA, 1971.

[82] G. Fox, M. Johnson, G. Lyzenga, S. Otto, J. Salmon, and D. Walker. *Solving Problems on Concurrent Processors:* Volume 1, *General Techniques and Regular Problems*. Prentice Hall, Englewood Cliffs, NJ. 1988.

[83] Mark A. Franklin. Parallel Solution of Ordinary Differential Equations. *IEEE Transactions on Computers*, 27(5):413–420, May 1978.

[84] U. Frisch, B. Hasslacher, and Y. Pomeau. Lattice Gas Automata for the Navier Stokes Equation. *Physical Review Letters*, 56(14):1505–1508, April 7, 1986.

[85] Richard M. Fujimoto. Parallel Discrete Event Simulation. *Communications of the ACM*, 33(10):31–53, October 1990.

[86] Ivan Futo and Tamas Gergely. *Artificial Intelligence in Simulation*. Ellis Horwood Limited, Chichester, West Sussex, England, John Wiley & Sons, New York, 1990.

[87] Dedre Gentner and Donald R. Gentner. Flowing Waters or Teeming Crowds: Mental Models of Electricity. In Dedre Gentner and Albert L. Stevens, editors, *Mental Models*, pp. 99–129. Lawrence Erlbaum Associates, Hillsdale, NJ, 1983.

[88] Clark Glymour, Richard Scheines, Peter Spirtes, and Kevin Kelly. *Discovering Causal Structure*. Academic Press, New York, 1987.

[89] Suresh C. Goel and Kenneth D. Morris. Techniques for Extrapolation, Delay Compensation and Smoothing with Preliminary Results and an Evaluation Tool. In *Summary Report: Fifth Workshop on Standards for the Interoperability of Defense Simulations*, pp. A47–A66. Institute for Simulation and Training, Orlando, FL, September 1991.

[90] Harvey Gould and Jan Tobochnik. *An Introduction to Computer Simulation Methods: Applications to Physical Systems:* Part 1. *Simulation with Deterministic Processes*, Addison-Wesley, Reading, MA, 1988.

[91] Harvey Gould and Jan Tobochnik. *An Introduction to Computer Simulation Methods: Applications to Physical Systems:* Part 2, *Simulation with Random Processes*, Addison-Wesley, Reading, MA, 1988.

[92] Ian Graham. *Object Oriented Methods*. Addison-Wesley, Reading, MA, 1991.

[93] L. Greengard. *The Rapid Evaluation of Potential Fields in Particle Systems*. MIT Press, Cambridge, MA, 1988.

[94] L. Greengard and V. Rokhlin. A Fast Algorithm for Particle Simulations. *Journal of Computational Physics*, 73:325–348, 1987.

[95] Geoffrey Grimmett. *Percolation*. Springer-Verlag, New York, 1989.

[96] Anoop Gupta. *Parallelism in Production Systems*. Morgan Kaufmann, San Mateo, CA, 1987.

[97] Richard Haberman. *Mathematical Models: Mechanical Vibrations, Population Dynamics, and Traffic Flow*. Prentice Hall, Englewood Cliffs, NJ, 1977.

[98] Brian E. Harrington. XSimcode: A Simulation Model Translator. Master's thesis, University of Florida, 1991.

[99] Brian E. Harrington and Paul A. Fishwick. A Portable Process-Oriented Compiler for Event Driven Simulation. *Simulation*, 60(6):393–405, June 1993.

[100] Derek Hart and Tony Croft. *Modelling with Projectiles*. Halsted Press, New York, 1988.

[101] S. I. Hayakawa. *Language in Thought and Action*. 5th ed. Harcourt Brace Jovanovich, San Diego, CA, 1989.

[102] D. W. Heermann. *Computer Simulation Methods in Theoretical Physics*, 2nd ed. Springer-Verlag, New York, 1990.

[103] George E. Heidorn. English as a Very High Level Language for Simulation Programming. In *Proc. Symposium on Very High Level Languges*, volume 9, pp. 91–100. *SIGPLAN Notices*, April 1974.

[104] R. W. Hockney and J. W. Eastwood. *Computer Simulation Using Particles*. Adam Hilger, Philadelphia, 1988.

[105] P. Hogeweg and B. Hesper. A Model Study on Biomorphological Description. *Pattern Recognition*, 6:165–179, 1974.

[106] John E. Hopcroft and Jeffrey D. Ullman. *Introduction to Automata Theory, Languages and Computation*. Addison-Wesley, Reading, MA, 1979.

[107] L. Hopf. *Introduction to the Differential Equations of Physics*. Dover Publications, Mineola, NY, 1948. Translated by Walter Nef.

[108] Werner Horn, editor. *Causal AI Models: Steps toward Applications*. Hemisphere Publishing, New York, 1990.

[109] Ellis Horowitz and Sartaj Sahni. *Fundamentals of Data Structures in Pascal*, 4th ed. Computer Science Press, New York, 1990.

[110] Paul Horowitz and Winfield Hill. *The Art of Electronics*. Cambridge University Press, Cambridge, 1980.

[111] Gene H. Hostetter, Mohammed S. Santina, and Paul D'Carpio-Montalvo. *Analytical, Numerical, and Computational Methods for Science and Engineering*. Prentice Hall, Englewood Cliffs, NJ, 1991.

[112] Jhyfang Hu and Jerzy Rozenblit. Knowledge Acquisition Based on Explicit Representation. *Expert Systems with Applications*, 3:303–315, 1991.

[113] A. Hutchinson. *Labanotation*. Theatre Arts Books, New York, 1970.

[114] Dym Ivey. *Principles of Mathematical Modeling*. Academic Press, New York, 1980.

[115] John A. Jacquez. *Compartmental Analysis in Biology and Medicine*, 2nd ed. University of Michigan Press, Ann Arbor, MI, 1985.

[116] David Jefferson. Virtual Time. *ACM Transactions on Programming Languages and Systems*, 7(3):404–425, July 1985.

[117] Douglas W. Jones. An Empirical Comparison of Priority-Queue and Event-Set Implementations. *Communications of the ACM*, 29(4):300–311, April 1986.

[118] R. E. Kalman, P. L. Falb, and M. A. Arbib. *Topics in Mathematical Systems Theory*. McGraw-Hill, New York, 1962.

[119] Dean C. Karnopp, Donald L. Margolis, and Ronald C. Rosenberg. *System Dynamics*. John Wiley & Sons, New York, 1990.

[120] William J. Kaufmann and Larry L. Smarr. *Supercomputing and the Transformation of Science*. W. H. Freeman and Company, New York, 1993.

[121] Hartmut Keller, Horst Stolz, Andreas Ziegler, and Thomas Braeunl. Virtual Mechanics: Simulation and Animation of Rigid Body Systems. Technical report, University of Stuttgart, August 1993. Computer Science Report 8/93.

[122] Tag Gon Kim. Hierarchical Development of Model Classes in the DEVS-Scheme Simulation Environment. *Expert Systems with Applications*, 3:303–315, 1991.

[123] George J. Klir. *Architecture of Systems Problem Solving*. Plenum Press, New York, 1985.

[124] Huseyin Koçak. *Differential and Difference Equations through Computer Experiments*, 2nd ed. Springer-Verlag, New York, 1989.

[125] Steven E. Koonin. *Computational Physics*. Benjamin-Cummings, Redwood City, CA, 1986.

[126] G. Korn and J. Wait. *Digital Continuous-System Simulation*. Prentice Hall, Englewood Cliffs, NJ, 1978.

[127] Granino A. Korn. *Interactive Dynamic System Simulation*. McGraw-Hill, New York, 1989.

[128] Bart Kosko. *Neural Networks and Fuzzy Systems*. Prentice Hall, Englewood Cliffs, NJ, 1992.

[129] Robert Kowalski. *Logic for Problem Solving*. Elsevier/North-Holland, New York, 1979.

[130] W. Kreutzer. *System Simulation: Programming Languages and Styles*. Addison-Wesley, Reading, MA, 1986.

[131] Barratt Krome. *Logic and Design in Art, Science and Mathematics*. McGraw-Hill, New York, 1980.

[132] Averill M. Law and David W. Kelton. *Simulation Modeling and Analysis*, 2nd ed. McGraw-Hill, New York, 1991.

[133] P. David Lebling, Marc S. Blank, and Timothy A. Anderson. Zork: A Computerized Fantasy Simulation Game. *IEEE Computer*, April 1979.

[134] K. C. Lin. Dead-Reckoning in Distributed Interactive Simulation. In *Third Annual Conference of the Chinese-American Scholars Association of Florida*, pp. 1–5, June 1992.

[135] Aristid Lindenmeyer. Mathematical Models for Cellular Interaction in Development. *Journal of Theoretical Biology*, 18:280–315, 1968.

[136] M. H. MacDougall. *Simulating Computer Systems: Techniques and Tools*. MIT Press, Cambridge, 1987.

[137] Ernst Mach. The Economical Nature of Physics. In *Popular Scientific Lectures*, pp. 186–213. Open Court, La Salle, IL, 1894.

[138] M. C. Mackey and L. Glass. Oscillations and Chaos in Physiological Control Systems. *Science*, 197:287–289, 1977.

[139] Astronomy Magazine. Sky Almanac. *Astronomy*, 21(9):46–47, September 1993.

[140] Nadia Magnenat-Thalmann and Daniel Thalmann. *Computer Animation: Theory and Practice*. Springer-Verlag, New York, 1985.

[141] Nadia Magnenat-Thalmann and Daniel Thalmann. *Synthetic Actors in Computer-Generated 3D Films*. Springer-Verlag, New York, 1990.

[142] Peter C. Marzio. *Rube Goldberg, His Life and Work*. Harper & Row, New York, 1973.

[143] E. J. Mastascusa. *Computer-Assisted Network and System Analysis*. John Wiley & Sons, New York, 1988.

[144] Sven Erik Mattsson. Towards a New Standard for Modelling and Simulation Tools. In Torleif Iversen, editor, *SIMS '93, Applied Simulation in Industry: Proceedings of the 35th SIMS Simulation Conference*, pp. 1–10, Trondheim, Norway, June 1993. SIMS, Scandinavian Simulation Society, c/o SINTEF Automatic Control. Invited paper.

[145] Sven Erik Mattsson and Mats Andersson. Omola: An Object-Oriented Modeling Language. In M. Jamshidi and C. J. Herget, editors, *Recent Advances in Computer-Aided Control Systems*

Engineering, volume 9 of *Studies in Automation and Control*, pp. 291–310. Elsevier Science Publishers, New York, 1993.

[146] Bruce McDonald. Distributed Interactive Simulation: Operational Concept. Technical report, UCF Institute for Simulation and Training, Orlando, FL, September 1992.

[147] Gary McGath. *COMPUTE!'s Guide to Adventure Games*. COMPUTE! Publications, Inc., Greensboro, NC, 1984.

[148] John McLeod. PHYSBE: A PHYSiological BEnchmark Experiment. *Simulation*, 7(6):324, 1966.

[149] John R. Merrill. *Using Computers in Physics*. Houghton Mifflin, Boston, 1976.

[150] Alexander Miczo. *Digital Logic Testing and Simulation*. Harper & Row, New York, 1986.

[151] John H. Milsum. *Biological Control Systems Analysis*. McGraw-Hill, New York, 1966.

[152] J. Misra. Distributed Discrete Event Simulation. *ACM Computing Surveys*, 18(1):39–65, March 1986.

[153] Ben Moszkowski. A Temporal Logic for Multilevel Reasoning about Hardware. *IEEE Computer*, 18(2):10–19, February 1985.

[154] Ben Moszkowski. *Executing Temporal Logic Programs*. Cambridge University Press, Cambridge, 1986.

[155] J. D. Murray. *Mathematical Biology*. Springer-Verlag, New York, 1990.

[156] Richard E. Nance. The Time and State Relationships in Simulation Modeling. *Communications of the ACM*, 24(4):173–179, April 1981.

[157] Sanjai Narain and Jeff Rothenberg. Qualitative Modeling Using the Causality Relation. *Transactions of the Society for Computer Simulation*, 7(3):265–289, 1990.

[158] David Nicol and Richard Fujimoto. Parallel Simulation Today. Technical report, College of William and Mary, *Annals of Operation Research*, November 1994.

[159] Howard T. Odum. *Systems Ecology: An Introduction*. John Wiley & Sons, New York, 1983.

[160] Howard T. Odum. Simulation Models of Ecological Economics Developed with Energy Language Methods. *Simulation Journal*, 53(2):69–75, 1989.

[161] Katsuhiko Ogata. *Modern Control Engineering*. Prentice Hall, Englewood Cliffs, NJ, 1970.

[162] Tuncer I. Ören. Model-Based Activities: A Paradigm Shift. In T. I. Ören, B. P. Zeigler, and M. S. Elzas, editors, *Simulation and Model-Based Methodologies: An Integrative View*, pp. 3–40. Springer-Verlag, New York, 1984.

[163] Tuncer I. Ören. Dynamic Templates and Semantic Rules for Simulation Advisors and Certifiers. In Paul Fishwick and Richard Modjeski, editors, *Knowledge Based Simulation: Methodology and Application*, pp. 53–76. Springer-Verlag, New York, 1991.

[164] James M. Ortega and William G. Poole. *An Introduction to Numerical Methods for Differential Equations*. Pitman Publishing Co., Marshfield, MA, 1981.

[165] Louis Padulo and Michael A. Arbib. *Systems Theory: A Unified State Space Approach to Continuous and Discrete Systems*. W.B. Saunders, Philadelphia, 1974.

[166] James Peterson and Abraham Silberschatz. *Operating System Concepts*. Addison-Wesley, Reading, MA, 1988.

[167] James L. Peterson. Petri Nets. *Computing Surveys*, 9(3):223–252, September 1977.

[168] James L. Peterson. *Petri Net Theory and the Modeling of Systems*. Prentice Hall, Englewood Cliffs, NJ, 1981.

[169] Carl Petri. Fundamentals of a Theory of Asynchronous Information Flow. In *Information*

Processing 62: Proceedings of the 1962 IFIP Conference, pp. 386–390. North-Holland, Amsterdam, August 1962.

[170] Carl Petri. Kommunikation mit Automaten. Ph.D. thesis, University of Bonn, Germany, 1962.

[171] Ira Pohl. *C++ for C Programmers*. Benjamin-Cummings, Reading City, CA, 1989.

[172] Arthur Pope and Duncan C. Miller. The SIMNET Communications Protocol for Distributed Simulation. Technical report, Bold, Beranek and Neumann (BBN), Cambridge, MA, 1987.

[173] Hanns-Oskar A. Porr. High-Level Computer Animation Using Motion Libraries Created from Keyframes. Master's thesis, University of Florida, 1991.

[174] Herbert Praehofer. System Theoretic Foundations for Combined Discrete-Continuous System Simulation. Ph.D. thesis, Johannes Kepler University of Linz, 1991.

[175] Herbert Praehofer. Systems Theoretic Formalisms for Combined Discrete-Continuous System Simulation. *International Journal of General Systems*, 19(3):219–240, 1991.

[176] Herbert Praehofer, F. Auernig, and G. Reisinger. An Environment for DEVS-Based Multiformalism Simulation in Common Lisp/CLOSS. *Discrete Event Dynamic Systems: Theory and Applications*, 3:119–149, 1993.

[177] William H. Press, Brian P. Flannery, Saul A. Teukolsky, and William T. Vetterling. *Numerical Recipes: The Art of Scientific Programming*. Cambridge University Press, New York, 1988.

[178] A. A. B. Pritsker. *The GASP IV Simulation Language*. John Wiley & Sons, New York, 1974.

[179] A. A. B. Pritsker. *Introduction to Simulation and SLAM II*. Halsted Press, New York, 1986.

[180] Przemyslaw Prusinkiewicz and James Hanan. *Lindenmeyer Systems, Fractals and Plants*. Springer-Verlag, New York, 1989. *Lecture Notes in Biomathematics* (79).

[181] Przemyslaw Prusinkiewicz and Aristid Lindenmeyer. *The Algorithmic Beauty of Plants*. Springer-Verlag, New York, 1990.

[182] Charles J. Puccia and Richard Levins. *Qualitative Modeling of Complex Systems*. Harvard University Press, Cambridge, MA, 1985.

[183] Ashvin Radiya and Robert G. Sargent. A Logic Based Foundation of Discrete Event Modeling and Simulation, ACM Transactions on Modeling and Computer Simulation 4(1):3–51, January 1994.

[184] R. Raghuram. *Computer Simulation of Electronic Circuits*. John Wiley & Sons, New York, 1989.

[185] Hans Reichenbach. *The Philosophy of Space and Time*. Dover Publications, Mineola, New York, 1957. First published by General Publishing Company, Toronto, Ontario.

[186] George P. Richardson and A. L. Pugh. *Introduction to System Dynamics Modeling with DYNAMO*. MIT Press, Cambridge, MA, 1981.

[187] Rhonda Richter and Jean C. Walrand. Distributes Simulation of Discrete Event Systems. *Proceedings of the IEEE*, 77(1):99–113, January 1989.

[188] Fred S. Roberts. *Discrete Mathematical Models*. Prentice Hall, Englewood Cliffs, NJ, 1976.

[189] Nancy Roberts, David Andersen, Ralph Deal, Michael Garet, and William Shaffer. *Introduction to Computer Simulation: A Systems Dynamics Approach*. Addison-Wesley, Reading, MA, 1983.

[190] David Robertson, Alan Bundy, Robert Muetzelfeldt, Mandy Haggith, and Michael Uschold. *Eco-Logic: Logic-Based Approaches to Ecological Modeling*. MIT Press, Cambridge, MA, 1991.

[191] Ronald C. Rosenberg and Dean C. Karnopp. *Introduction to Physical System Dynamics*. McGraw-Hill, New York, 1983.

[192] Jerzy W. Rozenblit. Experimental Frame Specification Methodology for Hierarchical Simulation Modeling. *International Journal of General Systems*, 19(32):317–336, 1991.

[193] Jerzy W. Rozenblit and Bernard P. Zeigler. Knowledge-Based Simulation and Design Methodology: A Flexible Test Architecture Application. *Transactions of the Society for Computer Simulation*, 7(3), 1990.

[194] Rudy Rucker. *Cellular Automata Laboratory (CALAB)*. Autodesk, Inc., Sausalito, CA, 1989.

[195] James Rumbaugh, Michael Blaha, William Premerlani, Eddy Frederick, and William Lorenson. *Object-Oriented Modeling and Design*. Prentice Hall, Englewood Cliffs, NJ, 1991.

[196] Hanan Samet. *Applications of Spatial Data Structures: Computer Graphics, Image Processing, and GIS*. Addison-Wesley, Reading, MA, 1990.

[197] Hanan Samet. *The Design and Analysis of Spatial Data Structures*. Addison-Wesley, Reading, MA, 1990.

[198] H. M. Schey. *Div, Grad, Curl, and All That*. W.W. Norton and Co., New York, 1973.

[199] Lee W. Schruben. Simulation Modeling with Event Graphs. *Communications of the ACM*, 26(11), 1983.

[200] Jacob T. Schwartz. *Relativity in Illustrations*. Dover Publications, Mineola, New York, 1962.

[201] Herb Schwetman. Using CSIM to Model Complex Systems. In *1988 Winter Simulation Conference*, pp. 246–253, San Diego, CA, December 1988.

[202] R. Serra and G. Zanarini. *Complex Systems and Cognitive Processes*. Springer-Verlag, New York, 1990.

[203] Suleyman Sevinc. Learning Events from Simulation Model Runs: From Procedural Models to Relational Abstractions. *International Journal of General Systems*, 19(32):337–355, 1991.

[204] Robert E. Shannon. *Systems Simulation: The Art and Science*. Prentice Hall, Englewood Cliffs, NJ, 1975.

[205] Alan C. Shaw. *The Logical Design of Operating Systems*. Prentice Hall, Englewood Cliffs, NJ, 1974.

[206] J. Lowen Shearer, Arthur T. Murphy, and Herbert H. Richardson. *Introduction to System Dynamics*. Addison-Wesley, Reading, MA, 1967.

[207] Herbert A. Simon. *The Sciences of the Artificial*. MIT Press, Cambridge, MA, 1969.

[208] M. G. Singh, A. Y. Allidina, and B. K. Daniels. *Parallel Processing Techniques for Simulation*. Plenum Press, New York, 1986.

[209] John F. Sowa. *Conceptual Structures: Information Processing in Mind and Machine*. Addison-Wesley, Reading, MA, 1984.

[210] Donald F. Stanat and David F. McAllister. *Discrete Mathematics in Computer Science*. Prentice Hall, Englewood Cliffs, NJ, 1977.

[211] Anthony M. Starfield, Karl A. Smith, and Andrew L. Bleloch. *How to Model It: Problem Solving for the Computer Age*. McGraw-Hill, New York, 1990.

[212] Dietrich Stauffer. *Introduction to Percolation Theory*. Taylor & Francis, London, England, 1985.

[213] A. D. Steckhahn. *Industrial Applications for Microprocessors*. Reston Publishing Co., Reston, VA, 1982.

[214] D. Stirling and Suleyman Sevinc. Combined Simulation and Knowledge-Based Control of a Stainless Steel Rolling Mill. *Expert Systems with Applications*, 3:353–366, 1991.

[215] Bjarne Stroustrup. *The C++ Programming Language*, 2nd ed. Addison-Wesley, Reading, MA, 1991.

[216] Joseph Talvalage and Roger G. Hannam. *Flexible Manufacturing Systems in Practice: Applications, Design and Simulation*. Marcel Dekker, New York, 1988.

[217] Andrew S. Tanenbaum. *Computer Networks*. Prentice Hall, Englewood Cliffs, NJ, 1981.

[218] James G. Taylor. *Lanchester Models of Warfare*, Volumes I and II. Operations Research Society of America (ORSA), Baltimore, MD 1983.

[219] Jean Thoma. *Bond Graphs: Introduction and Application*. Pergamon Press, Elmsford, NY, 1975.

[220] Bob Thomas. *Disney's Art of Animation*. Hyperion, New York, 1991.

[221] Tommaso Toffoli and Norman Margolus. *Cellular Automata Machines: A New Environment for Modeling*, 2nd ed. MIT Press, Cambridge, MA, 1987.

[222] Peter S. Vail. *Computer Integrated Manufacturing*. PWS-Kent Publishing, Boston, 1988.

[223] John P. van Gigch. *System Design Modeling and Metamodeling*. Plenum Press, New York, 1991.

[224] Jiri Vlach and Kishore Singhal. *Computer Methods for Circuit Analysis and Design*. Van Nostrand Reinhold, New York, 1983.

[225] Steven Vogel. *Life's Devices: The Physical World of Animals and Plants*. Princeton University Press, Cambridge, MA, 1988.

[226] John Von Neumann. Theory of Self-Reproducing Automata. In Arthur W. Burks, editor, *Essays on Cellular Automata*. University of Illinois Press, Champaign, IL, 1970.

[227] Herbert F. Wang and Mary P. Anderson. *Introduction to Groundwater Modeling: Finite Difference and Finite Element Methods*. W.H. Freeman and Company, San Francisco, 1982.

[228] Gérard Weisbuch. *Complex Systems Dynamics*. Addison-Wesley, Reading, MA, 1991.

[229] Lawrence E. Widman, Kenneth A. Loparo, and Norman R. Nielsen. *Artificial Intelligence, Simulation and Modeling*. John Wiley & Sons, New York, 1989.

[230] Stephen Wolfram. *Theory and Applications of Cellular Automata*. World Scientific Publishing, Singapore, 1986 (includes selected papers from 1983–1986).

[231] William A. Woods. What's in a Link: Foundations for Semantic Networks. In Daniel Bobrow and Allan Collins, editors, *Representation and Understanding*, pp. 35–82. Academic Press, New York, 1975.

[232] Peter B. Worland. Parallel Methods for the Numerical Solution of Ordinary Differential Equations. *IEEE Transactions on Computers*, pp. 1045–1048, October 1976.

[233] Bernard P. Zeigler. Towards a Formal Theory of Modelling and Simulation: Structure Preserving Morphisms. *Journal of the Association for Computing Machinery*, 19(4):742–764, 1972.

[234] Bernard P. Zeigler. *Theory of Modelling and Simulation*. John Wiley and Sons, New York, 1976.

[235] Bernard P. Zeigler. *Multi-facetted Modelling and Discrete Event Simulation*. Academic Press, New York, 1984.

[236] Bernard P. Zeigler. Multifaceted Systems Modeling: Structure and Behavior at a Multiplicity

of Levels. In *Individual Development and Social Change: Explanatory Analysis*, pp. 265–293. Academic Press, New York, 1985.

[237] Bernard P. Zeigler. DEVS Representation of Dynamical Systems: Event-Based Intelligent Control. *Proceedings of the IEEE*, 77(1):72–80, January 1989.

[238] Bernard P. Zeigler. *Object Oriented Simulation with Hierarchical, Modular Models: Intelligent Agents and Endomorphic Systems*. Academic Press, New York, 1990.

[239] David Zeltzer. Motor Control Techniques for Figure Animation. *IEEE Computer Graphics and Applications*, 2(9):53–59, 1982.

Index

Abstraction:
   concept, 40
   homomorphism, 104–106, 289–94
     See Multimodeling
Actors, 115–16
Adaptive Methods:
   control,
     See Automatic Control
   integration interval,
     See Runge-Kutta-Fehlberg
Adventure games,
     See Role-Playing Models
AERO, 283
Algorithms:
   adaptive step size control, 202
   block model event scheduler, 163
   bond graph (junction-oriented), 222
   bond graph (object-oriented), 226
   cellular automata, 236
   conservative method, 341
   definition, 8
   dynamic hashing, 62
   event scheduling template, 168

   finite state machine, 88
   inverse method (RN), 31
   kinetics, 173
   lathe simulation with blocking, 170
   lathe simulation without blocking, 170
   linked list array, 56
   linked list FEL, 55
   markov model (chain), 99
   molecular dynamics (events), 263
   network structure in code, 159
   network structure in data, 160
   optimistic method, 346
   optimization, 122
   outset calculation, 129
   particle-particle, 268
   particle-mesh, 272
   percolation, 235
   Petri nets (timed), 130
   priority queue (heap), 59
   queuing network, 169
   randomizing set elements, 130
   real time synchronization, 360
   rejection method (RN), 31

Algorithms: (*continued*)
   ring N body, 355
   spatial partitioning, 350
   System Dynamics, 181
   time-slice execution, 52
   time-slice simulator (V1), 146
   time-slice simulator (V2), 150
   tree code, 272
Aggregation of models, 287–88, 294–96
Animation:
   dining philosophers, 127–35
   keyframe, 114–16
   quote, 85
Anti-messages, 347
Architectures, 334–36
ARPA, 369
Array, 56–59
Artificial life, 14
Astronomy, 112
Asynchronous, 88–89
   distributed simulation, 333
Attribute,
      See Object-oriented
Automatic control:
   fan, 136
   heating, 177
   industrial plant, 308–10
   monorail, 153–57
   traffic, 92–93
Autonomy, 46

Ball throwing:
   in concept models, 76
      See also Ballistics
Ballistics, 209–11
Behavioral modeling, 63–64
Binary search tree, 57–58
Black box modeling, 63–64
Blinkers,
      See Cellular automata
Block modeling:
   event scheduling, 163
   time slicing, 146
Blocking:
   queuing networks, 169–71
   conservative simulation, 341
   optimistic simulation, 347
   simulator, 401
Blood circulation, 186
Boiling liquids, 311–20

Boltzmann distribution, 244–45
Bond graph:
   concept, 221
   regimes, 221
   junction-oriented, 222–25
   object-oriented, 226–30
   simplification rules, 222
Business applications,
      See Financial applications
BWATER, 332,
      See also Boiling liquids

C++, 76–81, 83
CALAB, 282
Calendar queue, 68
      See also hashing
CAM, 282
Causality (among events), 338–39
Causal models, 83
      See also System Dynamics
CELLSIM, 282
CELLULAR, 282
Cellular automata:
   algorithm, 238–39
   example, 48
   parallel, 351
   overview, 236
   Life, 236–38
   rules, 236–37
Chaos:
   concept, 13
   Mackey-Glass, 217
   Lorenz system, 213–14
   Silnikov's orbit, 214
Chemical kinetics,
      See Kinetic graphs
Circular queue, 197–98
Class,
      See Object-oriented
Clock:
   FEL (with), 57
   simulator, 384–85
CLOS, 332
Cobweb cycles, 196
Collisions:
   gas molecules,
      See Gas dynamics
   particles, 242–43
   rigid bodies, 274–76
   simulator, 426

Combined modeling,
    See Multimodeling
Communications network:
    algorithm, 172–73
    fixed route simulator, 407–409
    optimal route simulator, 410–13
Compartmental modeling, 184
Complex systems, 281
Computational science, 12
Computer:
    analog, 3
    simulation, 1
Conceptual modeling, 69–70
Constraint modeling:
    concept, 193–95
    graph-based approach, 217
    equation-based approach, 195
Constructive simulation, 359
Control,
    See Automatic control
Convection, 250
Convolution, 236
Counter, 94
Coupling:
    state to state, 40
CPU/Disk simulation, 399–401
CSIM, 192

Damping, 152
Data: 29–40
Declarative modeling:
    concept, 48–50, 83–85
    predicate calculus, 51
Dead reckoning, 361–64
Decomposition (of models),
    See Aggregation
Delay:
    block, 199
    event scheduling,
        See Event
    in delay differential equation, 215–17
Delay differential equation:
    regulation of hematopoiesis, 216
    using circular queue, 216
DEQ, 215, 425–26
DEVS, 67, 83, 140
Difference equation:
    canonical form, 198–200
    circular queues, 197–98
    definition, 195

Fibonacci growth, 197
    using variables, 198–200
    simulator, 421–23
Differential equation:
    canonical form, 198–200
    ordinary, 200
    partial, 246–52
    simulator, 423–26
    See also DEQ
DEQ, 425
Diffusion:
    cellular automata, 236
    heat equation, 350
    partial differential equation, 246
    simulator, 426–27
    transport method, 248–52
Digital logic:
    example 161–65
    simulator, 380–82
Dining philosophers:
    FSA, 95–97
    markov model, 100–101
    multimodeling, 307
    Petri net, 123
    simulator, 378–79
Directionality,
    See Functional modeling
DIS,
    See Distributed interactive simulation
Distributed interactive simulation, 357–67
Distributed simulation,
    See Parallel simulation
Distributions,
    See Random variables
Dynamic hashing,
    See Hashing

Effort:
    See Bond graph
Engine, 90–92
Epidemiological model, 187–88
Equational modeling:
    difference equation, 197
    differential equation, 200
Electrical circuit:
    digital, 94–95, 161–65
    analog, 219–20
Energy:
    bond graph, 221–30
    Lagrangian, 212

Euler's method,
    See Forward Euler
Event:
    definition, 40–44
    examples, 41–43
    external, 41
    graphs,
        See FEA
    internal, 41
    scheduling, 53–63, 163, 372–74
    See also Finite event automata
Expert systems,
    See Behavioral modeling
External event,
    See Event

Fast multipole method, 282
FEA,
    See Finite event automata
FElt, 283
FEL,
    See Future event list
FFM,
    See Fast multipole method
Fibonacci growth, 197
Flow:
    water jugs, 107
    coupled water tanks, 157, 229–30
    fluid dynamics, 251
    System Dynamics method,
        See System Dynamics
Financial applications, 64–65
Finite event automata (FEA):
    definition, 113–14
    keyframe animation,
        See Animation
Finite state automata (FSA):
    definition, 88
    digital counter, 94–95
    dining philosophers, 95–97
    engine, 90
    fan controller, 136
    JK flip flop, 94
    role playing, 97
    simulator, 88–89, 374–78
    traffic control, 92
    water jugs, 106
Flip flop, 94
Flow chart, 101
Fluid mechanics, 251–52

Forest fire, 235
Forward Euler, 200
FSA,
    See Finite state automata
FTP, 21
Functional modeling:
    concept, 48–50, 143–45
    function-based approach, 143
    lambda calculus, 49
    variable-based approach, 173
Function-based approach:
    See Functional modeling
Future event list, 57, 339–41
Fuzzy:
    data, 38–40
    logic, 64, 68

Gas dynamics:
    fluid,
        See Fluid mechanics
    molecules, 260–62
    simulator, 426
GNU License, 25, 372
Gopher, 22
GPSS,
    See Mini-GPSS
Gravitational force, 210, 267
Grocery store, 166–68, 387–88
Groundwater modeling, 252
GSPN, 141

Harmonic oscillation:
    in monorail control, 153–57
    mass-spring, 151, 211, 224
    pendulum, 212
    spring: 151
Hashing, 60–63
Heap, 59–60
Heat equation,
    See Diffusion
Heterogeneous refinement, 299–301, 309,
    314–16
HIV infection, 187–88
Homoclinic orbit,
    See Silnikov's orbit
Homogeneous refinement, 299–301,
    312–14
Homomorphism, 104–105, 288–94
Human body modeling:
    See also Dining philosophers

Hybrid models:
  state-event graphs, 118–20
    See also Multimodeling

Industrial plant control, 308–10
Inertia, 152
Input, 45–46
Integration:
  in block models, 146
  in equations, 200–206
  capacitance, 157–58
Internal event,
    See Event
Interpolation in DIS, 364–65
Interval measure, 29
Invariance:
  events, 117
  states, 117
Inventory control, 120–23
Inverse method: 31
Ising model, 244–45
IST, 369

JK flip flop,
    See Flip Flop

Keyframe animation,
    See Animation
Kinetic graph, 173–75
Kirchoff's laws, 219–20

L-systems:
    See Lindenmeyer systems
Lsys, 282
Labanotation, 139, 140
Lagrangian, 212
Lattice gas, 240–44
Lazy cancellation, 349
Lindenmeyer systems:
  object aggregation, 253
  overview, 253
  plant models, 257–60
Linked list, 55–56
Linux,
    See Preface
LISP, 83
Live simulation, 358
Local virtual time,
    See Virtual time
Locality, principal of, 50–51
Lorentz contraction, 119

Lorenz system, 213–14
Lotka-Volterra equations,
    See Predator-prey model
Logistic equation, 196, 199

Magnetism,
    See Ising model
Manufacturing, 123–26, 169–71
Margolus neighborhood,
    See Neighborhood
Markov model:
  definition, 99–100
  dining philosophers, 101–102
  flowchart, 101
  matrix, 100
  object representation, 103–104
  rat maze, 137
  scheduling, 103
  simulator, 100, 374–78
Matching, 105
MC,
    See Monte Carlo
Message passing, 341
Method,
    See Object-oriented
Military models,
    See DIS
MIMD, 334–36
Mini-GPSS:
  execution, 413–15
  flowchart, 415–17
  overview, 413
  syntax, 413–15
MISD, 334–36
Model:
  definition: 27
Modeling:
  combined, 10
  definition, 2
  engineering, 5–6
  specification, 7
  textbooks, 24
Model engineering, 83
Molecular models,
    See Gas dynamics
Monte Carlo, 31, 37–38
Moore neighborhood,
    See Neighborhood
Multimodeling:
  abstraction, 288
  aggregation, 287

Multimodeling: (*continued*)
  definition, 27, 285
  discontinuity, 296
  homomorphism, 289–94
  phase space, 290, 294–96, 301–302,
    320–22
  refinement, 294–96, 297–301, 309,
    312–16
Multiplicative congruential generator, 29

N body algorithms (parallel):
  overview, 353
  ring, 354–55
  shared memory, 354
N body algorithms (serial):
  particle-particle, 268–70
  partition neighbor list, 270
  particle-mesh, 272
  treecode, 272–74
  Verlet neighbor list, 270
Natural language, 70–73, 83
Natural law, 193
Navier-Stokes equation, 243
Neighborhood (in Cellular automata):
  Von Neumann, 237
  Margolus, 239–40
  Moore, 237
NEST, 192
Network:
  communications, 172–73, 407–409
  functional, 159
  optimal route simulator, 410–13
Newton's laws of motion, 209–213
Newton-Ralphson, 297
Neural network, 64, 68
Nominal Measure, 29
Nondeterministic automata,
    See Markov model
Null messages, 345
Numerical methods:
  integration,
    See Integration
  root finding, 296

Object-oriented:
  C++,
    See C++
  database, 83
  declarative model, 85–86
  model design, 17–18
  symbology, 18

Ohm's law, 221
Olympus, 141
Omola, 332
Omsim, 332
Ordinary differential equation,
    See Differential equation
Operating system:
  scheduling, 103
Operators, 108–109
Optimal route simulator, 410–13
Optimization, 122
Ordinal measure, 29
Orbital mechanics,
    See Planetary orbital mechanics
Output, 45–46

Parallel computation, 52
Parallel simulation:
  architectures, 334–38
  asynchronous method, 341
  causality, 338–39
  conservative simulation, 341
  distributed interactive simulation, 357–67
  optimistic simulation, 347
  shared clock, 339–40
  shared FEL, 340–41
  spatial models, 350
  synchronous method, 339
Partial differential equation,
    See Differential equation
Particles,
    See N body algorithms
Partition:
  neighbor list, 271
  parallel, 352–53
Parameter, 46
PDU,
    See Protocol data unit
Percolation, 235
Pendulum, 211–13
Petri nets:
  colored, 125
  definition, 123
  manufacturing, 123–27
  object-oriented representation, 126–27
  simulator, 382–83
  stochastic, 125
  randomizing transitions, 130
  timed, 124–25, 130
  transition outset, 130
Phase, 86

PHASER, 233
Physically-based modeling, 5
Pictorial modeling, 74–76
Planetary orbital mechanics, 267–68
Population models:
   bacterial growth, 180
   buildings, 183–85
   Fibonacci growth, 197
   predator/prey, 206–209
   temporal logic, 109
Potential field, 271
Predator-prey model, 206–209
Prediction:
   dead reckoning, 361–64
   lookahead, 344
Preemption:
   conservative simulation, 341
   queuing models, 386
   routine, 399–401
Priority queue,
   See heap
Probability density function,
   See Random variables
Probability distributions,
   See Random variables
Procedural modeling,
   See Functional modeling
Production systems, 15, 104–108
Projects, 21
Prolog, 83, 141
Propagation problem, 150
Protocol data unit, 359, 365–67
Pulse process:
   method, 175
   simulator, 409

Quadtree, 274
Queuing:
   basic algorithm, 168
   blocking, 169–70, 339, 347
   grocery store, 166–68
   library routines, 386–87
   metrics, 166
   models, 165
   simulator, 401–407
   single server, 166, 389–99
   Xsimcode, 418–20

Ratio measure, 29
Random Number, 30

Random Variables:
   Bernoulli distribution, 34
   binomial distribution, 34
   continuous distribution, 34–36
   cumulative distribution, 31–32
   definition, 30
   discrete distribution, 32–34
   exponential distribution, 36
   generation, 30
   normal distribution, 36
   Poisson distribution, 34
   triangular distribution, 35
   uniform distribution, 32, 35
Random walk,
   See Diffusion
Reachability tree, 98–99
REAL, 192
Real time synchronization, 361
Reasoning, 28
Refinement of models, 294–96, 297–301,
   312–16
Rejection method, 31
Relativity, 119
Research, 20–12
Rigid body mechanics, 274
Robot operations, 125–27, 325–28
Role playing, 1, 97
Rollback, 347–49
Rule base, 106
Runge-Kutta (RK), 201–202
Runge-Kutta-Fehlberg (RKF),
   202–203

Schematic, 76
Script, 115–16
Semantic network, 75–76
Serial model execution, 51–63
Server modeling,
   See Queuing models
Shared:
   clock, 339–40
   FEL, 340–41
   memory, 334–36
Signal flow graphs, 173
Silnikov's orbit, 214
SIMD, 334–36, 351
SIM++, 192
Simula, 83, 192
Simulation:
   analog, 3

Simulation: (*continued*)
  definition, 1
  phases, 4
  scaled, 3
  types, 2
Simulation Digest, 25
SimPack Toolkit, 25, 371–72
Single server queue: 53–54, 389–99
SIMNET, 357–58
SISD, 334–36
SLAM, 24
Sketch, 74
Smalltalk, 83
SMPL, 192
Soft systems, 83
Space-time,
        See Hybrid models
Spatial model:
  concept, 233
  entity-based approach, 252
  space-based approach, 234
SPICE, 232
Spring model,
        See Harmonic oscillation
State:
  definition, 40
  estimation, 361
  examples, 42–44
  graph,
        See FEA
Statistics:
  simulation textbook, 11
Straggler message, 346
STRICOM, 369
Superstate, 47, 68, 104–106
Synchronous, 88–89
  distributed simulation, 339
System:
  definition, 46–48
  science, 67–68
System Dynamics:
  algorithm, 181
  causal graph, 176–78
  flow graph, 179
  symbology, 180

System science, 67–68
System theory, 67–68

tDOL, 281
Tea model, 41, 290–94
Temporal logic, 109–12, 380–82
Tempura, 141, 380–82
Terrain model, 359
Text,
        See Natural language
TIERRA, 283
Time:
  definition, 46
  slicing, 52, 146–50, 383
  stamp, 340
  window, 340
        See also Causality
Time Warp, 369
Traffic control, 92–94, 104–105
Transition function, 47–48
Tree model, 261

Variable-based approach,
        See Functional modeling
Vector:
  calculus, 247
  forces, 209–10
Verlet neighbor list,
        See N body algorithms
Virtual reality, 13–14
Virtual simulation, 359
Virtual time, 338
Von Neumann neighborhood,
        See Neighborhood

Water jugs, 107–109, 303–307
Water tanks, 158
Wave equation, 350
Wildcard,
        See Matching
World Wide Web (WWW), 22

XKEY, 140
XPP, 232
Xsimcode, 418–20